Brownian Motion, the Fredholm Determinant, and Time Series Analysis

Brownian motion is an important topic in various applied fields where the analysis of random events is necessary. Introducing Brownian motion from a statistical viewpoint, this detailed text examines the distribution of quadratic plus linear or bilinear functionals of Brownian motion and demonstrates the utility of this approach for time series analysis. It also offers the first comprehensive guide to deriving the Fredholm determinant and the resolvent associated with such statistics.

Presuming only a familiarity with standard statistical theory and the basics of stochastic processes, this book brings together a set of important statistical tools in one accessible resource for researchers and graduate students. Readers also benefit from online appendices, which provide probability density graphs and solutions to the chapter problems.

KATSUTO TANAKA is Professor Emeritus at Hitotsubashi University, Tokyo. He is a recipient of the Tjalling C. Koopmans Econometric Theory Prize (1996), the Japan Statistical Society Prize (1998), and the Econometric Theory Award (1999). He has previously authored *Time Series Analysis* (First Edition 1996, Second Edition 2017) and six statistics and econometrics books in Japanese.

INSTITUTE OF MATHEMATICAL STATISTICS MONOGRAPHS

IMS Monographs are concise research monographs of high quality on any branch of statistics or probability of sufficient interest to warrant publication as books. Some concern relatively traditional topics in need of up-to-date assessment. Others are on emerging themes. In all cases the objective is to provide a balanced view of the field.

In collaboration with the International Society for Bayesian Analysis (ISBA), selected volumes in the IMS Monographs series carry the "with ISBA" designation at the recommendation of the ISBA editorial representative.

Other Books in the Series (*with ISBA)

1. *Large-Scale Inference*, by Bradley Efron
2. *Nonparametric Inference on Manifolds*, by Abhishek Bhattacharya and Rabi Bhattacharya
3. *The Skew-Normal and Related Families*, by Adelchi Azzalini
4. *Case-Control Studies*, by Ruth H. Keogh and D. R. Cox
5. *Computer Age Statistical Inference*, by Bradley Efron and Trevor Hastie
6. *Computer Age Statistical Inference (Student Edition)*, by Bradley Efron and Trevor Hastie
7. *Stable Lévy Processes via Lamperti-Type Representations*, by Andreas E. Kyprianou and Juan Carlos Pardo
8. *The Conway–Maxwell–Poisson Distribution*, by Kimberly F. Sellers

Brownian Motion, the Fredholm Determinant, and Time Series Analysis

KATSUTO TANAKA
Hitotsubashi University, Tokyo

CAMBRIDGE
UNIVERSITY PRESS

CAMBRIDGE
UNIVERSITY PRESS

Shaftesbury Road, Cambridge CB2 8EA, United Kingdom

One Liberty Plaza, 20th Floor, New York, NY 10006, USA

477 Williamstown Road, Port Melbourne, VIC 3207, Australia

314–321, 3rd Floor, Plot 3, Splendor Forum, Jasola District Centre, New Delhi – 110025, India

103 Penang Road, #05–06/07, Visioncrest Commercial, Singapore 238467

Cambridge University Press is part of Cambridge University Press & Assessment, a department of the University of Cambridge.

We share the University's mission to contribute to society through the pursuit of education, learning and research at the highest international levels of excellence.

www.cambridge.org
Information on this title: www.cambridge.org/9781009566995

DOI: 10.1017/9781009567008

When citing this work, please include a reference to the DOI 10.1017/9781009567008

First published 2025

A catalogue record for this publication is available from the British Library

A Cataloging-in-Publication data record for this book is available from the Library of Congress

ISBN 978-1-009-56699-5 Hardback

Additional resources for this publication at www.cambridge.org/9781009566995.

Contents

Preface

The title of this book may sound strange and unacceptable because Brownian motion and the Fredholm determinant are topics that seem totally unrelated to each other. I first describe the reason why such a title arises.

Brownian motion (Bm) is one of the central topics in probability theory, on which many good books have been written from a mathematical point of view. It is now also an important topic in various applied fields where the analysis of random events is necessary. This book discusses Bm from a statistical viewpoint. In particular, it is often the case that Bm arises in the time series situation, and we need to consider distributions of its quadratic functionals, sometimes extended with linear or bilinear functionals or ratios of those functionals. Since it is hard in general to derive distribution functions of such functionals explicitly, we attempt to derive the associated characteristic function (c.f.). For purely quadratic functionals of Bm, the theorem established in Anderson and Darling (1952) tells us that the derivation of the c.f. of such statistics results in computing the Fredholm determinant (FD), which is an important quantity in the theory of homogeneous integral equations of Fredholm type. In this way, Bm encounters the FD, which is reflected by the book title. If linear or bilinear functionals are added to the quadratic Brownian functional, the notion of the resolvent in the theory of nonhomogeneous integral equations is also necessary to derive the c.f.s of such statistics. Once the c.f. is obtained, the distribution function can be computed by inverting it numerically.

This book is primarily intended to serve as a textbook or supplementary material for graduate courses in time series analysis, asymptotic statistics, stochastic processes, applied probability, and so on. It is hoped that the book will also be used as a reference by researchers in these fields. I have attempted to make the book as self-contained as possible, although readers are assumed to be familiar with the standard theory of statistics, including estimation and testing. A knowledge of the basic theory of stochastic processes is also presupposed.

The main text of the book is organized into Parts I and II. Part I consists of Chapters 1 through 3 and discusses theoretical aspects of the topics described above, whereas Part II consists of Chapters 4 through 8 and provides statistical applications of the theory presented in Part I. The book is complemented by Appendices A and B, available online at www.cambridge.org/9781009566995. Appendix A gives a set of solutions to the exercises embedded in most sections of each chapter, whereas Appendix B presents graphs of some statistical probability densities discussed in the main text.

After giving an overview of Part I, Chapter 1 provides an introductory discussion of our topics, starting with a statistical explanation of Bm and its quadratic functionals. Then we indicate by simple examples, most of which are testing problems, how to relate the c.f. of the functionals of Bm with the FD and the resolvent.

Chapter 2 begins with the definition of the FD in connection with the homogeneous integral equation of Fredholm type. Then we present a general method for computing the FD associated with purely quadratic functionals of Bm, which deals with the differential equation with some boundary conditions equivalent to the integral equation. Many examples are shown for computing the FD, classifying the corresponding differential equations into five categories. We also suggest, for special types of functionals of Bm, computationally efficient methods for obtaining the associated FD without dealing with differential equations. It is recognized that computerized algebra is quite useful for derivation of the FD. In this book I exclusively use the REDUCE computer algebra system, which is available free of charge.

Chapter 3 extends the class of functionals to quadratic plus linear or bilinear functionals of Bm. It turns out that the FD is not sufficient to obtain the c.f.s of such functionals, and we need to deal with the resolvent appearing in the theory of nonhomogeneous Fredholm integral equations. We explain the resolvent from a statistical point of view. It is recognized that the computation of the resolvent itself is unnecessary to compute the c.f. but that we only have to compute a function derived from the resolvent. We demonstrate by some statistical examples how to compute the c.f.s of those statistics.

Chapters 4 through 8, organized as Part II, take up statistical problems and illustrate how the FDs and resolvents are used, concentrating on one particular topic in each chapter. After giving an overview of Part II to make clear the role of the models used in these chapters, Chapter 4 considers goodness-of-fit tests and derives the limiting c.f.s of test statistics for testing if the underlying distribution follows the hypothesized distribution parameterized by a vector of parameters. It is seen that test statistics converge in distribution to a quadratic-plus-linear functional of Bm, for which the c.f. is obtained from the method discussed in Chapter 3.

Chapter 5 deals with state space models and discusses testing for the constancy of a parameter contained in the measurement equation. Since there is no uniformly best test, we employ the locally best invariant (LBI) test. Ensuring that the LBI statistics converge in distribution to quadratic functionals of Bm under the null and local alternatives, we derive the associated FD. We also compute and compare limiting local powers under the two situations where the initial value of the state equation is known or unknown.

Chapter 6 deals with the moving average (MA) model of order 1 with or without trend term, where two cases are considered for the intial value of the error term that drives the process. One is the unconditional case where the initial value follows the same distribution as the future errors, and the other is the conditional case where the initial value is set to 0. For both models we conduct the unit root test of the coefficient parameter. We apply the unbiased LBI (LBIU) test to the unconditional MA(1) model, whereas the LBI test is conducted on the conditional MA(1) model. The limiting local powers of the two tests are computed using the associated FD and are graphically compared.

Chapter 7 deals with the autoregressive model of order 1 with or without trend term. We conduct the unit root test and derive the limiting distribution of least squares estimator-based, LBI, and LBIU test statistics. It is found that the distributional behavior of test statistics is quite different from the statistics discussed in previous chapters. In fact, the Ornstein–Uhlenbeck (O–U) process emerges in the limiting distribution under the local alternative, which has never occurred in the previous problems studied. For this situation, in addition to our method for deriving the FD, we employ Girsanov's theorem, which is concerned with the transformation of measures induced by the O–U process, to compute the c.f. associated with the limiting distribution.

Chapter 8 extends Bm to fractional Bm (fBm) and considers the fractional unit root distribution based on fBm. fBm has many different properties from Bm, among which is the most striking fact that increments are dependent, unlike in Bm. Because of this property, the derivation of the associated FD still remains to be done. We suggest an approximation to the distribution using the fundamental martingale introduced in Norros, Valkeila, and Virtamo (1999). Also dealt with is the multiple unit root distribution based on multi-fold integrated Bm. We find some similarities between the two unit root distributions together with an interesting moment property.

There are many exercises posed at the end of most sections of each chapter. Most of the exercises are concerned with corroborating the results described in the main text, so that the reader can gain a better understanding of the details of the discussions.

Online Appendix A provides a complete set of solutions to these exercises and can be found at the following web page:

www.cambridge.org/9781009566995

Also available from this location is Appendix B, which presents graphs of some statistical probability densities associated with Bm and fBm dealt with in the main text. These are computed by numerically inverting the c.f.s we already derived.

This book originates from joint work with the late Professor Seiji Nabeya, who was a respected teacher of mine from my undergraduate days. I thank him posthumously for pleasant memories extending over many years.

Finally, I would like to mention that it would be my great pleasure if Brownian motion and the Fredholm determinant were combined successfully in the statistical way presented in this book.

Part I

Theory

Part I is composed of Chapters 1 through 3 and discusses the theory of the Fredholm determinant and the resolvent necessary to study statistical and distributional aspects of functionals of Brownian motion (Bm). Here is a summary of Part I.

Chapter 1 starts with introducing Bm $W(t)$ defined on $[0, 1]$. After describing various statistical properties of Bm and defining stochastic integrals in the mean square (m.s.) sense, we consider a quadratic functional of Bm given by

$$S = \int_0^1 \int_0^1 K(s,t)\, dW(s)\, dW(t), \tag{I.1}$$

where the kernel $K(s,t)$ is symmetric, continuous, and nearly definite in $[0, 1] \times [0, 1]$. The statistic S is a purely quadratic functional of Bm, and its various examples are presented. The *Fredholm determinant* (FD) is required to study the distribution of S, which is discussed in detail in Chapter 2. It is also mentioned that S arises in the weak limit of S_n given by

$$S_n = \frac{1}{n} \sum_{j=1}^{n} \sum_{k=1}^{n} B_n(j,k)\, \varepsilon_j\, \varepsilon_k = \frac{1}{n}\, \boldsymbol{\varepsilon}' B_n \boldsymbol{\varepsilon} \;\Rightarrow\; S,$$

where $\boldsymbol{\varepsilon} = (\varepsilon_1, \ldots, \varepsilon_n)'$, $\{\varepsilon_j\} \sim$ i.i.d.$(0, 1)$, and B_n is an $n \times n$ symmetric matrix whose (j,k)th element $B_n(j,k)$ satisfies

$$\lim_{n\to\infty} \max_{1\le j,k\le n} \left| B_n(j,k) - K\left(\frac{j}{n}, \frac{k}{n}\right) \right| = 0.$$

As a variant of the statistic S, we also deal with a positive-valued quadratic functional, which takes the form

$$S' = \int_0^1 X^2(t)\, dt \overset{\mathcal{D}}{=} \int_0^1 \int_0^1 \mathrm{Cov}(X(s), X(t))\, dW(s)\, dW(t), \tag{I.2}$$

where $X(t)$ is a zero-mean Gaussian process and $\stackrel{\mathcal{D}}{=}$ stands for distributional equivalence, which is justified by the Karhunen–Loève expansion.

An example of $X(t)$ is given by $X(t) = W(t) - tW(1)$, which is the Brownian bridge with $\mathrm{Cov}(X(s), X(t)) = \min(s,t) - st$. Thus we have

$$S' = \int_0^1 (W(t) - tW(1))^2 \, dt \stackrel{\mathcal{D}}{=} \int_0^1 \int_0^1 [\min(s,t) - st] \, dW(s) \, dW(t).$$

The statistic S' can be dealt with in the same way as S.

Chapter 2 examines the distributional properties of the statistic S in (I.1) and derives its *characteristic function* (c.f.). The Anderson and Darling (1952) theorem leads to

$$\mathrm{E}\left(e^{i\theta S}\right) = \lim_{n\to\infty} \mathrm{E}\left(e^{i\theta S_n}\right) = \lim_{n\to\infty} \left| I_n - \frac{2i\theta}{n} B_n \right|^{-1/2} = (D(2i\theta))^{-1/2}, \tag{I.3}$$

where $D(\lambda)$ is the FD of $K(s,t)$ with $D(0) = 1$. To define $D(\lambda)$ of a given $K(s,t)$, we introduce the homogeneous integral equation of Fredholm type

$$f(t) = \lambda \int_0^1 K(s,t) f(s) \, ds, \tag{I.4}$$

where λ is an eigenvalue of $K(s,t)$ and $f(t)$ is an eigenfunction corresponding to λ. The integral equation can be approximated by the algebraic system

$$\boldsymbol{f}_N = \frac{\lambda}{N} K_N \boldsymbol{f}_N \quad \Leftrightarrow \quad \left(I_N - \frac{\lambda}{N} K_N \right) \boldsymbol{f}_N = \boldsymbol{0},$$

where $\boldsymbol{f}_N = [(f(k/N))]$ is an $N \times 1$ vector and $K_N = [(K(j/N, k/N))]$ is an $N \times N$ symmetric matrix. We also define $D_N(\lambda) = |I_N - \lambda K_N/N|$. The zero of $D_N(\lambda)$ gives the reciprocal of the eigenvalues of K_N/N in matrix theory, so λ in the integral equation (I.4) plays a similar role. Then the FD of $K(s,t)$ is defined and expressed as

$$D(\lambda) = \lim_{N\to\infty} \left| I_N - \frac{\lambda}{N} K_N \right| = \prod_{n=1}^{\infty} \left(1 - \frac{\lambda}{\lambda_n} \right) = 1 + \sum_{n=1}^{\infty} \frac{(-1)^n d_n}{n!} \lambda^n, \tag{I.5}$$

where $\lambda_1, \lambda_2, \ldots$ are the eigenvalues of $K(s,t)$ repeated as many times as their multiplicities, whereas

$$d_n = \int_0^1 \cdots \int_0^1 \begin{vmatrix} K(t_1,t_1) & K(t_1,t_2) & \cdot\ \cdot\ \cdot & K(t_1,t_n) \\ \vdots & \vdots & & \vdots \\ K(t_n,t_1) & K(t_n,t_2) & \cdot\ \cdot\ \cdot & K(t_n,t_n) \end{vmatrix} dt_1 \cdots dt_n. \tag{I.6}$$

In particular, we have

$$d_1 = \int_0^1 K(t,t)\,dt, \quad d_2 = \left(\int_0^1 K(t,t)\,dt\right)^2 - \int_0^1\int_0^1 K^2(s,t)\,ds\,dt.$$

We also have, using (I.5) and differentiating the c.f. $(D(2i\theta))^{-1/2}$,

$$\mathrm{E}(S) = \sum_{n=1}^{\infty} \frac{1}{\lambda_n} = \int_0^1 K(t,t)\,dt = d_1, \tag{I.7}$$

$$\mathrm{Var}(S) = 2\sum_{n=1}^{\infty} \frac{1}{\lambda_n^2} = 2\int_0^1\int_0^1 K^2(s,t)\,ds\,dt = 2(d_1^2 - d_2). \tag{I.8}$$

However, it proves to be quite hard in general to compute $D(\lambda)$ following the formulas in (I.5).

We present a general method for deriving the FD, which transforms the integral equation (I.4) to an equivalent differential equation with some boundary conditions. Then, solving the differential equation, we find a candidate for the FD from the boundary conditions and ensure that the candidate is really the FD by checking that it satisfies the sufficient conditions described in the text. Many statistical examples are given to demonstrate how to derive the FD by classifying the type of the differential equation into five categories, where we also deal with ratio statistics of quadratic functionals of Bm.

For some special types of kernels, we suggest computationally efficient methods for obtaining the FD without dealing with differential equations. These are based on Darling's formula (Darling, 1955) and applied to the kernels

$$K_1(s,t) = \min(s,t) - st - \sum_{j=1}^{m} \psi_j(s)\psi_j(t),$$

$$K_2(s,t) = 1 - \max(s,t) - \sum_{j=1}^{m} \psi_j(s)\psi_j(t),$$

where $\psi_j(t)$ $(j=1,\ldots,m)$ are continuously differentiable with $\psi_j(0) = \psi_j(1) = 0$ for $K_1(s,t)$ and $\psi_j(1) = 0$ for $K_2(s,t)$.

Chapter 3 extends the purely quadratic functionals S and S' to quadratic plus linear or bilinear functionals of Bm, for which a method for computing the corresponding c.f. is suggested.

As an extension of S to the quadratic plus linear functional, we deal with

$$T = \int_0^1\int_0^1 K(s,t)\,dW(s)\,dW(t) + 2a\int_0^1 K(0,t)\,dW(t) + b, \tag{I.9}$$

where a and b are constants. It is shown that

$$\mathrm{E}\left(e^{i\theta T}\right) = (D(\lambda))^{-1/2} \exp\left[\frac{\lambda}{2}\left\{a^2(\Gamma(0,0;\lambda) - K(0,0)) + b\right\}\right], \quad \text{(I.10)}$$

where $\lambda = 2i\theta$, $D(\lambda)$ is the FD of $K(s,t)$, and $\Gamma(0,0;\lambda)$ is the resolvent $\Gamma(s,t;\lambda)$ of $K(s,t)$ evaluated at $(s,t) = (0,0)$.

Given $K(s,t)$ and $g(t)$, the resolvent $\Gamma(s,t;\lambda)$ is defined by introducing the nonhomogeneous integral equation

$$f(t) = g(t) + \lambda \int_0^1 K(s,t) f(s)\, ds. \quad \text{(I.11)}$$

Then the solution $f(t)$ to this equation is expressed in the following way:

$$f(t) = g(t) + \lambda \int_0^1 \Gamma(s,t;\lambda) g(s)\, ds,$$

where $\Gamma(s,t;\lambda)$ satisfies

$$\Gamma(s,t;\lambda) = K(s,t) + \lambda \int_0^1 \Gamma(s,u;\lambda) K(u,t)\, du. \quad \text{(I.12)}$$

We can obtain $\Gamma(0,0;\lambda)$ in (I.10) by solving for $\Gamma(0,t;\lambda)$ in (I.12).

As an extension of S' to the quadratic plus linear functional, we deal with

$$\begin{aligned} T' &= \int_0^1 (X(t) + m(t))^2\, dt \\ &= \int_0^1 \left(X^2(t) + 2m(t)X(t) + m^2(t)\right) dt, \end{aligned} \quad \text{(I.13)}$$

where $\{X(t)\}$ is a zero-mean Gaussian process with $\mathrm{Cov}(X(s), X(t)) = K(s,t)$, and $m(t)$ is a continuous and deterministic function. It is shown that

$$\begin{aligned} \mathrm{E}\left(e^{i\theta T'}\right) = (D(\lambda))^{-1/2} \exp\bigg[& \frac{\lambda}{2} \int_0^1 m^2(t)\, dt \\ & + \frac{\lambda^2}{2} \int_0^1 \int_0^1 \Gamma(s,t;\lambda) m(s) m(t)\, ds\, dt \bigg], \end{aligned} \quad \text{(I.14)}$$

where $\lambda = 2i\theta$ and $D(\lambda)$ is the FD of $K(s,t)$. The computation of the resolvent $\Gamma(s,t;\lambda)$ for any (s,t) is difficult, but we can avoid this. In fact, it follows from (I.12) that

$$\begin{aligned} h(t;\lambda) &= \int_0^1 \Gamma(s,t;\lambda) m(s)\, ds \\ &= \int_0^1 K(s,t) m(s)\, ds + \lambda \int_0^1 K(s,t) h(s;\lambda)\, ds, \end{aligned} \quad \text{(I.15)}$$

which yields

$$\int_0^1 \int_0^1 \Gamma(s,t;\lambda)m(s)m(t)\,ds\,dt = \int_0^1 h(t;\lambda)m(t)\,dt. \tag{I.16}$$

Thus, if we can solve for $h(t;\lambda)$ in (I.15), we can compute the c.f. of T' in (I.14) using the right-hand side of (I.16) without computing $\Gamma(s,t;\lambda)$.

Various examples of T and T' are presented in the text. A common feature in the statistics T and T' is asymptotic normality as the linear term dominates, which is not anticipated for the statistics S in (I.1) and S' in (I.2).

We also consider the quadratic plus bilinear functionals of Bm given by

$$U = \int_0^1 \int_0^1 K(s,t)\,dW(s)\,dW(t) + 2aZ\int_0^1 K(0,t)\,dW(t) + bZ^2,$$

$$U' = \int_0^1 (X(t) + Zm(t))^2,$$

where $Z \sim \mathrm{N}(0,1)$ and is independent of $\{W(t)\}$ and $\{X(t)\}$. The computation of the c.f.s of U and U' can be done easily by the conditional argument using the results for $\mathrm{E}\left(e^{i\theta T}\right)$ and $\mathrm{E}\left(e^{i\theta T'}\right)$. For example, we have

$$\begin{aligned}
\mathrm{E}\left(e^{i\theta U}\right) &= \mathrm{E}\left[\mathrm{E}\left(e^{i\theta U}\,\middle|\,Z\right)\right] \\
&= (D(\lambda))^{-1/2}\mathrm{E}\left[\exp\left\{Z^2\frac{\lambda}{2}\left(a^2(\Gamma(0,0;\lambda) - K(0,0)) + b\right)\right\}\right] \\
&= (D(\lambda))^{-1/2}\left[1 - \lambda\left\{a^2(\Gamma(0,0;\lambda) - K(0,0)) + b\right\}\right]^{-1/2},
\end{aligned}$$

where $\lambda = 2i\theta$. We can compute $\mathrm{E}\left(e^{i\theta U'}\right)$ similarly. Unlike T and T', asymptotic normality does not hold for U and U' because the bilinear term of Bm is always coupled with the random variable Z.

Note: Once we obtain the c.f. of our statistic, we can compute its distribution function and probability density by inverting the c.f. numerically. Simlarly we can compute percentage points of the distribution. These procedures are important in applications when we conduct tests and examine power properties, which will be carried out in Part II.

1

Quadratic Functionals of Brownian Motion

Brownian motion (Bm) has been a central topic, not only in probability theory but also in various applied fields. Throughout this book we deal with Bm from a statistical viewpoint. In particular, we deal with quadratic functionals of Bm or ratios of those functionals, and consider their distributions, where functionals are expressed by the Riemann–Stieltjes double integral with respect to Bm. Since it is hard in general to derive distribution functions of such functionals explicitly, we attempt to derive the associated characteristic function (c.f.). For this purpose we use the theory of integral equations of Fredholm type, among which is the Fredholm determinant. The notion of the resolvent is also necessary if we deal with quadratic plus linear or bilinear functionals of Bm. In this chapter we give an introductory discussion on these. In particular, we indicate by some theorems and simple examples how to relate the c.f. with the Fredholm determinant and the resolvent.

1.1 Brownian Motion and Some Statistical Properties

Brownian motion (Bm) plays a fundamental role in subsequent discussions. We thus start by defining Bm. Let $(\Omega, \mathcal{F}, \mathcal{P})$ be a probability space on which Bm is defined.

Definition 1.1 (Brownian motion) The stochastic process $\{W(t)\}$ for $t \in [0, 1]$ is called Bm if the following conditions are satisfied.

(1) $P(W(0) = 0) = 1$.
(2) Increments $W(t_1) - W(t_0), W(t_2) - W(t_1), \ldots, W(t_n) - W(t_{n-1})$ are independent for any positive integer n and time points $0 \le t_0 < t_1 < \cdots < t_n \le 1$.

(3) $W(t) - W(s) \sim \mathrm{N}(0,\ t-s)$ for any $0 \le s < t \le 1$, where $\mathrm{N}(\mu,\ \sigma^2)$ stands for a normal distribution with mean μ and variance σ^2.

Note that the parameter t denotes time and extends over the unit interval $[0, 1]$ throughout this book, unless otherwise stated. Note also that we shall always use $\{W(t)\}$ for Bm without mentioning its definition. It follows from (3) above that Bm is a zero-mean Gaussian process, which means that the finite-dimensional distributions of $W(t_1), \ldots, W(t_n)$ for each collection $0 \le t_1 < \cdots < t_n \le 1$ are multivariate normal with means 0 for all n. Thus it suffices to know only the covariance $\mathrm{Cov}(W(s),\ W(t))$ to determine the distributions. We have from (2) above, for $s < t$,

$$\begin{aligned}\mathrm{Cov}(W(s),\ W(t)) &= \mathrm{E}(W(s)\,W(t)) = \mathrm{E}\left[W(s)\,(W(t) - W(s) + W(s))\right]\\ &= \mathrm{E}\left[W(s)\,(W(t) - W(s))\right] + \mathrm{E}(W^2(s)) = s,\end{aligned}$$

so that $\mathrm{Cov}(W(s),\ W(t)) = \min(s,t)$. It also follows from (2) that the increment $W(t) - W(s)$ for $t > s$ is independent of the past $W(u)$ for $u \le s$. More generally $W(t)$ is independent of $\mathcal{F}_s$, the σ-field generated by $W(u)$ for $u \le s$, which implies that Bm is a *martingale* satisfying

$$\mathrm{E}(|W(t)|) = \sqrt{\frac{2t}{\pi}} < \infty, \quad \mathrm{E}(W(t)|\mathcal{F}_s) = W(s) \quad (s \le t).$$

The mean square (m.s.) continuity of Bm is another consequence from the definition. It holds that

$$\lim_{h\to 0} \mathrm{E}\left[(W(t+h) - W(t))^2\right] = \lim_{h\to 0} |h| = 0,$$

which says that $W(t)$ is continuous at t in the m.s. sense. This is abbreviated as

$$\underset{h\to 0}{\mathrm{l.i.m.}}\ W(t+h) = W(t),$$

where "l.i.m." stands for limit in mean square. The second moment is necessary for the m.s. limit to exist. It follows that $W(t)$ is not m.s. differentiable in the sense that

$$\underset{h\to 0}{\mathrm{l.i.m.}}\ \frac{W(t+h) - W(t)}{h}$$

does not exist (Exercise 1.1.1 in this section).

Bm is also continuous with probability 1. In fact, it holds that

$$\mathrm{E}\left[(W(t) - W(s))^4\right] = 3\,(t-s)^2,$$

which implies that $W(t)$ is continuous with probability 1 because of Kolmogorov's continuity criterion (Liptser and Shiryaev, 2001a, p.23).

Some other properties (a) through (d) of Bm follow (Exercises 1.1.2, 1.1.3, 1.1.4).

(a) If $\{X(t)\}$ is a zero-mean Gaussian process with covariance function given by $\mathrm{Cov}(X(s), X(t)) = \min(s,t)$, then $\{X(t)\}$ is Bm.
(b) $W(t)$ is of unbounded variation, that is, for any $0 \le t_0 < t_1 < \cdots < t_n \le 1$, and $\Delta_n = \max_{1 \le k \le n}(t_k - t_{k-1})$,

$$\lim_{\Delta_n \to 0} \sum_{k=1}^{n} \mathrm{E}\left(|W(t_k) - W(t_{k-1})|\right) = \infty.$$

(c) The quadratic variation is finite and bounded away from 0. More specifically, for any $0 \le a = t_0 < t_1 < \cdots < t_n = b \le 1$, and $\Delta_n = \max_{1 \le k \le n}(t_k - t_{k-1})$,

$$\lim_{\Delta_n \to 0} \sum_{k=1}^{n} \mathrm{E}\left(|W(t_k) - W(t_{k-1})|^2\right) = b - a.$$

(d) The stochastic process $X(t)$ defined in the m.s. sense by

$$X(t) = \sum_{n=1}^{\infty} \frac{\sqrt{2}\,\sin\left(n - \frac{1}{2}\right)\pi\, t}{\left(n - \frac{1}{2}\right)\pi} Z_n \tag{1.1}$$

is Bm, where $\{Z_n\}$ is a sequence of independent N(0, 1) random variables, which is denoted as $\{Z_n\} \sim \mathrm{NID}(0,1)$. The expansion in (1.1) is called the *Karhunen–Loève expansion*, which will be explained in detail in Chapter 2.

If $W(t)$ is conditioned on $P(W(1) = 0) = 1$, the resulting process is called the *Brownian bridge* (Bb), which may be defined as follows:

Definition 1.2 (Brownian bridge) The stochastic process $\{\tilde{W}(t)\}$ for $t \in [0,1]$ is called the Bb if the following condition is satisfied.

(1) $\{\tilde{W}(t)\}$ is a zero-mean Gaussian process with covariance function given by $\mathrm{Cov}(\tilde{W}(s), \tilde{W}(t)) = \min(s,t) - st$.

The increments of the Bb are not independent, unlike Bm. In fact,

$$\mathrm{Cov}\left(\tilde{W}(t_2) - \tilde{W}(t_1), \tilde{W}(t_4) - \tilde{W}(t_3)\right) = -(t_2 - t_1)(t_4 - t_3) < 0$$

for $0 \le t_1 < t_2 < t_3 < t_4 \le 1$. Some other related properties (e), (f), and (g) of the Bb follow (Exercises 1.1.5 and 1.1.6).

(e) The stochastic process $\{W(t) - tW(1)\}$ is the Bb.

(f) $\mathrm{Cov}(W(s), W(t)|W(1) = 0) = \min(s, t) - st$.

(g) The stochastic process $X(t)$ defined in the m.s. sense by

$$X(t) = \sum_{n=1}^{\infty} \frac{\sqrt{2} \sin n\pi t}{n\pi} Z_n \tag{1.2}$$

is the Bb, where $\{Z_n\} \sim \mathrm{NID}(0, 1)$. This is the Karhunen–Loève expansion for the Bb.

We note in passing that Bm is a continuous-time version of the *random walk*. More specifically, suppose that the random walk is defined by

$$y_j = y_{j-1} + \varepsilon_j = \sum_{i=1}^{j} \varepsilon_i \qquad (j = 1, \ldots, n), \tag{1.3}$$

where $y_0 = 0$ and $\{\varepsilon_i\}$ is a sequence of independent and identically distributed random variables with mean 0 and variance σ^2, which is abbreviated as $\{\varepsilon_i\} \sim$ i.i.d.$(0, \sigma^2)$. Define also the *partial sum process*

$$\begin{aligned} Y_n(t) &= \frac{1}{\sqrt{n}\sigma} y_{j-1} + n\left(t - \frac{j-1}{n}\right) \frac{1}{\sqrt{n}\sigma} \varepsilon_j \\ &= \frac{1}{\sqrt{n}\sigma} y_{[nt]} + (nt - [nt]) \frac{1}{\sqrt{n}\sigma} \varepsilon_{[nt]+1}, \end{aligned} \tag{1.4}$$

where $(j-1)/n \le t \le j/n$ $(j = 1, \ldots, n)$, and $[x]$ denotes the largest integer not exceeding x. Note that $Y_n(t)$ is continuous on $[0, 1]$ by connecting $y_{j-1}/(\sqrt{n}\sigma)$ and $y_j/(\sqrt{n}\sigma)$ $(j = 1, \ldots, n)$. Then it holds that, as $n \to \infty$,

$$\{Y_n(t)\} \Rightarrow \{W(t)\}, \tag{1.5}$$

where $\Rightarrow$ stands for the *weak convergence*. This is the result established by Donsker (1951) and is referred to as *Donsker's theorem* or the *functional central limit theorem* (FCLT) because the weak convergence is concerned with the function space $C[0, 1]$, where $C[0, 1]$ is the space of all real-valued continuous functions defined on $[0, 1]$. Billingsley (1999) discusses details on the FCLT and weak convergence.

Our main subject in this book is to deal with functionals of Bm. In particular we consider functionals expressed by integrals. For this purpose we describe the stochastic integrals associated with Bm in the next section.

Exercises for Section 1.1

Exercise 1.1.1 Show that Bm is not m.s. differentiable.

Exercise 1.1.2 Prove that, if $\{X(t)\}$ is a zero-mean Gaussian process with covariance function given by $\mathrm{Cov}(X(s), X(t)) = \min(s, t)$, then $\{X(t)\}$ is Bm.

Exercise 1.1.3 Prove that

$$\lim_{\Delta_n \to 0} \sum_{k=1}^{n} \mathrm{E}\left(|W(t_k) - W(t_{k-1})|\right) = \infty,$$

$$\lim_{\Delta_n \to 0} \sum_{k=1}^{n} \mathrm{E}\left(|W(t_k) - W(t_{k-1})|^2\right) = b - a$$

for $a = t_0 < t_1 < \cdots < t_n = b$, and $\Delta_n = \max_{1 \le k \le n} (t_k - t_{k-1})$.

Exercise 1.1.4 Show that the stochastic process $X(t)$ defined in (1.1) is Bm, using the formula

$$\sum_{n=1}^{\infty} \frac{\cos\left(n - \frac{1}{2}\right)\pi x}{\left(n - \frac{1}{2}\right)^2 \pi^2} = \frac{1}{2}(1 - |x|) \qquad (|x| \le 2).$$

Exercise 1.1.5 Prove that Bm $W(t)$ conditioned on $W(1) = 0$ is the Bb. Prove also that the stochastic process $\{W(t) - tW(1)\}$ is a Bb that is independent of $W(1)$.

Exercise 1.1.6 Show that the stochastic process $X(t)$ defined in (1.2) is the Bb, using the formula

$$\sum_{n=1}^{\infty} \frac{\cos n\pi x}{n^2 \pi^2} = \frac{1}{4}(|x| - 1)^2 - \frac{1}{12} \qquad (|x| \le 2).$$

1.2 Mean Square Integral Associated with Brownian Motion

Stochastic integrals associated with Bm play an important role in the development of subsequent discussions. We define such integrals in the m.s. sense. For this purpose we introduce the space L_2, which is the space of random variables defined on a common probability space with finite second moment. It is known that the space L_2 is complete, so that, if $X_n \in L_2$ and $X = \operatorname*{l.i.m.}_{n \to \infty} X_n$ exists, X also belongs to L_2 (Loève, 1977, p.163). Moreover, X_n converges in the m.s. sense if and only if $\mathrm{E}(X_m X_n)$ converges as $m, n \to \infty$ in any manner. We also note that "l.i.m." and "expectation" commute in the following sense. If $\operatorname*{l.i.m.}_{n \to \infty} X_n = X$ and $\operatorname*{l.i.m.}_{n \to \infty} Y_n = Y$, it holds (Exercise 1.2.1) that

$$\lim_{n \to \infty} \mathrm{E}(X_n) = \mathrm{E}(X), \qquad \lim_{m,n \to \infty} \mathrm{E}(X_m Y_n) = \mathrm{E}(XY). \tag{1.6}$$

Using the above m.s. properties, we present stochastic integrals of two types in this section. Integrals of other types will be presented in the subsequent two sections.

1.2.1 Mean Square Riemann Integral

Let a stochastic process $\{X(t)\}$ belong to L_2. Then consider

$$Y = \int_0^1 X(t)\,dt = \underset{\substack{n\to\infty\\ \Delta_n\to 0}}{\text{l.i.m.}}\, Y_n, \quad Y_n = \sum_{i=1}^{n} X(t_i')(t_i - t_{i-1}), \tag{1.7}$$

where $\Delta_n = \max_{1\le i\le n}(t_i - t_{i-1})$ and $t_i' \in [t_{i-1}, t_i)$ $(i = 1, \ldots, n)$. The integral in (1.7) is called the *m.s. Riemann integral* of $X(t)$, and $X(t)$ is said to be m.s. Riemann integrable if $Y = \underset{n\to\infty}{\text{l.i.m.}}\, Y_n$ exists. It follows that a sufficient condition for the m.s. integral in (1.7) to be well defined is that $X(t)$ is m.s. continuous (Exercise 1.2.2). It can also be shown (Exercise 1.2.3) that, if $X(t)$ is m.s. continuous,

$$\mathrm{E}\left(\int_0^1 X(t)\,dt\right) = \int_0^1 \mathrm{E}(X(t))\,dt,$$

$$\mathrm{E}\left[\left(\int_0^1 X(t)\,dt\right)^2\right] = \int_0^1\int_0^1 \mathrm{E}(X(s)X(t))\,ds\,dt.$$

The following are three simple examples of the m.s. Riemann integral:

$$I_1 = \int_0^1 W(t)\,dt, \quad I_2 = \int_0^1 |W(t)|\,dt, \quad I_3 = \int_0^1 e^{W(t)}\,dt,$$

where I_1 is a linear functional of $W(t)$ so that it is normal because "l.i.m." and "normality" commute (Exercise 1.2.4), whereas I_2 and I_3 are not normal. We have (Exercise 1.2.5)

$$\mathrm{E}(I_1) = 0, \qquad \mathrm{E}(I_2) = \frac{2}{3}\sqrt{\frac{2}{\pi}}, \qquad \mathrm{E}(I_3) = 2(\sqrt{e} - 1),$$

$$\mathrm{E}(I_1^2) = \frac{1}{3}, \qquad \mathrm{E}(I_2^2) = \frac{3}{8}, \qquad \mathrm{E}(I_3^2) = \frac{2}{3}e^2 - \frac{8}{3}\sqrt{e} + 2.$$

The following are three examples of the m.s. Riemann integral of quadratic functions of Bm:

$$I_4 = \int_0^1 W^2(t)\,dt, \quad I_5 = \int_0^1 \tilde{W}^2(t)\,dt, \quad I_6 = \int_0^1 \bar{W}^2(t)\,dt,$$

where $\tilde{W}(t)$ is the Bb, whereas

$$\bar{W}(t) = W(t) - \int_0^1 W(s)\,ds$$

is called *demeaned Bm*. We have (Exercise 1.2.6)

$$\mathrm{E}(I_4) = \frac{1}{2}, \quad \mathrm{E}(I_4^2) = \frac{7}{12}, \quad \mathrm{E}(I_5) = \mathrm{E}(I_6) = \frac{1}{6}, \quad \mathrm{E}(I_5^2) = \mathrm{E}(I_6^2) = \frac{1}{20}.$$

It is noted that I_5 and I_6 have the same mean and variance. It will be shown later that the two have the same distribution.

Some other properties (a), (b), and (c) of the m.s. Riemann integral follow (Soong, 1973; Kuo, 2006; Klebaner, 2012).

(a) If $f(W(t))$ is m.s. continuous, $f(W(t))$ is m.s. Riemann integrable.
(b) The m.s. Riemann integral of $f(W(t))$, if it exists, is unique.
(c) If $f(W(t))$ is m.s. integrable, $X(t) = \int_0^t f(W(s))\,ds$ is m.s. differentiable with derivative given by $\dot{X}(t) = f(W(t))$.

1.2.2 Mean Square Riemann–Stieltjes Integral

Let $g(t)$ be a deterministic function of $t \in [0, 1]$. Then we define

$$A = \int_0^1 g(t)\,dW(t) = \underset{\substack{n\to\infty \\ \Delta_n\to 0}}{\text{l.i.m.}} \sum_{i=1}^{n} g(t_i')(W(t_i) - W(t_{i-1})), \tag{1.8}$$

$$B = \int_0^1 W(t)\,dg(t) = \underset{\substack{n\to\infty \\ \Delta_n\to 0}}{\text{l.i.m.}} \sum_{i=1}^{n} W(t_i')(g(t_i) - g(t_{i-1})), \tag{1.9}$$

where $t_i' \in [t_{i-1}, t_i)$ $(i = 1, \ldots, n)$. The integrals in (1.8) and (1.9) are called the *m.s. Riemann–Stieltjes integrals.* Normality is retained in these integrals with $\mathrm{E}(A) = \mathrm{E}(B) = 0$ (Exercise 1.2.7). Note that the Riemann–Stieltjes integral in (1.9) reduces to the Riemann integral in (1.7) when $g(t)$ is differentiable with $|g'(t)| < \infty$ on $[0, 1]$.

Because of the definition of the m.s. limit, the existence of the above integrals is determined by the existence of appropriate ordinary integrals, which leads us to

Theorem 1.3 *The m.s. Riemann–Stieltjes integrals in (1.8) and (1.9) are well defined if the following limits of the Riemann–Stieltjes double sums exist.*

$$\begin{aligned} E(A^2) &= \lim_{n\to\infty} \sum_{i=1}^{n}\sum_{j=1}^{n} g(t_i)g(t_j)E\left(\Delta W(t_i)\Delta W(t_j)\right) \\ &= \lim_{n\to\infty} \sum_{i=1}^{n} g^2(t_i)(t_i - t_{i-1}) \\ &= \int_0^1 g^2(t)\,dt, \end{aligned} \tag{1.10}$$

$$E(B^2) = \lim_{n\to\infty} \sum_{i=1}^{n} \sum_{j=1}^{n} E\left(W(t_i)W(t_j)\right) \Delta g(t_i)\Delta g(t_j)$$
$$= \lim_{n\to\infty} \sum_{i=1}^{n} \sum_{j=1}^{n} \min(t_i,t_j)\Delta g(t_i)\Delta g(t_j)$$
$$= \int_0^1 \int_0^1 \min(s,t)\, dg(s)\, dg(t), \tag{1.11}$$

where $\Delta W(t_i) = W(t_i) - W(t_{i-1})$ *and* $\Delta g(t_i) = g(t_i) - g(t_{i-1})$.

A sufficient condition for the existence of (1.8) is that $g(t)$ is continuous, whereas (1.9) exists if $g(t)$ is of bounded variation. Note that

$$A \sim \mathrm{N}\left(0, \int_0^1 g^2(t)\, dt\right), \quad B \sim \mathrm{N}\left(0, \int_0^1 \int_0^1 \min(s,t)\, dg(s)\, dg(t)\right). \tag{1.12}$$

Here are two examples of (1.8).

$$A_1 = \int_0^1 dW(t), \qquad A_2 = \int_0^1 (1-t)\, dW(t).$$

Then A_1 and A_2 are well defined with $g(t) = 1,\ 1-t$, respectively, in (1.8). Thus we have, from (1.10),

$$\mathrm{E}(A_1^2) = \mathrm{E}\left(\int_0^1 \int_0^1 dW(s)\, dW(t)\right) = \int_0^1 dt = 1,$$
$$\mathrm{E}(A_2^2) = \mathrm{E}\left(\int_0^1 \int_0^1 (1-s)(1-t) dW(s)\, dW(t)\right) = \int_0^1 (1-t)^2\, dt = \frac{1}{3}.$$

Here the Riemann–Stieltjes double integral appears, although its definition is deferred until the next section. In any case the above computation entails the following relation:

$$\mathrm{E}\left(dW(s)\, dW(t)\right) = \delta_{st}\, dt = \begin{cases} dt & (s=t), \\ 0 & (s \neq t). \end{cases}$$

An important and useful property associated with the two integrals in (1.8) and (1.9) is that the existence of one integral implies the other and vice versa, and they can be combined together in the following theorem for *integration by parts* (Soong, 1973, theorem 4.5.3; Kuo, 2006, p.9; Klebaner, 2012, p.12).

Theorem 1.4 *If either A in (1.8) or B in (1.9) exists, then both integrals exist, and*

$$\int_0^1 g(t)\,dW(t) = \big[g(t)W(t)\big]_0^1 - \int_0^1 W(t)\,dg(t). \tag{1.13}$$

For example, we have

$$A_3 = \int_0^1 dW(t) = [W(t)]_0^1 = W(1),$$

$$A_4 = \int_0^1 (1-t)\,dW(t) = [(1-t)W(t)]_0^1 - \int_0^1 (-1)W(t)\,dt = \int_0^1 W(t)\,dt.$$

As another example, let us consider, for $t \in [0,1]$,

$$F_g(t) = \int_0^t \frac{(t-s)^g}{g!}\,dW(s) \quad (g = 1,2,\dots), \quad F_0(t) = W(t). \tag{1.14}$$

This integral can be transformed into

$$\begin{aligned} F_g(t) &= \int_0^t \left(\int_s^t \frac{(u-s)^{g-1}}{(g-1)!}\,du\right) dW(s) = \int_0^t \left(\int_0^u \frac{(u-s)^{g-1}}{(g-1)!}\,dW(s)\right) du \\ &= \int_0^t F_{g-1}(u)\,du = \int_0^t \int_0^{t_1} \cdots \int_0^{t_{g-1}} W(t_g)\,dt_g\,dt_{g-1}\cdots dt_1. \end{aligned} \tag{1.15}$$

The stochastic process $\{F_g(t)\}$ is called *g-fold integrated Bm* because of this last expression, and it holds that

$$F_g(t) \sim \mathrm{N}\left(0, \frac{t^{2g+1}}{(2g+1)(g!)^2}\right) \quad (g = 0,1,\dots), \tag{1.16}$$

which means that the variation becomes smaller quickly as g gets large. Note that, unlike the Bm $F_0(t) = W(t)$, $F_g(t)$ for $g \geq 1$ is m.s. g-times continuously differentiable, and it holds that

$$\frac{dF_g(t)}{dt} = F_{g-1}(t), \qquad \int_0^t F_g(s)\,dF_g(s) = \frac{1}{2}F_g^2(t) \qquad (g \geq 1).$$

Figure 1.1 draws sample paths of $F_0(t)$, $F_1(t)$, and $F_2(t)$ with variations adjusted allowing for (1.16) so that the three sample paths can be compared on the same sheet without the vertical axis. These were generated by using the Karhunen–Loève expansion for Bm given in (1.1). It can be seen that the sample path of $F_2(t)$ is quite smooth and that of $F_1(t)$ is moderately smooth, whereas the Bm exhibits a zigzag fluctuation.

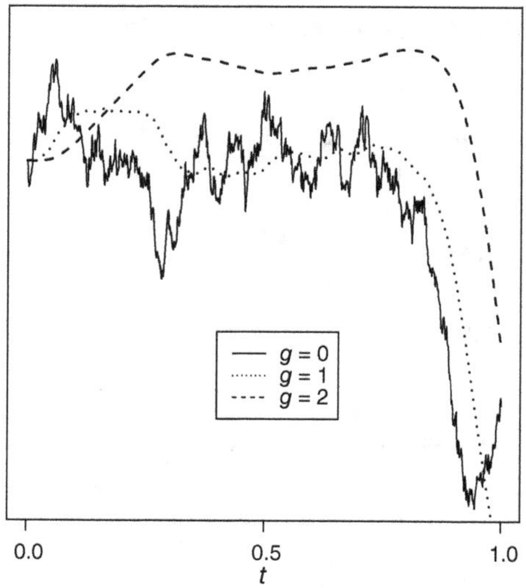

Figure 1.1 Sample paths of $\{F_g(t)\}$ for $g = 0, 1, 2$.

g-fold integrated Bm arises from the discrete-time process $\{y_j^{(g)}\}$ generated by

$$(1-L)^g\, y_j^{(g)} = \varepsilon_j \qquad (j = 1, \ldots, n), \tag{1.17}$$

where L is the *lag-operator* with the effect of lagging a variable $Ly_j = y_{j-1}, L^2 y_j = y_{j-2}$, and so on, whereas $\{\varepsilon_j\} \sim \text{i.i.d.}(0, \sigma^2)$. We assume that $y_{-k}^{(g)} = 0$ for $k = 0, 1, \ldots, g-1$. The process $\{y_j^{(g)}\}$ is called the I(g) process, where "I" stands for "integrated." Note that the random walk corresponds to the case of $g = 1$. It follows from (1.17) that

$$y_j^{(g)} = y_{j-1}^{(g)} + y_j^{(g-1)} = y_1^{(g-1)} + \cdots + y_j^{(g-1)}, \qquad y_j^{(0)} = \varepsilon_j.$$

Roughly speaking, the process $F_g(t)$ is related to $y_j^{(g)}$ in the following way. For Bm it holds that

$$W(t) = \int_0^t dW(u) \quad \Leftrightarrow \quad y_j^{(1)} = (1-L)^{-1}\varepsilon_j = \sum_{i=1}^{j} \varepsilon_i,$$

where $(1-L)^{-1}$ is interpreted as a cumulative sum operator. Similarly it holds that, for $F_1(t)$,

$$\int_0^t W(u)\, du = \int_0^t (t-u)\, dW(u) \;\Leftrightarrow\; y_j^{(2)} = (1-L)^{-2}\varepsilon_j \approx \sum_{i=1}^{j} (j-i)\varepsilon_i.$$

In general, we have, for $F_g(t)$,

$$\int_0^t \frac{(t-u)^g}{g!}\, dW(u) \Leftrightarrow y_j^{(g+1)} = (1-L)^{-(g+1)}\varepsilon_j \approx \frac{1}{g!}\sum_{i=1}^{j}(j-i)^g \varepsilon_i.$$

The above relationship between $F_g(t)$ and $y_j^{(g)}$ can be made exact in the following way. Let us put, for $g \geq 1$ and $(j-1)/n \leq t \leq j/n$,

$$\begin{aligned} Y_n^{(g)}(t) &= \frac{1}{n^{g-1/2}\sigma}\, y_{j-1}^{(g)} + n\left(t - \frac{j-1}{n}\right)\frac{1}{n^{g-1/2}\sigma}\, y_j^{(g-1)} \\ &= \frac{1}{n}\sum_{i=1}^{j-1} Y_n^{(g-1)}\left(\frac{i}{n}\right) + n\left(t - \frac{j-1}{n}\right)\frac{1}{n^{g-1/2}\sigma}\, y_j^{(g-1)}. \end{aligned}$$

Then it was shown by Chan and Wei (1988) that

$$\left\{Y_n^{(g)}(t)\right\} \Rightarrow \{F_{g-1}(t)\} \quad (g = 1, 2, \ldots), \tag{1.18}$$

which is an extended version of Donsker's FCLT in (1.5).

Exercises for Section 1.2

Exercise 1.2.1 If $\underset{n\to\infty}{\text{l.i.m.}}\, X_n = X$ and $\underset{n\to\infty}{\text{l.i.m.}}\, Y_n = Y$, prove that

$$\underset{n\to\infty}{\text{l.i.m.}}\, (aX_n + bY_n) = aX + bY, \qquad \lim_{m,n\to\infty} \mathrm{E}(X_m Y_n) = \mathrm{E}(XY),$$

where a and b are constants.

Exercise 1.2.2 Show that the m.s. integral $\int_0^1 X(t)\, dt$ is well defined if $X(t)$ is m.s. continuous.

Exercise 1.2.3 Show that, if $X(t)$ is m.s. continuous, it holds that

$$\mathrm{E}\left(\int_0^1 X(t)\, dt\right) = \int_0^1 \mathrm{E}(X(t))\, dt,$$

$$\mathrm{E}\left[\left(\int_0^1 X(t)\, dt\right)^2\right] = \int_0^1\int_0^1 \mathrm{E}(X(s)X(t))\, ds\, dt.$$

Exercise 1.2.4 Prove that, if $\{X_n(t)\}$ is a Gaussian process and $\underset{n\to\infty}{\text{l.i.m.}}\, X_n(t)$ exists, then $\{X(t)\}$ is also a Gaussian process.

Exercise 1.2.5 Compute the means and second moments of

$$I_1 = \int_0^1 W(t)\, dt, \quad I_2 = \int_0^1 |W(t)|\, dt, \quad I_3 = \int_0^1 e^{W(t)}\, dt.$$

Exercise 1.2.6 Compute the means and second moments of

$$I_4 = \int_0^1 W^2(t)\,dt, \quad I_5 = \int_0^1 (W(t) - tW(1))^2\,dt,$$

$$I_6 = \int_0^1 \left(W(t) - \int_0^1 W(s)\,ds\right)^2 dt.$$

Exercise 1.2.7 Prove that $\int_0^1 g(t)\,dW(t) \sim \mathrm{N}\left(0, \int_0^1 g^2(t)\,dt\right)$.

1.3 Statistics of Quadratic Functionals of Brownian Motion

Here we extend the m.s. Riemann–Stieltjes integral discussed in the previous section to the corresponding double integral, which will be frequently used in subsequent discussions. Let the partition $p_{m,n}$ of the unit square $[0,1] \times [0,1]$ be

$$p_{m,n} \; : \; 0 = s_0 < s_1 < \cdots < s_m = 1; \; 0 = t_0 < t_1 < \cdots < t_n = 1,$$

and put

$$\Delta_{m,n} = \max_{1 \le i, j \le n} \left(s_i - s_{i-1},\; t_j - t_{j-1}\right).$$

Then we consider, for a symmetric function $K(s,t)$ defined in $[0,1] \times [0,1]$,

$$\begin{aligned} X &= \int_0^1 \int_0^1 K(s,t)\,dW(s)\,dW(t) \\ &= \underset{\substack{m,n \to \infty \\ \Delta_{m,n} \to 0}}{\text{l.i.m.}} \sum_{i=1}^{m} \sum_{j=1}^{n} K\left(s_i', t_j'\right) \Delta W(s_i)\, \Delta W(t_j), \end{aligned} \tag{1.19}$$

where $s_i' \in [s_{i-1}, s_i)$, $t_j' \in [t_{j-1}, t_j)$, and $\Delta W(s_i) = W(s_i) - W(s_{i-1})$. The integral in (1.19) is called the *m.s. Riemann–Stieltjes double integral*, whereas X is referred to as a *quadratic functional* of Bm. In Section 1.2.1 we dealt with the m.s. Riemann integral of the quadratic functions of Bm. There is a close relationship between the two integrals, which will be explained later.

A sufficient condition for the existence of the integral in (1.19) is that $K(s,t)$ is continuous in $[0,1] \times [0,1]$. If $K(s,t)$ is symmetric and continuous, it holds (Exercise 1.3.1) that

$$\begin{aligned} \mathrm{E}(X) &= \int_0^1 K(t,t)\,dt, \\ \mathrm{Var}(X) &= 2\int_0^1 \int_0^1 K^2(s,t)\,ds\,dt = 4\int_0^1 \int_0^t K^2(s,t)\,ds\,dt. \end{aligned} \tag{1.20}$$

Moreover, the c.f. of X in (1.19) is given by

$$\mathrm{E}\left(e^{i\theta X}\right) = \mathrm{E}\left[\exp\left\{i\theta \int_0^1 \int_0^1 K(s,t)\,dW(s)\,dW(t)\right\}\right] = (D(2i\theta))^{-1/2}, \tag{1.21}$$

where $D(\lambda)$ is called the *Fredholm determinant* (FD) of $K(s,t)$. This result originates from Anderson and Darling (1952), whose proof will be given in Theorem 2.3 under an additional assumption on $K(s,t)$. It holds that the FD of $K(s,t)$ is the same as that of $K(1-s, 1-t)$ because of *Mercer's theorem* (Theorem 2.2). Thus we have

$$\int_0^1 \int_0^1 K(s,t)\,dW(s)\,dW(t) \stackrel{\mathcal{D}}{=} \int_0^1 \int_0^1 K(1-s, 1-t)\,dW(s)\,dW(t), \tag{1.22}$$

where $\stackrel{\mathcal{D}}{=}$ stands for distributional equivalence. The results described in (1.20) will also be derived easily via the properties of the FD (see Theorem 2.4).

Let us deal with some examples of (1.19). For this purpose we put, for $j = 1, 2, \ldots,$

$$X_j = \int_0^1 \int_0^1 K_j(s,t)\,dW(s)\,dW(t) \stackrel{\mathcal{D}}{=} \int_0^1 \int_0^1 K_j(1-s, 1-t)\,dW(s)\,dW(t). \tag{1.23}$$

Consider first X_1 with

$$K_1(s,t) = \sum_{k=1}^{n} g_k(s) g_k(t),$$

where $g_k(t)$ $(k = 1, \ldots, n)$ are continuous and linearly independent on [0, 1]. In this case $K_1(s,t)$ is said to be *degenerate*. To obtain the c.f. of X_1, we have

$$X_1 = \sum_{k=1}^{n} \int_0^1 \int_0^1 g_k(s) g_k(t)\,dW(s)\,dW(t) = \sum_{k=1}^{n} Y_k^2,$$

where

$$Y_k = \int_0^1 g_k(t)\,dW(t) \sim \mathrm{N}\left(0, \int_0^1 g_k^2(t)\,dt\right),$$

$$\mathrm{Cov}(Y_k, Y_\ell) = \int_0^1 g_k(t)\,g_\ell(t)\,dt.$$

Each Y_k^2 is a constant multiple of a $\chi^2(1)$ random variable, where $Y_1, \ldots, Y_n$ are not independent. We can compute the c.f. of X_1 following

$$\mathrm{E}\left(e^{i\theta X_1}\right) = |I_n - 2i\theta\Sigma|^{-1/2} = \prod_{k=1}^{n} (1 - 2i\theta\lambda_k)^{-1/2},$$

where I_n is the $n \times n$ identity matrix and Σ is the covariance matrix of $\boldsymbol{Y} = (Y_1, \dots, Y_n)'$, whereas $\lambda_1, \dots, \lambda_n$ are the eigenvalues of Σ, which are all positive. It holds that

$$X_1 \stackrel{\mathcal{D}}{=} \sum_{k=1}^{n} \lambda_k Z_k^2, \quad \boldsymbol{Z} = (Z_1, \dots, Z_n) \sim \mathrm{N}(\boldsymbol{0}, I_n).$$

According to the result in (1.21), the FD of $K_1(s,t)$ is given by

$$D_1(\lambda) = \left| I_n - \lambda \Sigma \right| = \prod_{k=1}^{n} \left(1 - \lambda \lambda_k\right), \tag{1.24}$$

where it is noted that $\lambda_1, \lambda_2, \dots, \lambda_n$ are the solutions to $D_1\left(\lambda^{-1}\right) = 0$.

As a simple example of X_1, let us consider X_2 with

$$K_2(s,t) = 1 + st, \quad g_1(t) = 1, \quad g_2(t) = t. \tag{1.25}$$

Then we obtain (Exercise 1.3.2)

$$\begin{aligned}
\mathrm{E}(X_2) &= \frac{4}{3}, \quad \mathrm{Var}(X_2) = \frac{29}{9}, \\
\mathrm{E}\left(e^{i\theta X_2}\right) &= (D_2(2i\theta))^{-1/2} = \left(1 - \frac{8}{3} i\theta - \frac{1}{3}\theta^2\right)^{-1/2}, \\
X_2 &\stackrel{\mathcal{D}}{=} \lambda_1 Z_1^2 + \lambda_2 Z_2^2 \quad \left(\lambda_1, \lambda_2 = \frac{2}{3} \pm \frac{\sqrt{13}}{6}\right),
\end{aligned}$$

where $D_2(\lambda) = 1 - 4\lambda/3 + \lambda^2/12$, which is the FD of $K_2(s,t)$, whereas λ_1 and λ_2 are the solutions to $D_2\left(\lambda^{-1}\right) = 0$.

As another example, consider X_3 with

$$K_3(s,t) = 1 + st + s^2t^2, \quad g_1(t) = 1, \quad g_2(t) = t, \quad g_3(t) = t^2. \tag{1.26}$$

We have

$$\begin{aligned}
\mathrm{E}(X_3) &= \frac{23}{15}, \quad \mathrm{Var}(X_3) = \frac{1199}{300}, \\
\mathrm{E}\left(e^{i\theta X_3}\right) &= (D_3(2i\theta))^{-1/2} = \left(1 - \frac{46}{15} i\theta - \frac{127}{180}\theta^2 + \frac{1}{270} i\theta^3\right)^{-1/2}, \\
X_3 &\stackrel{\mathcal{D}}{=} \lambda_1 Z_1^2 + \lambda_2 Z_2^2 + \lambda_3 Z_3^2,
\end{aligned}$$

where $D_3(\lambda) = 1 - 23\lambda/15 + 127\lambda^2/720 - \lambda^3/2160$, which is the FD of $K_3(s,t)$, whereas λ_1, λ_2, and λ_3 are the solutions to $D_3\left(\lambda^{-1}\right) = 0$.

In the above examples, the number of solutions to $D_j(\lambda) = 0$ $(j = 1, 2, 3)$ is finite. For these cases the functions $K_j(s,t)$ $(j = 1, 2, 3)$ are degenerate.

The next example deals with the case where the function $K(s,t)$ is *non-degenerate*, that is, the case where the number of zeros of the FD is infinite. Consider

$$K_4(s,t) = \sum_{n=1}^{\infty} \frac{f_n(s)\, f_n(t)}{\lambda_n}, \tag{1.27}$$

where λ_n is positive for all n and the sum converges absolutely and uniformly in the unit square. It is also assumed that $\{f_n(t)\}$ is an *orthonormal* sequence of functions on [0, 1] that satisfies

$$\int_0^1 f_m(t)\, f_n(t)\, dt = \delta_{mn} = \begin{cases} 1 & (m = n), \\ 0 & (m \neq n). \end{cases}$$

Then we can change the order of the sum and integration to get

$$X_4 = \int_0^1 \int_0^1 \sum_{n=1}^{\infty} \frac{f_n(s)\, f_n(t)}{\lambda_n}\, dW(s)\, dW(t) = \sum_{n=1}^{\infty} \frac{1}{\lambda_n} Z_n^2,$$

where

$$Z_n = \int_0^1 f_n(t)\, dW(t) \sim \mathrm{NID}(0,1), \quad \mathrm{E}(X_4) = \sum_{n=1}^{\infty} \frac{1}{\lambda_n}, \quad \mathrm{Var}(X_4) = \sum_{n=1}^{\infty} \frac{2}{\lambda_n^2}.$$

The c.f. of X_4 is obtained as

$$\mathrm{E}\left(e^{i\theta X_4}\right) = \mathrm{E}\left[\exp\left\{ i\theta \sum_{n=1}^{\infty} \frac{1}{\lambda_n} Z_n^2 \right\}\right] = \prod_{n=1}^{\infty} \left(1 - \frac{2i\theta}{\lambda_n}\right)^{-1/2}.$$

According to the result in (1.21), the FD of $K_4(s,t)$ is given by

$$D_4(\lambda) = \prod_{n=1}^{\infty} \left(1 - \frac{\lambda}{\lambda_n}\right),$$

where the solutions to $D_4(\lambda) = 0$ are λ_n $(n = 1, 2, \ldots)$.

As a specific example of $K_4(s,t)$, let us take up

$$K_5(s,t) = \sum_{n=1}^{\infty} \frac{2 \cos\left(n - \frac{1}{2}\right)\pi s \, \cos\left(n - \frac{1}{2}\right)\pi t}{\left(n - \frac{1}{2}\right)^2 \pi^2} = 1 - \max(s,t), \tag{1.28}$$

where $\left\{\sqrt{2}\cos\left(n-\frac{1}{2}\right)\pi t\right\}$ is orthonormal on $[0,1]$, and this last equality is ensured by Mercer's theorem (Theorem 2.2). Then we have, for the corresponding X_5,

$$\mathrm{E}(X_5) = \sum_{n=1}^{\infty} \frac{1}{\left(n-\frac{1}{2}\right)^2 \pi^2} = \int_0^1 (1-t)\,dt = \frac{1}{2},$$

$$\mathrm{Var}(X_5) = \sum_{n=1}^{\infty} \frac{2}{\left(n-\frac{1}{2}\right)^4 \pi^4} = 4\int_0^1\int_0^t (1-\max(s,t))^2\,ds\,dt = \frac{1}{3}.$$

The c.f. of X_5 is given by

$$\mathrm{E}\left(e^{i\theta X_5}\right) = \prod_{n=1}^{\infty}\left(1-\frac{2i\theta}{\left(n-\frac{1}{2}\right)^2\pi^2}\right)^{-1/2} = \left(\cos\sqrt{2i\theta}\right)^{-1/2}, \qquad (1.29)$$

where this last equality comes from the infinite product expansion formula for the cos function:

$$\cos z = \prod_{n=1}^{\infty}\left(1-\frac{z^2}{\left(n-\frac{1}{2}\right)^2\pi^2}\right). \qquad (1.30)$$

Thus the FD of $K_5(s,t) = 1-\max(s,t)$ is given by

$$D_5(\lambda) = \cos\sqrt{\lambda} = \prod_{n=1}^{\infty}\left(1-\frac{\lambda}{\left(n-\frac{1}{2}\right)^2\pi^2}\right),$$

where the solutions to $D_5(\lambda)=0$ are $(n-1/2)^2\pi^2$ $(n=1,2,\ldots)$.

We note here that

$$\begin{aligned}
&\int_0^1\int_0^1 K_5(s,t)\,dW(s)\,dW(t)\\
&= \int_0^1\int_0^1 [1-\max(s,t)]\,dW(s)\,dW(t)\\
&= \int_0^1\left(W(1) - t\int_0^t dW(s) - \int_t^1 s\,dW(s)\right)dW(t)\\
&= \int_0^1\left(\int_t^1 W(s)\,ds\right)dW(t) = \int_0^1 W^2(t)\,dt.
\end{aligned}$$

This relation may be established in a reversed direction in a general form as

$$\int_0^1 g(t)\, W^2(t)\, dt = \int_0^1 g(t) \left(\int_0^t \int_0^t dW(u)\, dW(v) \right) dt$$
$$= \int_0^1 \int_0^1 \left(\int_{\max(u,v)}^1 g(t)\, dt \right) dW(u)\, dW(v),$$

where it is justified to interchange the order of integration (Exercise 1.3.3). For example, we have

$$\int_0^1 t^{2k}\, W^2(t)\, dt = \int_0^1 \int_0^1 \frac{1}{2k+1} \left[1 - (\max(s,t))^{2k+1} \right] dW(s)\, dW(t),$$

where $k > -1/2$.

It is noted that, because of the distributional equivalence described in (1.22), the distribution of X_6 with

$$K_6(s,t) = K_5(1-s, 1-t) = \sum_{i=1}^{\infty} \frac{2 \sin\left(n - \frac{1}{2}\right)\pi s \, \sin\left(n - \frac{1}{2}\right)\pi t}{\left(n - \frac{1}{2}\right)^2 \pi^2}$$
$$= 1 - \max(1-s, 1-t) = \min(s,t) \tag{1.31}$$

is the same as that of X_5. Moreover, for X_6, it holds (Exercise 1.3.4) that

$$X_6 = \int_0^1 \int_0^1 \min(s,t)\, dW(s)\, dW(t) = \int_0^1 (W(1) - W(t))^2\, dt.$$

Thus it follows that

$$\int_0^1 W^2(t)\, dt = \int_0^1 \int_0^1 [1 - \max(s,t)]\, dW(s)\, dW(t)$$
$$\stackrel{\mathcal{D}}{=} \int_0^1 \int_0^1 \min(s,t)\, dW(s)\, dW(t)$$
$$= \int_0^1 (W(1) - W(t))^2\, dt, \tag{1.32}$$

which shows a close relationship between the m.s. Riemann integral of squared functions of Bm and the m.s. Riemann–Stieltjes double integral with respect to Bm. More generally, it will be shown in Theorem 2.11 that any zero-mean Gaussian process $\{X(t)\}$ defined on $[0, 1]$ satisfies

$$\int_0^1 X^2(t)\, dt \stackrel{\mathcal{D}}{=} \int_0^1 \int_0^1 \mathrm{Cov}\,(X(s), X(t))\, dW(s)\, dW(t). \tag{1.33}$$

Let us consider another example of the nondegenerate case for which

$$K_7(s,t) = \sum_{n=1}^{\infty} \frac{2 \sin n\pi s \, \sin n\pi t}{n^2\pi^2} = \min(s,t) - st, \tag{1.34}$$

where this last equality also comes from Mercer's theorem. Note here that $K_7(s,t) = K_7(1-s, 1-t)$, unlike $K_5(s,t) = 1 - \max(s,t)$. Then it is shown (Exercise 1.3.5) that the corresponding X_7 has the following properties:

$$\mathrm{E}(X_7) = \sum_{n=1}^{\infty} \frac{1}{n^2\pi^2} = \frac{1}{6}, \quad \mathrm{Var}(X_7) = \sum_{n=1}^{\infty} \frac{2}{n^4\pi^4} = \frac{1}{45},$$

$$\mathrm{E}\left(e^{i\theta X_7}\right) = \prod_{n=1}^{\infty} \left(1 - \frac{2i\theta}{n^2\pi^2}\right)^{-1/2} = \left(\frac{\sin\sqrt{2i\theta}}{\sqrt{2i\theta}}\right)^{-1/2}, \tag{1.35}$$

where this last equality comes from the infinite product expansion formula for the sin function:

$$\frac{\sin z}{z} = \prod_{n=1}^{\infty} \left(1 - \frac{z^2}{n^2\pi^2}\right). \tag{1.36}$$

Thus the FD of $K_7(s,t) = \min(s,t) - st$ is given by

$$D_7(\lambda) = \frac{\sin\sqrt{\lambda}}{\sqrt{\lambda}} = \prod_{n=1}^{\infty} \left(1 - \frac{\lambda}{n^2\pi^2}\right),$$

where the solutions to $D_7(\lambda) = 0$ are given by $n^2\pi^2$ $(n = 1, 2, \ldots)$. It is also shown (Exercise 1.3.6) that

$$\int_0^1 \left(W(t) - \int_0^1 W(s)\,ds\right)^2 dt = \int_0^1 W^2(t)\,dt - \left(\int_0^1 W(s)\,ds\right)^2$$
$$= \int_0^1 \int_0^1 [\min(s,t) - st]\,dW(s)\,dW(t),$$

where the integrand on the left-hand side is the square of demeaned Bm.

The next example gives the same FD as $K_7(s,t)$, which is given by

$$K_8(s,t) = \sum_{n=1}^{\infty} \frac{2 \cos n\pi s \, \cos n\pi t}{n^2\pi^2} = \frac{1}{3} - \max(s,t) + \frac{s^2 + t^2}{2}, \tag{1.37}$$

where this last equality comes from Mercer's theorem. Using the relation described in (1.33), we can relate the distribution of X_7 with that of X_8 as follows (Exercise 1.3.7):

$$\begin{aligned}
&\int_0^1\int_0^1 K_7(s,t)\,dW(s)\,dW(t)\\
&= \int_0^1\int_0^1 [\min(s,t) - st]\,dW(s)\,dW(t) = \int_0^1 \left(W(t) - \int_0^1 W(s)\,ds\right)^2 dt\\
&\stackrel{\mathcal{D}}{=} \int_0^1 (W(t) - tW(1))^2\,dt = \int_0^1\int_0^1 K_8(s,t)\,dW(s)\,dW(t), \qquad (1.38)
\end{aligned}$$

where $\{W(t) - tW(1)\}$ is the Bb.

The next example is concerned with the quadratic functional of the g-fold integrated Bm $\{F_g(t)\}$ discussed in the last section. Consider here the case of $g = 1$. Then we have

$$\begin{aligned}
X_9 &= \int_0^1 F_1^2(t)\,dt = \int_0^1 \left(\int_0^t (t-s)\,dW(s)\right)^2 dt\\
&= \int_0^1\int_0^1 K_9(s,t)\,dW(s)\,dW(t) \qquad (1.39)\\
&\stackrel{\mathcal{D}}{=} X_{10} = \int_0^1\int_0^1 K_{10}(s,t)\,dW(s)\,dW(t), \qquad (1.40)
\end{aligned}$$

where, for $s \le t$,

$$K_9(s,t) = \frac{1}{6}(1-t)^2(t+2-3s), \qquad (1.41)$$

$$K_{10}(s,t) = \mathrm{Cov}\,(F_1(s), F_1(t)) = -\frac{1}{6}s^3 + \frac{1}{2}s^2 t. \qquad (1.42)$$

The c.f. of this statistic cannot be obtained by the approach discussed so far. However, it can be obtained by the method we shall use later, which yields

$$\mathrm{E}\left(e^{i\theta X_9}\right) = \mathrm{E}\left(e^{i\theta X_{10}}\right) = \left[\frac{1}{2}\left(1 + \cos(2i\theta)^{1/4}\cosh(2i\theta)^{1/4}\right)\right]^{-1/2}.$$

Thus the FD of $K_9(s,t)$ is equal to that of $K_{10}(s,t)$ and is given by

$$D_9(\lambda) = D_{10}(\lambda) = \frac{1}{2}\left(1 + \cos\lambda^{1/4}\cosh\lambda^{1/4}\right).$$

The computation of the c.f. becomes more complicated as g becomes large, which we discuss in detail in Chapter 2.

The class of quadratic functionals of Bm is quite large. We shall examine many quadratic functionals in later chapters.

Exercises for Section 1.3

Exercise 1.3.1 Suppose that

$$X = \int_0^1 \int_0^1 K(s,t)\,dW(s)\,dW(t),$$

where $K(s,t)$ is symmetric and continuous in $[0,1] \times [0,1]$. Then prove that

$$\mathrm{E}(X) = \int_0^1 K(t,t)\,dt,$$

$$\mathrm{Var}(X) = 2\int_0^1 \int_0^1 K^2(s,t)\,ds\,dt = 4\int_0^1 \int_0^t K^2(s,t)\,ds\,dt.$$

Exercise 1.3.2 For X_2 and X_3 defined in (1.23) with $K_2(s,t) = 1 + st$ and $K_3(s,t) = 1 + st + s^2t^2$, show that

$$\mathrm{E}(X_2) = \frac{4}{3}, \quad \mathrm{Var}(X_2) = \frac{29}{9}, \quad \mathrm{E}\left(e^{i\theta X_2}\right) = \left(1 - \frac{8}{3}i\theta - \frac{1}{3}\theta^2\right)^{-1/2},$$

$$\mathrm{E}(X_3) = \frac{23}{15}, \quad \mathrm{Var}(X_3) = \frac{1199}{300},$$

$$\mathrm{E}\left(e^{i\theta X_3}\right) = \left(1 - \frac{46}{15}i\theta - \frac{127}{180}\theta^2 + \frac{1}{270}i\theta^3\right)^{-1/2}.$$

Exercise 1.3.3 Justify the following computation:

$$\int_0^1 g(t)W^2(t)\,dt = \int_0^1 \int_0^1 \left(\int_{\max(s,t)}^1 g(u)\,du\right) dW(s)\,dW(t),$$

where $g(t)$ is continuous on $[0,1]$.

Exercise 1.3.4 Prove the following equality by showing that (i) the left-hand side leads to the right-hand side and (ii) the right-hand side leads to the left-hand side:

$$\int_0^1 \int_0^1 \min(s,t)\,dW(s)\,dW(t) = \int_0^1 (W(1) - W(t))^2\,dt.$$

Exercise 1.3.5 For X_7 defined in (1.23) with

$$K_7(s,t) = \sum_{n=1}^{\infty} \frac{2\sin n\pi s\,\sin n\pi t}{n^2\pi^2},$$

show that

$$\mathrm{E}(X_7) = \sum_{n=1}^{\infty} \frac{1}{n^2\pi^2} = \frac{1}{6}, \quad \mathrm{Var}(X_7) = \sum_{n=1}^{\infty} \frac{2}{n^4\pi^4} = \frac{1}{45},$$

$$\mathrm{E}\left(e^{i\theta X_7}\right) = \prod_{n=1}^{\infty} \left(1 - \frac{2i\theta}{n^2\pi^2}\right)^{-1/2} = \left(\frac{\sin\sqrt{2i\theta}}{\sqrt{2i\theta}}\right)^{-1/2}.$$

Exercise 1.3.6 Show that

$$\int_0^1 \left(W(t) - \int_0^1 W(s)\,ds\right)^2 dt = \int_0^1 \int_0^1 [\min(s,t) - st]\,dW(s)\,dW(t).$$

Exercise 1.3.7 Show that

$$\begin{aligned}&\int_0^1 (W(t) - tW(1))^2\,dt \\ &= \int_0^1 \int_0^1 \left[\frac{1}{3} - \max(s,t) + \frac{s^2+t^2}{2}\right] dW(s)\,dW(t) \\ &\overset{\mathcal{D}}{=} \int_0^1 \int_0^1 [\min(s,t) - st]\,dW(s)\,dW(t).\end{aligned}$$

1.4 Mean Square Itô Integral

Here we discuss the m.s. *Itô integral* in connection with quadratic functionals of Bm. Let $\{X(t)\}$ ($\in L_2$) be a stochastic process that is m.s. continuous and independent of $W(v) - W(u)$ for all $0 \le t \le u \le v \le 1$. Let the partition of $[0,1]$ be $p_n : 0 = t_0 < t_1 < \cdots < t_n = 1$, and put $\Delta_n = \max_{1\le i\le n}(t_i - t_{i-1})$. Then the m.s. Itô integral of $X(t)$ with respect to $W(t)$ is defined by

$$Y = \int_0^1 X(t)\,dW(t) = \underset{\substack{n\to\infty \\ \Delta_n\to 0}}{\mathrm{l.i.m.}} \sum_{i=1}^{n} X(t_{i-1})(W(t_i) - W(t_{i-1})). \tag{1.43}$$

The Itô integral defined in this way exists and is unique. For the proof, see Jazwinski (1970, theorem 4.2), Soong (1973, theorem 5.2.1), Kuo (2006, theorem 4.7.1), and Klebaner (2012, theorem 4.3). It evidently holds that

$$\mathrm{E}(Y) = 0, \quad \mathrm{Var}(Y) = \int_0^1 \mathrm{E}(X^2(t))\,dt.$$

We notice in (1.43) that the value of $X(t)$ taken in the interval $[t_{i-1}, t_i)$ is $X(t_{i-1})$, unlike in the m.s. Riemann or Riemann–Stieltjes integral. If we take t_i' ($\ne t_{i-1}$) in $[t_{i-1}, t_i)$, then the m.s. limit is different, which will be shown later.

According to the definition of the Itô integral, the following relations can be established (Exercise 1.4.1):

$$\int_0^1 X(t)\,(dW(t))^j = \begin{cases} \displaystyle\int_0^1 X(t)\,dt & (j = 2), \\ 0 & (j \geq 3). \end{cases} \tag{1.44}$$

In particular, putting $X(t) \equiv 1$, we formally have

$$(dW(t))^j = \begin{cases} dt & (j = 2), \\ 0 & (j \geq 3). \end{cases} \tag{1.45}$$

To examine some other properties, consider the stochastic process defined by

$$Y(t) = \int_0^t X(s)\,dW(s), \tag{1.46}$$

which is well defined for each $t \in [0, 1]$. It is clear that $\{Y(t)\}$ is m.s. continuous because, for $0 \leq t < t + h \leq 1$,

$$\begin{aligned} \mathrm{E}\left[(Y(t+h) - Y(t))^2\right] &= \mathrm{E}\left[\left(\int_t^{t+h} X(s)\,dW(s)\right)^2\right] \\ &= \int_t^{t+h} \mathrm{E}(X^2(s))\,ds \\ &\leq h \max_{t \leq s \leq t+h} \mathrm{E}(X^2(s)). \end{aligned}$$

The m.s. differentiability of $Y(t)$ is not ensured, but we express (1.46) as

$$Y(t) = \int_0^t X(s)\,dW(s) \quad \Leftrightarrow \quad dY(t) = X(t)\,dW(t). \tag{1.47}$$

The differential expression on the right is called the *stochastic differential*, which is always taken to imply the left-hand integral. It also holds that $\{Y(t)\}$ is a zero-mean martingale, that is, $\mathrm{E}(Y(t)) = 0$ and, for $s \leq t$,

$$\mathrm{E}[Y(t)|Y(s)] = Y(s) + \mathrm{E}[Y(t) - Y(s)|Y(s)] = Y(s).$$

As an example, let us consider

$$Y_1(t) = \int_0^t W(s)\,dW(s). \tag{1.48}$$

We cannot apply Theorem 1.4 for integration by parts to $Y_1(t)$. If it is applied, we have $Y_1(t) = W^2(t)/2$, which is evidently incorrect because $\{Y_1(t)\}$ is a zero-mean martingale, whereas $\mathrm{E}(W^2(t)/2) = t/2 \neq 0$. Let the partition of $[0,t]$ be $p_m(t)\ :\ 0 = t_0 < t_1 < \cdots < t_m = t$, and put $\Delta_m(t) = \max\limits_{1\le i\le m}(t_i - t_{i-1})$. Then, as $m\to\infty$ and $\Delta_m(t)\to 0$,

$$\begin{aligned}
Y_{m1}(t) &= \sum_{i=1}^{m} W(t_{i-1})(W(t_i) - W(t_{i-1})) \\
&= -\frac{1}{2}\left[\sum_{i=1}^{m}(W(t_i) - W(t_{i-1}))^2 - \sum_{i=1}^{m} W^2(t_i) + \sum_{i=1}^{m} W^2(t_{i-1})\right] \\
&= \frac{1}{2}W^2(t) - \frac{1}{2}\sum_{i=1}^{m}(W(t_i) - W(t_{i-1}))^2 \\
&\Rightarrow \frac{1}{2}W^2(t) - \frac{1}{2}t,
\end{aligned}$$

where the last result follows from the quadratic variation property of Bm described in Section 1.1. Thus it holds that

$$Y_1(t) = \int_0^t W(s)\,dW(s) = \underset{\substack{m\to\infty \\ \Delta_m(t)\to 0}}{\text{l.i.m.}}\ Y_{m1}(t) = \frac{1}{2}(W^2(t) - t), \tag{1.49}$$

and formally we have

$$d\left(\frac{1}{2}(W^2(t) - t)\right) = W(t)\,dW(t) \quad \Leftrightarrow \quad d(W^2(t)) = 2W(t)\,dW(t) + dt. \tag{1.50}$$

This is a simplified version of the Itô calculus discussed below. In connection with the random walk $\{y_j\}$ defined in (1.3), it holds (Exercise 1.4.2) that, as $n\to\infty$,

$$\frac{1}{n\sigma^2}\sum_{j=1}^{n} y_{j-1}\varepsilon_j \Rightarrow \frac{1}{2}(W^2(1) - 1) = \int_0^1 W(t)\,dW(t), \tag{1.51}$$

where $y_j = y_{j-1} + \varepsilon_j$, $y_0 = 0$ and $\{\varepsilon_j\} \sim$i.i.d.$(0,\sigma^2)$. Because of the *continuous mapping theorem* (CMT) and joint weak convergence (Billingsley, 1999, chapter 2; Phillips, 1987a), it also holds that

$$S_n = \frac{\frac{1}{n}\sum_{j=2}^{n} y_{j-1}\varepsilon_j}{\frac{1}{n^2}\sum_{j=2}^{n} y_{j-1}^2} \Rightarrow \frac{\int_0^1 W(t)\,dW(t)}{\int_0^1 W^2(t)\,dt}. \tag{1.52}$$

The distribution of this last random variable is called the *unit root distribution* because

$$S_n = n\left(\frac{\sum_{j=2}^n y_{j-1}y_j}{\sum_{j=2}^n y_{j-1}^2} - 1\right) = n(\hat{\rho} - 1),$$

where $\hat{\rho}$ is the least squares estimator (LSE) of ρ for the model $y_j = \rho y_{j-1} + \varepsilon_j$, with the true ρ being unity.

It was mentioned above that the m.s. limit of the sum that defines the Itô integral crucially depends on the choice of the value of the integrand on each interval. For example, we have (Exercise 1.4.3)

$$\underset{\substack{m\to\infty \\ \Delta_m(t)\to 0}}{\text{l.i.m.}} \sum_{i=1}^m W(\tau_{i-1})(W(t_i) - W(t_{i-1})) = \frac{1}{2}(W^2(t) - t) + \lambda t,$$

where $\tau_{i-1} = (1-\lambda)t_{i-1} + \lambda t_i$ for $0 \le \lambda \le 1$. The case of $\lambda = 0$ corresponds to the Itô integral, whereas the integral with $\lambda = 1/2$ is called the *Stratonovich integral*. When $\lambda = 0$, the above quantity becomes a zero-mean martingale.

We now discuss the *Itô calculus*. Let $\{X(t)\}$ be a stochastic process on $[0, 1]$ that belongs to L_2, and consider the integral equation of the form

$$X(t) = X(0) + \int_0^t \mu(X(s), s)\, ds + \int_0^t \sigma(X(s), s)\, dW(s). \qquad (1.53)$$

Notice that there are two types of m.s. integrals. One is the Riemann integral and the other the Itô integral. Whether these integrals can be defined in the m.s. sense, and, more importantly, whether this integral equation has a unique m.s. solution $X(t)$, can be answered in the affirmative by the following theorem (Jazwinski, 1970, theorem 4.5; Kuo, 2006, theorem 10.3.5).

Theorem 1.5 *Suppose that*

(i) $X(0)$ is any random variable that belongs to L_2 and is independent of $W(t) - W(s)$ for all $0 \le s \le t \le 1$.

(ii) There is a positive constant K such that

$$|\mu(x,t) - \mu(y,t)| \le K|x-y|, \quad |\sigma(x,t) - \sigma(y,t)| \le K|x-y|,$$

$$|\mu(x,s) - \mu(x,t)| \le K|s-t|, \quad |\sigma(x,s) - \sigma(x,t)| \le K|s-t|,$$

$$|\mu(x,t)| \le K(1+x^2)^{1/2}, \quad |\sigma(x,t)| \le K(1+x^2)^{1/2}.$$

Then (1.53) has a unique m.s. continuous solution $X(t)$ on $[0, 1]$ such that $X(t) - X(0)$ is independent of $W(v) - W(u)$ for all $0 \le t \le u \le v \le 1$.

This theorem ensures that the two integrals appearing in (1.53) are well defined in the m.s. sense. Note also that the solution process $\{X(t)\}$ is not m.s. differentiable in general. Nonetheless we write (1.53) as

$$dX(t) = \mu(X(t), t)\, dt + \sigma(X(t), t)\, dW(t). \tag{1.54}$$

Here $dX(t)$ is called the *stochastic differential* of $X(t)$, and we call this equation the Itô *stochastic differential equation* (SDE), which is always understood in terms of the integral equation (1.53).

The idea of stochastic differentials can further be developed for functions of $X(t)$ and t. Namely, we can consider another SDE that $f(X(t), t)$ satisfies on the basis of the following theorem (Jazwinski, 1970, lemma 4.2; Kuo, 2006, theorem 7.4.3; Klebaner, 2012, theorem 4.16).

Theorem 1.6 (Itô's formula) *Suppose that $X(t)$ has the stochastic differential (1.54), where the conditions (i) and (ii) in Theorem 1.5 are satisfied. Let $f(x, t)$ denote a continuous function in $(-\infty, \infty) \times [0, 1]$ with continuous partial derivatives $f_x = \partial f(x, t)/\partial x$, $f_{xx} = \partial^2 f(x, t)/\partial x^2$, and $f_t = \partial f(x, t)/\partial t$. Assume further that $f_t(X(t), t)$ and $f_{xx}(X(t), t)\, \sigma^2(X(t), t)$ are m.s. Riemann integrable. Then $f(X(t), t)$ has the stochastic differential*

$$df(X(t), t) = f_x(X(t), t)\, dX(t) + \left(f_t(X(t), t) + \frac{1}{2} f_{xx}(X(t), t)\sigma^2(X(t), t) \right) dt. \tag{1.55}$$

Some important implications of Itô's formula follow.

(a) If we expand $df(X(t), t)$ formally as

$$df \sim f_x\, dX(t) + f_t\, dt + \frac{1}{2} (f_{xx}\, d^2X(t) + 2 f_{xt}\, dX(t)\, dt + f_{tt}\, d^2 t),$$

then (1.55) implies that we may put $d^2X(t) = \sigma^2(X(t), t)\, dt$, $dX(t)\, dt = 0$ and $d^2 t = 0$.

(b) The integral equation corresponding to (1.55) may be written as

$$\int_a^b f_x(X(t), t)\, dX(t) = f(X(b), b) - f(X(a), a) - \int_a^b \left(f_t(X(t), t) + \frac{1}{2} f_{xx}(X(t), t)\, \sigma^2(X(t), t) \right) dt,$$

where $0 \le a \le t \le b \le 1$. The integral on the left-hand side has never appeared before, and may be called the extended Itô integral. The above relation tells us that the extended Itô integral can be computed via the

Riemann integral on the right-hand side. In particular, for $f(x,t) = x^2$, we have

$$\int_0^t X(s)\,dX(s) = \frac{1}{2}(X^2(t) - X^2(0)) - \frac{1}{2}\int_0^t \sigma^2(X(s),s)\,ds. \quad (1.56)$$

(c) For any integer $n \geq 2$, the following stochastic differential can be derived from (1.55) as

$$d(W^n(t)) = nW^{n-1}(t)\,dW(t) + \frac{n(n-1)}{2}W^{n-2}(t)\,dt. \quad (1.57)$$

In particular, when $n = 2$, we have

$$d(W^2(t)) = 2W(t)\,dW(t) + dt,$$

which was earlier presented in (1.50) as a simplified version of the Itô calculus. The relation (1.57) is equivalent to

$$\int_0^t W^n(s)\,dW(s) = \frac{1}{n+1}W^{n+1}(t) - \frac{n}{2}\int_0^t W^{n-1}(s)\,ds, \quad (1.58)$$

which reduces to (1.49) when $n = 1$. We see that the formulas (1.56) and (1.58) enable us to convert (extended) Itô integrals into the usual Riemann integrals.

(d) Itô's formula is useful for solving SDEs in conjunction with the uniqueness property of the solution described in Theorem 1.5, which will be exemplified shortly.

Two examples follow in connection with quadratic functionals associated with the Itô integral.

Example 1.7 Let us consider the SDE

$$d\left(g(t)W^2(t)\right) = 2g(t)W(t)\,dW(t) + (g'(t)W^2(t) + g(t))\,dt,$$

which yields (Exercise 1.4.4)

$$\int_0^1 g(t)W(t)\,dW(t) = \frac{1}{2}\int_0^1\int_0^1 g(\max(s,t))\,dW(s)\,dW(t) - \frac{1}{2}\int_0^1 g(t)\,dt.$$

In particular, we have

$$\int_0^1 t^k\, W(t)\,dW(t) = \int_0^1\int_0^1 \frac{1}{2}(\max(s,t))^k\,dW(s)\,dW(t) - \frac{1}{2(k+1)}, \quad (1.59)$$

where $k \geq 0$.

Example 1.8 Consider the SDE

$$dX(t) = \gamma\, X(t)\, dt + dW(t) \quad \Leftrightarrow \quad X(t) = e^{\gamma t} X(0) + e^{\gamma t} \int_0^t e^{-\gamma s}\, dW(s), \tag{1.60}$$

which is called the *Ornstein–Uhlenbeck (O–U) process* defined on [0, 1], where $X(0)$ is independent of $\{W(t)\}$. Then it follows from (1.56) that

$$\int_0^t X(s)\, dX(s) = \frac{1}{2}\,(X^2(t) - X^2(0) - t), \tag{1.61}$$

which is of the nature of the Itô integral. The O–U process has the following moments:

$$\mathrm{E}(X(t)) = e^{\gamma t}\mathrm{E}(X(0)),$$
$$\mathrm{Cov}(X(s),\, X(t)) = e^{\gamma(s+t)} \left[\mathrm{Var}(X(0)) + \frac{1 - e^{-2\gamma \min(s,t)}}{2\gamma} \right].$$

It turns out that, if γ is positive, $\{X(t)\}$ is explosive and $\mathrm{Var}(X(t))$ increases with t. On the other hand, if γ is negative and $X(0) \sim \mathrm{N}(0, -1/(2\gamma))$, then $\{X(t)\}$ is stationary with $\mathrm{E}(X(t)) = 0$ and $\mathrm{Cov}(X(s),\, X(t)) = -e^{\gamma|s-t|}/(2\gamma)$ so that $\mathrm{Var}(X(t)) = -1/(2\gamma)$ (Exercise 1.4.5).

The O–U process $\{X(t)\}$ in (1.60) arises from the *near-random walk* defined by

$$y_j = \rho\, y_{j-1} + \varepsilon_j \qquad (j = 1, \ldots, n), \quad y_0 = 0, \tag{1.62}$$

where $\rho = 1 + \gamma/n$ and $\{\varepsilon_j\} \sim \mathrm{i.i.d.}(0, \sigma^2)$. In fact, if we construct the partial sum process

$$X_n(t) = \frac{1}{\sqrt{n}\sigma}\, y_{[nt]} + (nt - [nt])\frac{y_{[nt]+1} - y_{[nt]}}{\sqrt{n}\sigma}, \tag{1.63}$$

it holds (Bobkoski, 1983; Phillips, 1987b) that

$$\{X_n(t)\} \Rightarrow \{X(t)\}, \quad X(t) = e^{\gamma t} \int_0^t e^{-\gamma s}\, dW(s). \tag{1.64}$$

In particular, the CMT gives

$$\frac{1}{n\sigma^2}\, y_n^2 = X_n^2(1) \Rightarrow X^2(1) = \left(\int_0^1 e^{\gamma(1-s)}\, dW(s) \right)^2. \tag{1.65}$$

Let us consider the asymptotic distribution of the LSE $\hat{\rho}$ of ρ in (1.62), where

$$n(\hat{\rho}-1) = n\left(\frac{\sum_{j=2}^{n} y_{j-1}y_j}{\sum_{j=2}^{n} y_{j-1}^2} - 1\right) = \frac{\frac{1}{n}\sum_{j=2}^{n} y_{j-1}(y_j - y_{j-1})}{\frac{1}{n^2}\sum_{j=2}^{n} y_{j-1}^2}.$$

The FCLT and CMT lead us to

$$\begin{aligned}
\frac{1}{n\sigma^2}\sum_{j=2}^{n} y_{j-1}(y_j - y_{j-1}) &= -\frac{1}{2n\sigma^2}\left(\sum_{j=2}^{n}(y_j - y_{j-1})^2 - \sum_{j=2}^{n} y_j^2 + \sum_{j=2}^{n} y_{j-1}^2\right) \\
&= -\frac{1}{2n\sigma^2}\left(\sum_{j=2}^{n}\left(\frac{\gamma}{n}y_{j-1} + \varepsilon_j\right)^2 - y_n^2 + y_1^2\right) \\
&\Rightarrow \frac{1}{2}(X^2(1) - 1) = \int_0^1 X(t)\,dX(t),
\end{aligned}$$

$$\frac{1}{n^2\sigma^2}\sum_{j=2}^{n} y_j^2 = \frac{1}{n}\sum_{j=2}^{n} X_n^2\left(\frac{j}{n}\right) \Rightarrow \int_0^1 X^2(t)\,dt.$$

Thus, because of joint weak convergence (Phillips, 1987b), we have

$$n(\hat{\rho}-1) \Rightarrow R = \frac{\int_0^1 X(t)\,dX(t)}{\int_0^1 X^2(t)\,dt}. \tag{1.66}$$

The distribution of R in (1.66) is called the *near-unit root distribution*, which will be examined in detail in Chapter 7.

To describe an extended version of Itô's formula, we introduce

Definition 1.9 (Vector Brownian motion) A multiple stochastic process $\boldsymbol{W}(t) = (W_1(t), \ldots, W_q(t))'$ is q-dimensional Bm that satisfies

(1) $P(\boldsymbol{W}(0) = \boldsymbol{0}) = 1$.
(2) $\boldsymbol{W}(t_1) - \boldsymbol{W}(t_0)$, $\boldsymbol{W}(t_2) - \boldsymbol{W}(t_1)$, $\ldots$, $\boldsymbol{W}(t_n) - \boldsymbol{W}(t_{n-1})$ are independent for any positive integer n and time points $t_0 < t_1 < \cdots < t_n$.
(3) $\boldsymbol{W}(t) - \boldsymbol{W}(s) \sim \mathrm{N}(\boldsymbol{0}, (t-s)I_q)$ for any $0 \le s < t \le 1$.

We now extend Itô's formula to the vector case (Jazwinski, 1970, lemma 4.2; Kuo, 2006, p.107; Klebaner, 2012, p.119).

Theorem 1.10 (An extended version of Itô's formula) *Let the p-dimensional stochastic process $\{\boldsymbol{Y}(t)\}$ be the unique solution of the Itô SDE :*

$$d\boldsymbol{Y}(t) = \boldsymbol{\mu}(\boldsymbol{Y}(t), t)dt + G(\boldsymbol{Y}(t), t)\, d\boldsymbol{W}(t), \tag{1.67}$$

where $\boldsymbol{\mu}$ is $p\times 1$ and G is $p\times q$. Let $g(\boldsymbol{y}, t)$ be a real-valued function of $\boldsymbol{y}$: $p\times 1$ and $t \in [0, 1]$ with continuous partial derivatives $g_y = \partial g/\partial \boldsymbol{y}$: $p \times 1$, $g_{yy} = \partial^2 g/(\partial \boldsymbol{y}\partial \boldsymbol{y}')$: $p \times p$, and $g_t = \partial g/\partial t$. Assume further that g_t and $GG'g_{yy}$ are m.s. Riemann integrable. Then $g(\boldsymbol{Y}(t), t)$ has the stochastic differential

$$dg = g_y' d\boldsymbol{Y}(t) + \left(g_t + \frac{1}{2}\,\mathrm{tr}(GG'g_{yy})\right) dt. \tag{1.68}$$

Some applications of Theorem 1.10 follow. Putting $g(\boldsymbol{Y}(t), t) = \boldsymbol{W}'(t)H\boldsymbol{W}(t)$ with H being a $q \times q$ symmetric matrix, it follows from (1.68) that

$$d(\boldsymbol{W}'(t)H\boldsymbol{W}(t)) = 2\boldsymbol{W}'(t)Hd\boldsymbol{W}(t) + \mathrm{tr}\,(H)\, dt,$$

which yields

$$S_1 = \int_0^1 \boldsymbol{W}'(t)H\, d\boldsymbol{W}(t) = \frac{1}{2}\left(\boldsymbol{W}'(1)H\boldsymbol{W}(1) - \mathrm{tr}\,(H)\right). \tag{1.69}$$

Thus it is easy to derive the c.f. of S_1 (Exercise 1.4.6). Note that, if $H = I_q$, we have

$$\begin{aligned}\int_0^1 \boldsymbol{W}'(t)\, d\boldsymbol{W}(t) &= \sum_{k=1}^{q} \int_0^1 W_k(t)\, dW_k(t)\\ &= \sum_{k=1}^{q} \frac{1}{2}\left(W_k^2(1) - 1\right) = \frac{1}{2}\left(\boldsymbol{W}'(1)\boldsymbol{W}(1) - q\right),\end{aligned}$$

which may be established from the scalar Itô integral in (1.49).

As another simple case of (1.69), consider the case of $q = 2$, where

$$S_2 = \int_0^1 \boldsymbol{W}'(t)\begin{pmatrix} 0 & 1\\ 1 & 0\end{pmatrix} d\boldsymbol{W}(t) = \int_0^1 (W_1(t)\, dW_2(t) + W_2(t)\, dW_1(t)). \tag{1.70}$$

Then we have (Exercise 1.4.7)

$$S_2 = \frac{1}{2}\boldsymbol{W}'(1)\begin{pmatrix} 0 & 1\\ 1 & 0\end{pmatrix}\boldsymbol{W}(1) = W_1(1)W_2(1) \stackrel{\mathcal{D}}{=} \frac{1}{2}\left(W_1^2(1) - W_2^2(1)\right).$$

If the matrix H is not symmetric, the formula (1.69) is invalid. As an example, consider the case of $q = 2$, where

$$S_3 = \int_0^1 \mathbf{W}'(t) \begin{pmatrix} 0 & 1/2 \\ -1/2 & 0 \end{pmatrix} d\mathbf{W}(t)$$
$$= \frac{1}{2} \int_0^1 \Big(W_1(t)\, dW_2(t) - W_2(t)\, dW_1(t) \Big), \tag{1.71}$$

which is called *Lévy's stochastic area.* If we apply the right-hand side of (1.69) to this integral, it reduces to 0, which is evidently incorrect.

To derive the distribution of S_3, we note that the conditional distribution of S_3 given $\{W_1(t)\}$ is normal with mean 0. Since it holds that

$$\mathrm{E}(W_2(s)dW_2(t)) = \begin{cases} 0 & (s < t), \\ dt & (s \geq t), \end{cases}$$

we have

$$\mathrm{Var}\Big(S_3 | \{W_1(t)\}\Big)$$
$$= \frac{1}{4}\mathrm{E}\Big[\int_0^1 \int_0^1 \Big(W_1(s)\, dW_2(s) - W_2(s)\, dW_1(s)\Big)$$
$$\times \Big(W_1(t)\, dW_2(t) - W_2(t)\, dW_1(t)\Big) \Big| \{W_1(t)\} \Big]$$
$$= \frac{1}{4} \Big[\int_0^1 W_1^2(t)\, dt - 2 \int\int_{0 \leq s \leq t \leq 1} W_1(s)\, ds\, dW_1(t)$$
$$+ \int_0^1 \int_0^1 \min(s,t)\, dW_1(s)\, dW_1(t) \Big]$$
$$= \frac{1}{4} \int_0^1 \Big[W_1^2(t) - 2(W_1(1) - W_1(t))W_1(t) + (W_1(1) - W_1(t))^2 \Big] dt$$
$$= \int_0^1 \Big(W_1(t) - \frac{1}{2} W_1(1) \Big)^2 dt$$
$$\stackrel{\mathcal{D}}{=} \int_0^1 \int_0^1 \frac{1}{4} \Big[1 - 2|s - t| \Big] dW_1(s) dW_1(t). \tag{1.72}$$

Thus it is possible to obtain, using the conditional argument and the result in (1.21),

$$\begin{aligned}
\mathrm{E}\left(e^{i\theta S_3}\right) &= \mathrm{E}\left[\mathrm{E}\left(e^{i\theta S_3}\Big|\{W_1(t)\}\right)\right] = \mathrm{E}\left[\exp\left\{-\frac{\theta^2}{2}\,\mathrm{Var}\,(S_3|\{W_1(t)\})\right\}\right] \\
&= \mathrm{E}\left[\exp\left\{-\frac{\theta^2}{2}\int_0^1\left(W_1(t)-\frac{1}{2}W_1(1)\right)^2 dt\right\}\right] \\
&= \mathrm{E}\left[\exp\left\{-\frac{\theta^2}{2}\int_0^1\int_0^1\frac{1}{4}\Big[1-2|s-t|\Big]\,dW_1(s)\,dW_1(t)\right\}\right] \\
&= \left(D(-\theta^2)\right)^{-1/2} = \left(\cosh\frac{\theta}{2}\right)^{-1},
\end{aligned}$$

where $D(\lambda)=\left(\cos(\sqrt{\lambda}/2)\right)^2$ is the FD of $K(s,t)=(1-2|s-t|)/4$, and this last result will be proved in Section 2.4. Assuming this result, it can be shown that

$$S_3 \stackrel{\mathcal{D}}{=} \frac{1}{4}\sum_{n=1}^{\infty}\frac{Z_{n1}^2+Z_{n2}^2-Z_{n3}^2-Z_{n4}^2}{\left(n-\frac{1}{2}\right)\pi},$$

where $\left(Z_{n1},\ Z_{n2},\ Z_{n3},\ Z_{n4}\right)'\sim \mathrm{NID}(\mathbf{0},\ I_4)$. It is interesting to note that S_2 and S_3 resemble each other closely, but the two distributions are totally different. The distribution of S_2 is that of the weighted sum of a finite number of $\chi^2(1)$ random variables, whereas that of S_3 is the weighted sum of an infinite number of $\chi^2(2)$ random variables.

If we consider a simplified version of S_3 defined by

$$S_4=\int_0^1 \mathbf{W}'(t)\begin{pmatrix}0 & 1/2\\ 0 & 0\end{pmatrix} d\mathbf{W}(t)=\frac{1}{2}\int_0^1 W_1(t)\,dW_2(t), \tag{1.73}$$

it can be shown (Exercise 1.4.8) that, for $\left(Z_{n1},\ Z_{n2}\right)'\sim \mathrm{NID}(\mathbf{0},\ I_2)$,

$$S_4 \stackrel{\mathcal{D}}{=} \frac{1}{4}\sum_{n=1}^{\infty}\frac{Z_{n1}^2-Z_{n2}^2}{\left(n-\frac{1}{2}\right)\pi}.$$

Thus we conclude that S_3 is a convolution of S_4, which gives

$$S_3 \stackrel{\mathcal{D}}{=} \frac{1}{2}\int_0^1\left(W_1(t)\,dW_2(t)+W_3(t)\,dW_4(t)\right),$$

where $\left(W_1(t),\ W_2(t),\ W_3(t),\ W_4(t)\right)'$ is four-dimensional Bm. The c.f. of S_4 is given by

$$\mathrm{E}\left(e^{i\theta S_4}\right) = \prod_{n=1}^{\infty}\left\{\left(1 - \frac{i\theta/2}{\left(n-\frac{1}{2}\right)\pi}\right)\left(1 + \frac{i\theta/2}{\left(n-\frac{1}{2}\right)\pi}\right)\right\}^{-1/2}$$

$$= \prod_{n=1}^{\infty}\left(1 + \frac{\theta^2/4}{\left(n-\frac{1}{2}\right)^2\pi^2}\right)^{-1/2} = \left(\cosh\frac{\theta}{2}\right)^{-1/2},$$

so that the c.f. of S_3 is given by

$$\mathrm{E}\left(e^{i\theta S_3}\right) = \left[\mathrm{E}\left(e^{i\theta S_4}\right)\right]^2 = \left(\cosh\frac{\theta}{2}\right)^{-1}. \tag{1.74}$$

Exercises for Section 1.4

Exercise 1.4.1 Prove that

$$\int_0^1 X(t)\,(dW(t))^j = \begin{cases} \displaystyle\int_0^1 X(t)\,dt & (j=2), \\ 0 & (j \geq 3). \end{cases}$$

Exercise 1.4.2 Prove that

$$\frac{1}{n\sigma^2}\sum_{j=1}^{n} y_{j-1}\varepsilon_j \Rightarrow \frac{1}{2}(W^2(1)-1) = \int_0^1 W(t)\,dW(t),$$

where $y_j = y_{j-1} + \varepsilon_j$, $y_0 = 0$ and $\{\varepsilon_j\} \sim \text{i.i.d.}(0, \sigma^2)$.

Exercise 1.4.3 Show that

$$\underset{\substack{m\to\infty \\ \Delta_m(t)\to 0}}{\text{l.i.m.}} \sum_{i=1}^{m} W(t_i')(W(t_i) - W(t_{i-1})) = \frac{1}{2}(W^2(t) - t) + \lambda t,$$

where $0 = t_0 < t_1 < \cdots < t_m = t$, $\Delta_m(t) = \max_{1\leq i\leq m}(t_i - t_{i-1})$ and $t_i' = (1-\lambda)t_{i-1} + \lambda t_i$ for $0 \leq \lambda \leq 1$.

Exercise 1.4.4 Show that

$$\int_0^1 g(t)W(t)\,dW(t)$$
$$= \frac{1}{2}\int_0^1\int_0^1 g(\max(s,t))\,dW(s)\,dW(t) - \frac{1}{2}\int_0^1 g(t)\,dt,$$

where $g(t)$ is an ordinary Riemann integrable function.

Exercise 1.4.5 For the O–U process $dX(t) = \gamma X(t)\,dt + dW(t)$ with $X(0) = 0$, prove the following relations:

$$\int_0^1 X^2(t)\,dt = \int_0^1\int_0^1 \frac{e^{\gamma(2-s-t)} - e^{\gamma|s-t|}}{2\gamma}\,dW(s)\,dW(t)$$
$$\stackrel{\mathcal{D}}{=} \int_0^1\int_0^1 \frac{e^{\gamma(s+t)} - e^{\gamma|s-t|}}{2\gamma}\,dW(s)\,dW(t).$$

Exercise 1.4.6 Derive the c.f. of

$$S_1 = \int_0^1 \boldsymbol{W}'(t) H\,d\boldsymbol{W}(t),$$

where $\boldsymbol{W}(t)$ is q-dimensional Bm, and H is a $q \times q$ symmetric matrix.

Exercise 1.4.7 Show that

$$S_2 = \int_0^1 \Big(W_2(t)\,dW_1(t) + W_1(t)\,dW_2(t)\Big) \stackrel{\mathcal{D}}{=} \frac{1}{2}\Big(W_1^2(1) - W_2^2(1)\Big),$$

where $\boldsymbol{W}(t) = \big(W_1(t), W_2(t)\big)'$ is two-dimensional Bm.

Exercise 1.4.8 Using (1.29) and the conditional argument, show that

$$\int_0^1 W_1(t)\,dW_2(t) \stackrel{\mathcal{D}}{=} \frac{1}{2}\sum_{n=1}^{\infty} \frac{Z_{n1}^2 - Z_{n2}^2}{\left(n - \frac{1}{2}\right)\pi},$$

where $\big(Z_{n1}, Z_{n2}\big)' \sim \mathrm{NID}(\mathbf{0}, I_2)$.

1.5 Statistical Examples and the Fredholm Determinant

In this section we explore various statistical examples where quadratic functionals of Bm are used, among which are (1) goodness-of-fit tests and Darling's formula, (2) LBI tests for parameter constancy, (3) LBIU tests for the moving average unit root, and (4) LBI and LBIU tests for the autoregressive unit root.

1.5.1 Goodness-of-Fit Tests and Darling's Formula

Let $X_1, \ldots, X_n$ be a random sample from the common distribution $G(x)$ assumed to be absolutely continuous. The testing problem considered here is to test if $G(x) = F(x;\xi)$, where $F(x;\xi)$ is a distribution function parameterized by ξ. For illustrative purposes we concentrate here on the case where ξ is a single parameter. The multi-parameter case will be examined in Chapters 3 and 4. Depending on whether ξ is specified or not, we deal with two cases.

Case 1.1 The testing problem considered here is

$$H_0: \ G(x) = F(x;\xi),$$

where ξ is a nuisance parameter to be estimated. This case was discussed in Darling (1955), where the test statistic is a modified version of the *Cramér–von Mises–Smirnov statistic* and is given by

$$W_n^2(\hat{\xi}) = n \int_{-\infty}^{\infty} \left(F_n(x) - F(x;\hat{\xi})\right)^2 dF(x;\hat{\xi}), \tag{1.75}$$

where $\hat{\xi}$ is a suitable estimator of ξ, and $F_n(x)$ is the empirical distribution function defined by

$$F_n(x) = \frac{1}{n}\sum_{j=1}^{n} \epsilon\left(x - X_j\right), \qquad \epsilon(x) = \begin{cases} 1 & (x \geq 0), \\ 0 & (x < 0). \end{cases}$$

For practical computation of $W_n^2(\hat{\xi})$, we rearrange the sample $X_1, \ldots, X_n$ to get $X_1' < \cdots < X_n'$, so that (Exercise 1.5.1)

$$W_n^2(\hat{\xi}) = \sum_{j=1}^{n} \left(F(X_j';\hat{\xi}) - \frac{2j-1}{2n}\right)^2 + \frac{1}{12n}.$$

Let us consider the asymptotic distribution of $W_n^2(\hat{\xi})$ under H_0. Suppose that $\hat{\xi}$ is the maximum likelihood estimator (MLE) of ξ that satisfies some regularity conditions. Here we exclude the super-efficient case such that $n\mathrm{E}[(\hat{\xi} - \xi)^2] \to 0$ as $n \to \infty$. Putting

$$g_\xi(t) = \left.\frac{\partial F(x;\xi)}{\partial \xi}\right|_{x=x(t;\xi)}, \qquad x(t;\xi) = \inf\{x : F(x;\xi) = t\},$$

Darling (1955) considered

$$\begin{aligned} W_n^2(\hat{\xi}) &= \int_{-\infty}^{\infty} \left(\sqrt{n}\,(F_n(x) - F(x;\xi)) - \sqrt{n}\,(\hat{\xi} - \xi)\frac{\partial F(x;\xi^*)}{\partial \xi}\right)^2 dF(x;\hat{\xi}) \\ &= \int_0^1 \left(Y_n(t) - \sqrt{n}\,(\hat{\xi} - \xi)\, g_\xi(t)\right)^2 dt + \delta_n \\ &= \int_0^1 Z_n^2(t)\, dt + \delta_n, \qquad \mathrm{plim}_{n\to\infty}\delta_n = 0, \end{aligned}$$

where ξ^* lies between ξ and $\hat{\xi}$ and plim stands for convergence in probability, whereas

$$Z_n(t) = Y_n(t) - \sqrt{n}\,(\hat{\xi} - \xi)\, g_\xi(t),$$

and

$$Y_n(t) = \sqrt{n}\,(H_n(t) - t), \qquad H_n(t) = \frac{1}{n}\sum_{j=1}^{n} \epsilon\big(t - F(X_j;\xi)\big).$$

Since $F(X_1;\xi), \ldots, F(X_n;\xi)$ are independent and uniformly distributed over [0, 1], it follows (Exercise 1.5.2) that

$$\mathrm{E}(H_n(t)) = t, \qquad \mathrm{Cov}(H_n(s), H_n(t)) = \frac{1}{n}\,(\min(s,t) - st),$$

so that

$$\mathrm{E}(Y_n(t)) = 0, \qquad \mathrm{Cov}(Y_n(s),\, Y_n(t)) = \min(s,t) - st.$$

Then it was shown by Darling (1955) that

$$W_n^2(\hat{\xi}) \Rightarrow \int_0^1 Z^2(t)\,dt \;=\; W^2(\xi),$$
$$W^2(\xi) \stackrel{\mathcal{D}}{=} \int_0^1\!\int_0^1 K(s,t)\,dW(s)\,dW(t), \tag{1.76}$$

where $\{Z(t)\}$ is a zero-mean Gaussian process and

$$K(s,t) = \mathrm{Cov}(Z(s),\, Z(t)) = \min(s,t) - st\, - \sigma^2(\xi)\, g_\xi(s)\, g_\xi(t), \tag{1.77}$$

$$\sigma^{-2}(\xi) = \mathrm{E}\left\{\left(\frac{\partial \log f(X_1;\xi)}{\partial \xi}\right)^2\right\} = \int_{-\infty}^{\infty} \left(\frac{\partial \log f(x;\xi)}{\partial \xi}\right)^2 f(x;\xi)\,dx,$$

with $f(x;\xi) = \partial F(x;\xi)/\partial x$.

It must be ensured that $\sigma(\xi)g_\xi(t)$ does not depend on ξ. If it does, the present test is useless because we cannot compute significance points. It was shown in Darling (1955) that, if ξ is a location or scale parameter, $\sigma(\xi)g_\xi(t)$ does not depend on ξ, so that the $W_n^2(\hat{\xi})$ test is parameter free. In fact, it holds (Exercise 1.5.3) that, when $F(x;\xi) = R(x-\xi)$ and $r(x) = R'(x)$,

$$\sigma(\xi)\, g_\xi(t) = -\left(\int_{-\infty}^{\infty} \frac{(r'(x))^2}{r(x)}\,dx\right)^{-1/2} r(R^{-1}(t)),$$

and, when $F(x;\xi) = R(x/\xi)$,

$$\sigma(\xi)\, g_\xi(t) = -\left(\int_{-\infty}^{\infty} \frac{(x\,r'(x))^2}{r(x)}\,dx - 1\right)^{-1/2} R^{-1}(t)\, r(R^{-1}(t)).$$

It also holds (Exercise 1.5.4) that, when $F(x;\xi) = (R(x))^{\xi}$ $(\xi > 0)$,

$$\sigma(\xi)\, g_\xi(t) = t\, \log t.$$

As a specific example, let us consider

$$F(x;\xi) = \frac{1}{1+e^{-(x-\xi)}},$$

which is the logistic distribution function with location parameter ξ. We have

$$\frac{\partial F(x;\xi)}{\partial \xi} = \frac{-e^{-(x-\xi)}}{\left(1+e^{-(x-\xi)}\right)^2}, \qquad F(x;\xi) = t \Leftrightarrow e^{-(x-\xi)} = \frac{1-t}{t},$$

which yields

$$g_\xi(t) = \left.\frac{\partial F(x;\xi)}{\partial \xi}\right|_{x=x(t;\xi)} = -t(1-t).$$

Since

$$f(x;\xi) = \frac{e^{-(x-\xi)}}{\left(1+e^{-(x-\xi)}\right)^2}, \qquad \frac{\partial \log f(x;\xi)}{\partial \xi} = 1 - \frac{2e^{-(x-\xi)}}{1+e^{-(x-\xi)}},$$

we have

$$\begin{aligned}
\sigma^{-2}(\xi) &= \int_{-\infty}^{\infty} \left(1 - \frac{2e^{-(x-\xi)}}{1+e^{-(x-\xi)}}\right)^2 \frac{e^{-(x-\xi)}}{\left(1+e^{-(x-\xi)}\right)^2}\, dx \\
&= \int_0^1 (2t-1)^2\, dt = \frac{1}{3}.
\end{aligned}$$

Thus the distribution of $W^2(\xi)$ in the present case is given by (1.76), and it follows from (1.77) that

$$\mathrm{Cov}(Z(s),\, Z(t)) = \min(s,t) - st - 3st(1-s)(1-t). \tag{1.78}$$

The computation of the c.f. of $W^2(\xi)$ can be carried out using Darling's formula (see Darling (1955, theorem 6.5) and (2.83)), which gives the FD of $\mathrm{Cov}(Z(s), Z(t))$ in (1.78) as

$$\begin{aligned}
D(\lambda) &= \frac{\sin\sqrt{\lambda}}{\sqrt{\lambda}} - 6\int_0^1 \left(\int_0^t g_\xi'(s)\cos\sqrt{\lambda}s\, ds\right) g_\xi'(t)\cos\sqrt{\lambda}(1-t)\, dt \\
&= \frac{12}{\lambda}\left(\frac{2\left(1-\cos\sqrt{\lambda}\right)}{\lambda} - \frac{\sin\sqrt{\lambda}}{\sqrt{\lambda}}\right).
\end{aligned} \tag{1.79}$$

Thus we obtain

$$\mathrm{E}\left(e^{i\theta W^2(\xi)}\right) = (D(2i\theta))^{-1/2} = \left[\frac{6}{i\theta}\left(\frac{1-\cos\sqrt{2i\theta}}{i\theta} - \frac{\sin\sqrt{2i\theta}}{\sqrt{2i\theta}}\right)\right]^{-1/2}. \tag{1.80}$$

As another example, consider

$$F(x;\xi) = \frac{1}{2} + \frac{1}{\pi}\tan^{-1}\frac{x}{\xi} \quad (\xi > 0),$$

which is the Cauchy distribution with scale parameter ξ. We have

$$\begin{aligned}\frac{\partial F(x;\xi)}{\partial \xi} &= -\frac{x}{\xi^2\pi}\frac{1}{1+x^2/\xi^2},\\ F(x;\xi) = t \ \Leftrightarrow\ \frac{x}{\xi} &= \tan\pi(t-1/2) = -\cot\pi t,\end{aligned}$$

which yields

$$g_\xi(t) = \left.\frac{\partial F(x;\xi)}{\partial \xi}\right|_{x=x(t;\xi)} = \frac{\cot\pi t}{\xi\pi}\frac{1}{1+\cot^2\pi t} = \frac{1}{2\xi\pi}\sin 2\pi t.$$

Since

$$f(x;\xi) = \frac{1}{\xi\pi}\frac{1}{1+x^2/\xi^2}, \qquad \frac{\partial \log f(x;\xi)}{\partial \xi} = -\frac{1}{\xi} + \frac{2x^2/\xi^3}{1+x^2/\xi^2},$$

we have

$$\begin{aligned}\sigma^{-2}(\xi) &= \int_{-\infty}^{\infty}\left(-\frac{1}{\xi} + \frac{2x^2/\xi^3}{1+x^2/\xi^2}\right)^2 \frac{1}{\xi\pi}\frac{1}{1+x^2/\xi^2}\,dx\\ &= \frac{1}{\xi^2} + \frac{8}{\xi^2\pi}\int_0^\infty\left(\frac{-u^2}{(1+u^2)^2} + \frac{u^4}{(1+u^2)^3}\right)du\\ &= \frac{1}{2\xi^2},\end{aligned}$$

where we have used the fact that

$$\int_0^\infty \frac{x^{a-1}}{(1+x^c)^b}\,dx = \frac{1}{c}\,B(b-a/c, a/c) \quad (a>0,\ b>a/c,\ c>0).$$

Here $B(p,q)$ is the beta function defined by

$$B(p,q) = \int_0^1 t^{p-1}(1-t)^{q-1}\,dt \quad (p>0,\ q>0).$$

Thus the distribution of $W^2(\xi)$ in the present case is given by (1.76) with

$$\mathrm{Cov}(Z(s),\ Z(t)) = \min(s,t) - st - \frac{1}{2\pi^2}\sin 2\pi s\,\sin 2\pi t. \tag{1.81}$$

Then, using Darling's formula in (2.83), the FD of $\mathrm{Cov}(Z(s), Z(t))$ in (1.81) is given by

$$
\begin{aligned}
D(\lambda) &= \frac{\sin\sqrt{\lambda}}{\sqrt{\lambda}} - 4\int_0^1 \left(\int_0^t \cos 2\pi s \, \cos\sqrt{\lambda} s \, ds\right) \cos 2\pi t \, \cos\sqrt{\lambda}(1-t)\, dt \\
&= \frac{\sin\sqrt{\lambda}}{\sqrt{\lambda}} \Big/ \left(1 - \frac{\lambda}{4\pi^2}\right).
\end{aligned}
\tag{1.82}
$$

Thus we now obtain

$$
\mathrm{E}\left(e^{i\theta W^2(\xi)}\right) = (D(2i\theta))^{-1/2} = \left[\frac{\sin\sqrt{2i\theta}}{\sqrt{2i\theta}} \Big/ \left(1 - \frac{i\theta}{2\pi^2}\right)\right]^{-1/2}. \tag{1.83}
$$

Case 1.2 Here we deal with the testing problem

$$
H_0 : G(x) = F(x;\xi_0) \quad \text{versus} \quad H_1 : G(x) = F(x;\xi), \quad \xi = \xi_0 + \frac{\gamma}{\sqrt{n}},
$$

where ξ_0 is a specified value and γ is a constant. This case was discussed in Durbin (1973). Using the Cramér–von Mises–Smirnov statistic

$$
W_n^2(\xi_0) = n\int_{-\infty}^{\infty} \left(F_n(x) - F(x;\xi_0)\right)^2 dF(x;\xi_0), \tag{1.84}
$$

the null hypothesis H_0 is rejected if $W_n^2(\xi_0)$ is sufficiently large.

Let us consider the limiting distribution of $W_n^2(\xi_0)$ as $n \to \infty$. We first deal with the distribution under H_0. Putting $G(x) = F(x;\xi_0) = t$, we have

$$
\begin{aligned}
W_n^2(\xi_0) &= n\int_{-\infty}^{\infty} \left(\frac{1}{n}\sum_{j=1}^n \epsilon(x - X_j) - F(x;\xi_0)\right)^2 dF(x;\xi_0) \\
&= n\int_0^1 \left(\frac{1}{n}\sum_{j=1}^n \epsilon\left(t - F(X_j;\xi_0)\right) - t\right)^2 dt \\
&= \int_0^1 Y_n^2(t)\, dt,
\end{aligned}
$$

where

$$
Y_n(t) = \sqrt{n}\,(H_n(t) - t), \qquad H_n(t) = \frac{1}{n}\sum_{j=1}^n \epsilon\left(t - F(X_j;\xi_0)\right).
$$

It follows that $\{Y_n(t)\} \Rightarrow \{Y(t)\}$, where $\{Y(t)\}$ is the Bb. Thus it holds that

$$\begin{aligned} W_n^2(\xi_0) \Rightarrow W^2(\xi_0) &= \int_0^1 Y^2(t)\,dt \\ &\stackrel{\mathcal{D}}{=} \int_0^1\int_0^1 [\min(s,t) - st]\,dW(s)\,dW(t), \end{aligned} \tag{1.85}$$

and the c.f. of $W^2(\xi_0)$ is given from (1.35) by

$$\mathrm{E}\left(e^{i\theta W^2(\xi_0)}\right) = (D(2i\theta))^{-1/2} = \left(\frac{\sin\sqrt{2i\theta}}{\sqrt{2i\theta}}\right)^{-1/2},$$

where $D(\lambda)$ is the FD of $K(s,t) = \min(s,t) - st$.

We next consider the limiting distribution of $W_n^2(\xi_0)$ under $H_1 : \xi = \xi_0 + \gamma/\sqrt{n}$. Expanding $F(x;\xi_0)$ around $\xi = \xi_0 + \gamma/\sqrt{n}$, we consider, in place of $W_n^2(\xi_0)$,

$$\begin{aligned} R_n^2(\xi_0) &= n\int_{-\infty}^{\infty}\left(F_n(x) - F(x;\xi) + \frac{\gamma}{\sqrt{n}}\left.\frac{\partial F(x;\xi)}{\partial \xi}\right|_{\xi=\xi_0}\right)^2 dF(x;\xi) \\ &= \int_0^1 \left(Y_n(t) + \gamma g_{\xi_0}(t)\right)^2 dt, \end{aligned} \tag{1.86}$$

where

$$Y_n(t) = \sqrt{n}(H_n(t) - t), \qquad H_n(t) = \frac{1}{n}\sum_{j=1}^{n} \epsilon\left(t - F(X_j;\xi)\right),$$

$$g_{\xi_0}(t) = \left.\frac{\partial F(x;\xi)}{\partial \xi}\right|_{x=x(t;\xi_0)}, \qquad x(t;\xi_0) = \inf\{x : F(x;\xi_0) = t\}.$$

It follows from Durbin (1973) that, under H_1 and some regularity conditions, the limiting distributions of $W_n^2(\xi_0)$ and $R_n^2(\xi_0)$ are the same and

$$R_n^2(\xi_0) \Rightarrow R^2(\xi_0) = \int_0^1 \left(Y(t) + \gamma g_{\xi_0}(t)\right)^2 dt, \tag{1.87}$$

where $\{Y(t)\}$ is the Bb. Unlike the previous statistics, the statistic $R^2(\xi_0)$ is a quadratic functional of a nonzero-mean Gaussian process $\{Y(t) + \gamma g_{\xi_0}(t)\}$. To derive the c.f. of such a functional, it is not sufficient to compute the FD of $K(s,t) = \mathrm{Cov}(Y(s), Y(t))$; we also need to deal with the *resolvent* associated with $K(s,t)$ and the function $m(t) = \gamma g_{\xi_0}(t)$.

Discussions on the resolvent are deferred until Chapter 3. Here we take up two examples only for demonstration purposes. Consider first the logistic distribution function $F(x;\xi) = 1/(1 + e^{-(x-\xi)})$ dealt with in Case 1.1. Then we have

$$g_{\xi_0}(t) = \left.\frac{\partial\left(1 + e^{-(x-\xi)}\right)^{-1}}{\partial \xi}\right|_{x=x(t;\xi_0)} = -t(1-t).$$

Thus (1.87) holds under H_1 with $g_{\xi_0}(t) = -t(1-t)$. The corresponding c.f. of $R^2(\xi_0)$, the derivation of which involves the resolvent and will be explained in Section 3.2, is given by

$$\begin{aligned} &\mathrm{E}\left(e^{i\theta R^2(\xi_0)}\right) \\ &= \mathrm{E}\left[\exp\left\{i\theta \int_0^1 (Y(t) - \gamma t(1-t))^2\, dt\right\}\right] \\ &= \left(\frac{\sin\sqrt{2i\theta}}{\sqrt{2i\theta}}\right)^{-1/2} \exp\left[\frac{\gamma^2}{\theta^2}\left(\frac{\cos\sqrt{2i\theta} - 1}{\sin\sqrt{2i\theta}/\sqrt{2i\theta}} + i\theta - \frac{\theta^2}{6}\right)\right]. \end{aligned} \tag{1.88}$$

It can be seen that $R^2(\xi_0)$ does not depend on ξ_0 and tends to normality as $|\gamma| \to \infty$.

As the second example, consider the Cauchy distribution $F(x;\xi) = 1/2 + \tan^{-1}(x/\xi)/\pi$ also dealt with in Case 1.1, where $\xi > 0$. We have

$$g_{\xi_0}(t) = \frac{\partial}{\partial \xi}\left.\left(\frac{1}{2} + \frac{1}{\pi}\tan^{-1}\frac{x}{\xi}\right)\right|_{x=x(t;\xi_0)} = \frac{1}{2\xi_0\pi}\sin 2\pi t,$$

and

$$\begin{aligned} \mathrm{E}\left(e^{i\theta R^2(\xi_0)}\right) &= \mathrm{E}\left(\exp\left\{i\theta \int_0^1 \left(Y(t) + \frac{\gamma}{2\xi_0\pi}\sin 2\pi t\right)^2 dt\right\}\right) \\ &= \left(\frac{\sin\sqrt{2i\theta}}{\sqrt{2i\theta}}\right)^{-1/2} \exp\left[\frac{\gamma^2}{\xi_0^2}\frac{i\theta}{4(2\pi^2 - i\theta)}\right]. \end{aligned} \tag{1.89}$$

The derivation of this c.f. also involves the resolvent and will be demonstrated in Section 3.2. Unlike $R^2(\xi_0)$ in (1.88), the present statistic $R^2(\xi_0)$ depends on ξ_0, although it also tends to normality as $|\gamma| \to \infty$.

Goodness-of-fit tests will be discussed in more detail in Chapter 4, dealing with the multi-parameter case.

1.5.2 LBI Tests for Parameter Constancy

Let us consider the following *state space model*:

$$y_j = \beta_j + \varepsilon_j, \quad \beta_j = \beta_{j-1} + \eta_j \qquad (j = 1, \ldots, n), \tag{1.90}$$

where

(i) $\{y_j\}$ is an observable sequence, whereas $\{\beta_j\}$ is an unobservable sequence starting from an unknown constant β_0;
(ii) $\{\varepsilon_j\} \sim \mathrm{NID}(0, \sigma_\varepsilon^2)$, $\{\eta_j\} \sim \mathrm{NID}(0, \sigma_\eta^2)$, and these are independent of each other.

The state space model is composed of two equations. One is a measurement equation that describes the relation between the observed variable y_j and the unobserved state variable β_j. The other is a state equation that reflects the dynamics of the state variable β_j.

Our concern is to test if β_j, the level of the process, is constant, that is, $\beta_j = \beta_0$ for all j. This is equivalent to testing

$$H_0 : \rho = \frac{\sigma_\eta^2}{\sigma_\varepsilon^2} = 0 \qquad \text{versus} \qquad H_1 : \rho > 0. \tag{1.91}$$

The present testing problem was first discussed in Nyblom and Mäkeläinen (1983) and Tanaka (1983), and was generalized by Nabeya and Tanaka (1988) and Nabeya (1989) to various cases, which we shall deal with in subsequent chapters.

Since $y_j = \beta_0 + \varepsilon_j + \eta_1 + \cdots + \eta_j$, the observation vector $\boldsymbol{y} = (y_1, \ldots, y_n)'$ follows

$$\boldsymbol{y} = \beta_0 \boldsymbol{e} + \boldsymbol{\varepsilon} + C\boldsymbol{\eta} \sim \mathrm{N}\big(\beta_0 \boldsymbol{e}, \sigma_\varepsilon^2 \Omega(\rho)\big), \tag{1.92}$$

where $\boldsymbol{e} = (1, \ldots, 1)' : n \times 1$, $\boldsymbol{\varepsilon} = (\varepsilon_1, \ldots, \varepsilon_n)'$, $\boldsymbol{\eta} = (\eta_1, \ldots, \eta_n)'$, and

$$\Omega(\rho) = I_n + \rho CC', \; C = \begin{pmatrix} 1 & & & 0 \\ \cdot & \cdot & & \\ \cdot & & \cdot & \\ \cdot & & & \cdot \\ 1 & \cdot & \cdot & 1 \end{pmatrix}, \; C^{-1} = \begin{pmatrix} 1 & & & 0 \\ -1 & \cdot & & \\ & \cdot & \cdot & \\ & & \cdot & \cdot \\ 0 & & -1 & 1 \end{pmatrix}. \tag{1.93}$$

Defining the parameter vector $\boldsymbol{\delta} = (\rho, \beta_0, \sigma_\varepsilon^2)'$, the log-likelihood for $\boldsymbol{y}$ is given by

$$L(\boldsymbol{\delta}) = -\frac{n}{2} \log 2\pi\sigma_\varepsilon^2 - \frac{1}{2} \log |\Omega(\rho)| - \frac{1}{2\sigma_\varepsilon^2} (\boldsymbol{y} - \beta_0 \boldsymbol{e})' \Omega^{-1}(\rho) (\boldsymbol{y} - \beta_0 \boldsymbol{e}). \tag{1.94}$$

The matrix C necessarily arises from the random walk process $\beta_j = \beta_{j-1} + \eta_j$ and may be called the *random walk-generating matrix*. Note that the (j,k)th element of CC' is $\min(j,k)$, and the eigenvalues of CC' were obtained by

Rutherford (1946) as

$$\lambda_j = \frac{1}{4}\left(\sin\frac{j-\frac{1}{2}}{2n+1}\pi\right)^{-2} \qquad (j = 1, \ldots, n).$$

For the present problem there exists no uniformly best test. Here we consider the *locally best invariant* (LBI) test (Ferguson, 1967). It follows from (1.92) that the testing problem (1.91) is invariant under the group of transformations: $\boldsymbol{y} \to a\boldsymbol{y} + b\boldsymbol{e}$ and $(\rho, \beta_0, \sigma_\varepsilon^2) \to (\rho, a\beta_0 + b, a^2\sigma_\varepsilon^2)$, where $a > 0$. Following Kariya (1980), choose an $n \times (n-1)$ matrix H such that $H'H = I_{n-1}$ and $HH' = I_n - \boldsymbol{e}\boldsymbol{e}'/n = M = M' = M^2$. Let us put $\boldsymbol{w} = H'\boldsymbol{y} = H'(\beta_0\boldsymbol{e} + \boldsymbol{\varepsilon} + C\boldsymbol{\eta})$. Noting that $M\boldsymbol{e} = HH'\boldsymbol{e} = \boldsymbol{0}$ and thus $H'\boldsymbol{e} = \boldsymbol{0}$, we have

$$\boldsymbol{w} \sim \mathrm{N}\left(\boldsymbol{0},\ \sigma_\varepsilon^2 H'\Omega(\rho)H\right), \tag{1.95}$$

and the statistic $s(\boldsymbol{w}) = \boldsymbol{w}/\sqrt{\boldsymbol{w}'\boldsymbol{w}}$ is a maximal invariant under the above group. Let $P_\rho(\cdot)$ be the distribution of $s(\boldsymbol{w})$, and put

$$f_n(s(\boldsymbol{w})|\rho) = |H'\Omega(\rho)H|^{-1/2}\left[\frac{\boldsymbol{w}'(H'\Omega(\rho)H)^{-1}\boldsymbol{w}}{\boldsymbol{w}'\boldsymbol{w}}\right]^{-(n-1)/2}, \tag{1.96}$$

which is the probability density of $P_\rho(\cdot)$ with respect to $P_0(\cdot)$. The rejection region of the LBI test is now obtained (Ferguson, 1967, p.235) as

$$\left.\frac{\partial \log f_n(s(\boldsymbol{w})|\rho)}{\partial\rho}\right|_{\rho=0} > \text{constant}, \tag{1.97}$$

which leads to the rejection region given by

$$R_{n1} = \frac{\boldsymbol{w}'H'CC'H\boldsymbol{w}}{\boldsymbol{w}'\boldsymbol{w}} = \frac{\boldsymbol{y}'MCC'M\boldsymbol{y}}{\boldsymbol{y}'M\boldsymbol{y}} > \text{constant}. \tag{1.98}$$

The derivation of the LBI test is somewhat complicated. It can be shown (Tanaka, 1996, chapter 9) that the LBI test for the present problem is equivalent to the *Lagrange multiplier* (LM) *test*, which is easier to derive. In fact, the rejection region of the LM test is given by

$$\left.\frac{\partial L(\boldsymbol{\delta})}{\partial\rho}\right|_{H_0} > \text{constant}, \tag{1.99}$$

where the partial derivative is evaluated at the MLE $\tilde{\boldsymbol{\delta}} = (\tilde{\rho}, \tilde{\beta}_0, \tilde{\sigma}_\varepsilon^2)'$ of $\boldsymbol{\delta}$ under H_0. Note that $\tilde{\rho} = 0$. We also have

$$\tilde{\beta}_0 = \frac{1}{n}\sum_{j=1}^{n} y_j = \frac{1}{n}\boldsymbol{e}'\boldsymbol{y}, \quad \tilde{\sigma}_\varepsilon^2 = \frac{1}{n}\sum_{j=1}^{n}(y_j - \bar{y})^2 = \frac{1}{n}\boldsymbol{y}'M\boldsymbol{y}.$$

Then, using

$$\frac{d\log|\Omega(\rho)|}{d\rho} = \mathrm{tr}\left(\Omega^{-1}(\rho)\frac{d\Omega(\rho)}{d\rho}\right), \qquad \frac{d\Omega^{-1}(\rho)}{d\rho} = -\Omega^{-1}(\rho)\frac{d\Omega(\rho)}{d\rho}\Omega^{-1}(\rho),$$

we have

$$\begin{aligned} \left.\frac{\partial L(\boldsymbol{\delta})}{\partial \rho}\right|_{H_0} &= -\frac{1}{2}\mathrm{tr}(CC') + \frac{1}{2\tilde{\sigma}_\varepsilon^2}\, \boldsymbol{y}'MCC'M\boldsymbol{y} \\ &= -\frac{n(n+1)}{4} + \frac{n}{2}\frac{\boldsymbol{y}'MCC'M\boldsymbol{y}}{\boldsymbol{y}'M\boldsymbol{y}}. \end{aligned} \tag{1.100}$$

Thus it is seen that the LM test is equivalent to the LBI test. An interesting feature of the present test is that $\partial L(\boldsymbol{\delta})/\partial\rho|_{H_0}$ does not tend to normality, which will be seen below. This contrasts with the standard situation.

Let us derive the asymptotic distribution of the LBI statistic R_{n1} in (1.98) under $H_0 : \rho = 0$. For this purpose we can assume that $\sigma_\varepsilon^2 = 1$ without any loss of generality and consider

$$S_{n1} = \frac{1}{n} R_{n1} = \frac{1}{n}\frac{\boldsymbol{y}'MCC'M\boldsymbol{y}}{\boldsymbol{y}'M\boldsymbol{y}}. \tag{1.101}$$

Under H_0, it holds that $\boldsymbol{y} = \beta_0\boldsymbol{e} + \boldsymbol{\varepsilon}$ and

$$\frac{1}{n}\,\boldsymbol{y}'M\boldsymbol{y} = \frac{1}{n}\,\boldsymbol{\varepsilon}'M\boldsymbol{\varepsilon} \;\to\; 1 \quad \text{in probability,} \tag{1.102}$$

$$\begin{aligned} \frac{1}{n^2}\,\boldsymbol{y}'MCC'M\boldsymbol{y} &= \frac{1}{n^2}\,\boldsymbol{\varepsilon}'MCC'M\boldsymbol{\varepsilon} = \frac{1}{n}\,\boldsymbol{\varepsilon}'B_{n1}\boldsymbol{\varepsilon} \\ &= \frac{1}{n}\sum_{j=1}^{n}\sum_{k=1}^{n} B_{n1}(j,k)\varepsilon_j\varepsilon_k, \end{aligned} \tag{1.103}$$

where $B_{n1} = MCC'M/n$, and $B_{n1}(j,k)$ is the (j,k)th element of B_{n1} given by

$$\begin{aligned} B_{n1}(j,k) &= \frac{1}{n}\min(j,k) - \frac{1}{2n^2}\,j(2n-j+1) - \frac{1}{2n^2}\,k(2n-k+1) \\ &\quad + \frac{1}{6n^3}\,n(n+1)(2n+1). \end{aligned} \tag{1.104}$$

To derive the limiting distribution of the quadratic form in (1.103), we can use the following theorem. Its extended version is given in Nabeya and Tanaka (1988) and will be proved in Chapter 5 (see Theorem 5.1).

Theorem 1.11 *Consider*

$$S_n = \frac{1}{n}\boldsymbol{\varepsilon}'B_n\boldsymbol{\varepsilon} = \frac{1}{n}\sum_{j=1}^{n}\sum_{k=1}^{n} B_n(j,k)\varepsilon_j\varepsilon_k,$$

where $\{\varepsilon_j\} \sim i.i.d.(0,1)$ *and* B_n *is an* $n \times n$ *symmetric and nonnegative definite matrix. Assume that, for the* (j,k)*th element* $B_n(j,k)$ *of* B_n*, there exists a symmetric and continuous function* $K(s,t)$ *satisfying*

$$\lim_{n\to\infty} \max_{1\le j,k\le n} \left| B_n(j,k) - K\left(\frac{j}{n},\frac{k}{n}\right)\right| = 0. \tag{1.105}$$

Then it holds that, as $n \to \infty$,

$$S_n \Rightarrow S = \int_0^1\int_0^1 K(s,t)\,dW(s)\,dW(t). \tag{1.106}$$

Note that, because B_n is nonnegative definite and the condition (1.105) is imposed, the statistic S is nonnegative. This can be weakened to the extent that the matrix B_n is indefinite so that S is not necessarily nonnegative to deal with more general cases, which will be discussed in Chapter 5.

Let us return to (1.103) and (1.104). It holds that the matrix $B_{n1} = MCC'M/n$ with $B_{n1}(j,k)$ in the (j,k)th place is nonnegative definite and the function

$$K_1(s,t) = \min(s,t) - s - t + \frac{s^2+t^2}{2} + \frac{1}{3} = \frac{1}{3} - \max(s,t) + \frac{s^2+t^2}{2}$$

is the uniform limit of $B_{n1}(j,k)$ in the sense of (1.105). Thus Theorem 1.11 leads us to establish that

$$\begin{aligned}\frac{1}{n^2}\,y'MCC'My &= \frac{1}{n}\sum_{j=1}^n\sum_{k=1}^n B_{n1}(j,k)\varepsilon_j\varepsilon_k \\ &\Rightarrow \int_0^1\int_0^1 K_1(s,t)\,dW(s)\,dW(t). \end{aligned}\tag{1.107}$$

Therefore it follows that

$$S_{n1} = \frac{1}{n}\frac{y'MCC'My}{y'My} \Rightarrow S_1 = \int_0^1\int_0^1 K_1(s,t)\,dW(s)\,dW(t). \tag{1.108}$$

The c.f. of S_1 is given from (1.35) and (1.38) by $(D_1(2i\theta))^{-1/2}$, where $D_1(\lambda) = \sin\sqrt{\lambda}/\sqrt{\lambda}$ is the FD of $K_1(s,t)$.

In the present case there is a simpler way of deriving the limiting distribution of S_{n1}. Noting that

$$\begin{aligned}\frac{1}{n^2}\,y'MCC'My &= \frac{1}{n^2}\,\boldsymbol{\varepsilon}'MCC'M\boldsymbol{\varepsilon} \overset{\mathcal{D}}{=} \frac{1}{n^2}\,\boldsymbol{\varepsilon}C'MC\boldsymbol{\varepsilon} \\ &= \frac{1}{n}\boldsymbol{\varepsilon}'B^\star_{n1}\boldsymbol{\varepsilon} = \frac{1}{n}\sum_{j=1}^n\sum_{k=1}^n B^\star_{n1}(j,k),\end{aligned}$$

where $B_{n1}^{\star} = C'MC/n$ and $B_{n1}^{\star}(j,k)$ is the (j,k)th element of $B_{n1}^{\star}$ defined by

$$B_{n1}^{\star}(j,k) = \frac{1}{n}\min(n+1-j, n+1-k) - \frac{1}{n^2}(n-j+1)(n-k+1), \qquad (1.109)$$

we can find the uniform limit $K_1^{\star}(s,t)$ of $B_{n1}^{\star}(j,k)$ as

$$K_1^{\star}(s,t) = \min(1-s, 1-t) - (1-s)(1-t) = \min(s,t) - st.$$

Thus it follows that

$$S_{n1} \Rightarrow S_1 \stackrel{\mathcal{D}}{=} \int_0^1 \int_0^1 [\min(s,t) - st]\, dW(s)\, dW(t). \qquad (1.110)$$

In the present model (1.92), the initial value β_0 was unknown. If β_0 is assumed to be known, in which case we can put $\beta_0 = 0$, then the above results hold true by replacing $K_1(s,t)$ by $\min(s,t)$ (Exercise 1.5.5).

Nabeya and Tanaka (1988) extended the present model to

$$y_j = x_j\beta_j + \boldsymbol{z}_j'\boldsymbol{\gamma} + \varepsilon_j, \quad \beta_j = \beta_{j-1} + \eta_j \quad (j = 1, \ldots, n),$$

where x_j and $\boldsymbol{z}_j$ are scalar and vector deterministic regressors, respectively. We shall deal with this extended case in Chapter 5.

It is also interesting to examine the power of the LBI test. For this purpose we consider the distribution of the LBI test statistic under the local alternative

$$H_1(c,\kappa) : \rho = \frac{c}{n^{\kappa}},$$

where c and κ are positive constants. We shall discuss the local alternative in Chapter 5.

1.5.3 LBIU Tests for the Moving Average Unit Root

In this section we deal with the first-order moving average (MA(1)) model given by

$$y_j - \mu = \varepsilon_j - \alpha\,\varepsilon_{j-1} \qquad (j = 1, \ldots, n), \qquad (1.111)$$

where μ is an unknown mean and α $(0 < \alpha \leq 1)$ is the coefficient parameter, whereas $\{\varepsilon_j\}$ $(j = 0, 1, \ldots, n)$ is an error sequence following NID$(0, \sigma_\varepsilon^2)$.

Our concern here is to consider testing

$$H_0 : \alpha = 1 \qquad \text{versus} \qquad H_1 : \alpha < 1, \qquad (1.112)$$

which was initially discussed in Tanaka (1990b) when $\mu = 0$. The MA(1) model (1.111) is said to be noninvertible under H_0. The log-likelihood for $\boldsymbol{y} = (y_1, y_2, \ldots, y_n)'$ is

$$L(\boldsymbol{\delta}) = -\frac{n}{2}\log 2\pi\sigma_\varepsilon^2 - \frac{1}{2}\log|\Omega(\alpha)| - \frac{1}{2\sigma_\varepsilon^2}(\boldsymbol{y}-\mu\boldsymbol{e})'\Omega^{-1}(\alpha)(\boldsymbol{y}-\mu\boldsymbol{e}), \tag{1.113}$$

where $\boldsymbol{\delta} = (\alpha, \mu, \sigma_\varepsilon^2)'$, $\boldsymbol{e} = (1, \ldots, 1)' : n \times 1$, and

$$\Omega(\alpha) = \begin{pmatrix} 1+\alpha^2 & -\alpha & & & 0 \\ -\alpha & 1+\alpha^2 & \cdot & & \\ & \cdot & \cdot & \cdot & \\ & & \cdot & \cdot & -\alpha \\ 0 & & & -\alpha & 1+\alpha^2 \end{pmatrix}.$$

Proceeding in the same way as in the last section, the testing problem (1.112) is shown to be invariant under the group of transformations: $\boldsymbol{y} \to a\boldsymbol{y} + b\boldsymbol{e}$ and $(\alpha,\ \mu,\ \sigma_\varepsilon^2) \to (\alpha,\ a\mu+b,\ a^2\sigma_\varepsilon^2)$, where $a > 0$. Let us put $\boldsymbol{w} = H'\boldsymbol{y}$, where H is the $n \times (n-1)$ matrix such that $H'H = I_{n-1}$ and $HH' = I_n - \boldsymbol{e}\boldsymbol{e}'/n = M$. Then we have

$$\boldsymbol{w} \sim \mathrm{N}\left(\boldsymbol{0},\ \sigma_\varepsilon^2 H'\Omega(\alpha)H\right), \tag{1.114}$$

and the probability density of the maximal invariant $s(\boldsymbol{w}) = \boldsymbol{w}/\sqrt{\boldsymbol{w}'\boldsymbol{w}}$ is given by

$$f_n(s(\boldsymbol{w})|\alpha) = |H'\Omega(\alpha)H|^{-1/2}\left[\frac{\boldsymbol{w}'(H'\Omega(\alpha)H)^{-1}\boldsymbol{w}}{\boldsymbol{w}'\boldsymbol{w}}\right]^{-(n-1)/2}. \tag{1.115}$$

Using

$$\left.\frac{d\Omega(\alpha)}{d\alpha}\right|_{\alpha=1} = \Omega(1) = \Omega, \qquad \left.\frac{d^2\Omega(\alpha)}{d\alpha^2}\right|_{\alpha=1} = 2I_n, \tag{1.116}$$

we obtain

$$\left.\frac{\partial \log f_n(s(\boldsymbol{w})|\alpha)}{\partial\alpha}\right|_{\alpha=1} = 0,$$
$$\left.\frac{\partial^2 \log f_n(s(\boldsymbol{w})|\alpha)}{\partial\alpha^2}\right|_{\alpha=1} = c_1 + \frac{n-1}{2}\,\frac{\boldsymbol{y}'H(H'\Omega H)^{-2}H'\boldsymbol{y}}{\boldsymbol{y}'H(H'\Omega H)^{-1}H'\boldsymbol{y}},$$

where c_1 is a constant. Then we can conduct the *unbiased locally best invariant* (LBIU) *test* (Ferguson, 1967, p.237), which rejects H_0 when

$$\left.\frac{\partial^2 \log f_n(s(\boldsymbol{w})|\alpha)}{\partial\alpha^2}\right|_{\alpha=1} > c_2 \quad \Leftrightarrow \quad R_{n2} = \frac{\boldsymbol{y}'H(H'\Omega H)^{-2}H'\boldsymbol{y}}{\boldsymbol{y}'H(H'\Omega H)^{-1}H'\boldsymbol{y}} > c_3, \tag{1.117}$$

where c_2 and c_3 are constants. We can rewrite the LBIU statistic R_{n2} so that it does not depend on H. For this purpose we have (Rao, 1973, p.77)

Lemma 1.12 *Let $R = (P \;\; Q)$ be an $n \times n$ orthogonal matrix, where P and Q are $n \times (n-p)$ and $n \times p$ matrices, respectively. Then, for any $n \times n$ nonsingular symmetric matrix A, it holds that*

$$P(P'AP)^{-1}P' = A^{-1}N = N'A^{-1} = N'A^{-1}N, \tag{1.118}$$

$$P(P'AP)^{-2}P' = N'A^{-2}N, \tag{1.119}$$

where $N = I_n - Q(Q'A^{-1}Q)^{-1}Q'A^{-1}$.

Proof Let us consider

$$G = \begin{pmatrix} P'A \\ \\ Q' \end{pmatrix}, \quad GR = \begin{pmatrix} P'A \\ \\ Q' \end{pmatrix} (P \quad Q) = \begin{pmatrix} P'AP & P'AQ \\ \\ 0 & I_p \end{pmatrix}.$$

It follows from the assumption that the $n \times n$ matrix G is nonsingular. Then, noting that $P'Q = 0$ and $Q'P = 0$, the first equality in (1.118) is established because

$$G\left[P(P'AP)^{-1}P' - \left(A^{-1} - A^{-1}Q(Q'A^{-1}Q)^{-1}Q'A^{-1}\right)\right]$$
$$= G\left[P(P'AP)^{-1}P' - A^{-1}N\right] = 0.$$

It can be checked that $A^{-1}N = N'A^{-1} = N'A^{-1}N$, which proves (1.118). Since

$$P(P'AP)^{-2}P' = P(P'AP)^{-1}P'P(P'AP)^{-1}P' = N'A^{-1}A^{-1}N = N'A^{-2}N,$$

the relation (1.119) is also established. □

Note that the above lemma holds if P and Q are replaced by PF_1 and QF_2, respectively, where F_1 and F_2 are any nonsigular matrices of size $(n-p) \times (n-p)$ and $p \times p$, respectively. Using this fact, the LBIU statistic R_{n2} in (1.117) can be expressed (Exercise 1.5.6) as

$$R_{n2} = \frac{\boldsymbol{y}'N'\Omega^{-2}N\boldsymbol{y}}{\boldsymbol{y}'N'\Omega^{-1}N\boldsymbol{y}}, \tag{1.120}$$

where $N = I_n - \boldsymbol{e}(\boldsymbol{e}'\Omega^{-1}\boldsymbol{e})^{-1}\boldsymbol{e}'\Omega^{-1}$, $N'\Omega^{-1}N = \Omega^{-1}N = N'\Omega^{-1}$, and

$$\Omega^{-1} = \left[(CC')^{-1} + \boldsymbol{e}_n\boldsymbol{e}_n'\right]^{-1} = CC' - \frac{1}{n+1}C\boldsymbol{e}\boldsymbol{e}'C'$$
$$= \left[(C'C)^{-1} + \boldsymbol{e}_1\boldsymbol{e}_1'\right]^{-1} = C'C - \frac{1}{n+1}C'\boldsymbol{e}\boldsymbol{e}'C. \tag{1.121}$$

Here $\boldsymbol{e}_n = (0, \ldots, 0, 1)' : n \times 1$, $\boldsymbol{e}_1 = (1, 0, \ldots, 0)' : n \times 1$, and C is the random walk-generating matrix defined in (1.93). Note that the (j,k)th element of Ω^{-1} is $\min(j,k) - jk/(n+1)$, and the eigenvalues of Ω^{-1} are given (Anderson, 1971, theorem 6.5.5) by

$$\lambda_j = \frac{1}{4}\left(\sin\frac{j\pi}{2(n+1)}\right)^{-2} \qquad (j = 1, \ldots, n).$$

The present LBIU test can also be derived on the basis of the log-likelihood $L(\boldsymbol{\delta})$ defined in (1.113). Using the relations in (1.116) and $\mathrm{tr}(\Omega^{-1}) = n(n+2)/6$, we obtain

$$\left.\frac{\partial L(\boldsymbol{\delta})}{\partial \alpha}\right|_{H_0} = 0, \qquad \left.\frac{\partial^2 L(\boldsymbol{\delta})}{\partial \alpha^2}\right|_{H_0} = -\frac{n(n+5)}{6} + n\frac{\boldsymbol{y}'N'\Omega^{-2}N\boldsymbol{y}}{\boldsymbol{y}'N'\Omega^{-1}N\boldsymbol{y}}, \tag{1.122}$$

where the derivatives are evaluated at the MLE of $\boldsymbol{\delta}$ under H_0. Then it is seen that the resulting statistic is the same as the LBIU statistic R_{n2} in (1.117) and (1.120). It is also recognized that $\alpha = 1$ gives a local maximum of $L(\boldsymbol{\delta})$ when $\partial^2 L(\boldsymbol{\delta})/\partial\alpha^2|_{H_0} < 0$, which will be discussed in Chapter 6.

Let us consider the asymptotic distribution of R_{n2} as $n \to \infty$ under H_0. We can assume that $\sigma_\varepsilon^2 = 1$. Putting $\boldsymbol{\zeta} \sim \mathrm{N}(\mathbf{0}, I_n)$, we have

$$\begin{aligned}
\frac{1}{n}\,\boldsymbol{y}'N'\Omega^{-1}N\boldsymbol{y} &\stackrel{\mathcal{D}}{=} \frac{1}{n}\,\boldsymbol{\zeta}'\Omega^{1/2}N'\Omega^{-1}N\Omega^{1/2}\boldsymbol{\zeta} \stackrel{\mathcal{D}}{=} \frac{1}{n}\,\boldsymbol{\zeta}'\Omega^{-1/2}N\Omega N'\Omega^{-1/2}\boldsymbol{\zeta} \\
&= \frac{1}{n}\,\boldsymbol{\zeta}'\tilde{M}\boldsymbol{\zeta} \;\to\; 1 \quad \text{in probability,} \\
\frac{1}{n^2}\,\boldsymbol{y}'N'\Omega^{-2}N\boldsymbol{y} &\stackrel{\mathcal{D}}{=} \frac{1}{n^2}\,\boldsymbol{\zeta}'\Omega^{1/2}N'\Omega^{-2}N\Omega^{1/2}\boldsymbol{\zeta} \stackrel{\mathcal{D}}{=} \frac{1}{n^2}\,\boldsymbol{\zeta}'\Omega^{-1}N\Omega N'\Omega^{-1}\boldsymbol{\zeta} \\
&= \frac{1}{n}\,\boldsymbol{\zeta}'\,B_{n2}\,\boldsymbol{\zeta},
\end{aligned}$$

where

$$\begin{aligned}
\tilde{M} &= I_n - \Omega^{-1/2}\boldsymbol{e}(\boldsymbol{e}'\Omega^{-1}\boldsymbol{e})^{-1}\boldsymbol{e}'\Omega^{-1/2} = \tilde{M}' = \tilde{M}^2, \\
B_{n2} &= \frac{1}{n}\,\Omega^{-1}N = \frac{1}{n}\left(\Omega^{-1} - \Omega^{-1}\boldsymbol{e}(\boldsymbol{e}'\Omega^{-1}\boldsymbol{e})^{-1}\boldsymbol{e}'\Omega^{-1}\right).
\end{aligned}$$

Here $\mathrm{tr}(\tilde{M}) = n-1$ and Ω^{-1} is given in (1.121), whereas

$$\Omega^{-1}\boldsymbol{e} = C'\boldsymbol{d} - \frac{n}{2}C'\boldsymbol{e}, \quad \boldsymbol{e}'\Omega^{-1}\boldsymbol{e} = \frac{n(n+1)(n+2)}{12},$$

where $\boldsymbol{d} = (1, 2, \ldots, n)'$. Then we have

$$B_{n2}(j,k) = \frac{1}{n}\left(\min(j,k) - \frac{jk}{n+1} - \frac{3jk(n-j+1)(n-k+1)}{n(n+1)(n+2)}\right),$$

which has the uniform limit

$$K_2(s,t) = \min(s,t) - st - 3st(1-s)(1-t). \tag{1.123}$$

It follows from Theorem 1.11 that

$$S_{n2} = \frac{1}{n} R_{n2} = \frac{1}{n} \frac{\boldsymbol{y}' N' \Omega^{-2} N \boldsymbol{y}}{\boldsymbol{y}' N' \Omega^{-1} N \boldsymbol{y}} \Rightarrow S_2 = \int_0^1 \int_0^1 K_2(s,t)\, dW(s)\, dW(t). \tag{1.124}$$

It is interesting to note that the function $K_2(s,t)$ in (1.123) was earlier presented in connection with goodness-of-fit tests in Section 1.5.1, and its FD is given in (1.79) together with the c.f. of S_2 in (1.80).

In this section we have dealt with the MA(1) model with unknown mean. If the mean is known and assumed to be 0, the analysis is easier (Exercise 1.5.7).

In Chapter 6 we shall deal with more general models and derive the asymptotic distributions of the LBIU statistics under the local alternative

$$H_1(c,\kappa) : \alpha = 1 - \frac{c}{n^{\kappa}},$$

where c and κ are positive constants. We shall also deal with the conditional model where the initial value ε_0 of the error process $\{\varepsilon_j\}$ is set at 0. The LBI test is derived for that model, and the power performance of the LBI test is compared with that of the LBIU test discussed in this section.

1.5.4 LBI and LBIU Tests for the Autoregressive Unit Root

Let us consider the following model:

$$y_j = \boldsymbol{\beta}' \boldsymbol{x}_j + u_j, \quad u_j = \rho\, u_{j-1} + \varepsilon_j \qquad (j = 1, \ldots, n), \tag{1.125}$$

where $\boldsymbol{\beta}$ and $\boldsymbol{x}_j$ are $p \times 1$ coefficient and regression vectors, respectively, and $\{u_j\}$ is an error term that follows the first-order autoregressive (AR(1)) process with $\{\varepsilon_j\} \sim \text{i.i.d.}(0, \sigma_\varepsilon^2)$. The AR(1) coefficient parameter ρ is assumed to take any value, and the initial value u_0 of $\{u_j\}$ is assumed to be $u_0 = 0$.

Our concern here is to conduct the test

$$H_0 : \rho = 1 \quad \text{versus} \quad H_1 : \rho < 1 \quad \text{or} \quad H_1 : \rho \neq 1. \tag{1.126}$$

Dickey and Fuller (1979) first suggested the unit root test using the following three models:

$$y_j = \begin{cases} \rho\, y_{j-1} + \varepsilon_j, \\ \\ \rho\, y_{j-1} + \alpha + \varepsilon_j, \\ \\ \rho\, y_{j-1} + \beta + \gamma\, j, \end{cases} \tag{1.127}$$

where it is assumed that ρ is close to unity and the true value of (α, γ) is $(0, 0)$, while β is arbitrary. This seemingly curious restriction may be understood from the model in (1.125). In fact, when $\boldsymbol{x}_j = (1, j)'$ in (1.125), we have

$$y_j = \beta_1 + \beta_2 j + u_j = \rho y_{j-1} + \beta_1(1 - \rho) + \beta_2\rho + \beta_2(1 - \rho) j + \varepsilon_j,$$

and it is recognized that the above restriction is imposed on this model when $\rho = 1$.

The unit root test applied to the model (1.125) is based on $n(\hat{\rho} - 1)$, where $\hat{\rho}$ is the LSE of ρ computed from

$$\hat{\rho} = \frac{\sum_{j=2}^{n} \hat{u}_{j-1}\hat{u}_j}{\sum_{j=2}^{n} \hat{u}_{j-1}^2}, \quad \hat{u}_j = y_j - \hat{\boldsymbol{\beta}}'\boldsymbol{x}_j. \tag{1.128}$$

Here $\hat{\boldsymbol{\beta}}$ is the LSE of $\boldsymbol{\beta}$ obtained from the regression of $\{y_j\}$ on $\{\boldsymbol{x}_j\}$, where we consider three cases for $\boldsymbol{x}_j$:

$$\text{Case 1: } \boldsymbol{x}_j = 0, \quad \text{Case 2: } \boldsymbol{x}_j = 1, \quad \text{Case 3: } \boldsymbol{x}_j = (1, j)'.$$

Note that $\hat{\rho}$ computed for the three cases is asymptotically equivalent to the LSE computed from the three models in (1.127). Then it can be shown (Phillips, 1987a; Nabeya and Tanaka, 1990a) that, as $n \to \infty$ under H_0,

$$n(\hat{\rho} - 1) \Rightarrow \frac{\int_0^1 G(t)\, dW(t)}{\int_0^1 G^2(t)\, dt}, \tag{1.129}$$

where

$$G(t) = \begin{cases} W(t), & \text{(Case 1)} \\ W(t) - \int_0^1 W(s)\, ds, & \text{(Case 2)} \\ W(t) + (6t - 4)\int_0^1 W(s)\, ds - (12t - 6)\int_0^1 sW(s)\, ds. & \text{(Case 3)} \end{cases}$$

We already presented an expression for the limit in distribution of $n(\hat{\rho} - 1)$ for Case 1, which is the unit root distribution given in (1.52). Note that $G(t)$ in Case 2 is referred to as *demeaned Bm*, which comes from

$$\frac{1}{\sqrt{n}\sigma_\varepsilon}\hat{u}_j = \frac{1}{\sqrt{n}\sigma_\varepsilon}\left(u_j - \frac{1}{n}\sum_{k=1}^{n} u_k\right) \quad \Leftrightarrow \quad W(t) - \int_0^1 W(s)\, ds,$$

whereas $G(t)$ in Case 3 is *detrended Bm*, which comes from (Exercise 1.5.8)

$$\begin{aligned}\frac{1}{\sqrt{n}\sigma_\varepsilon}\hat{u}_j &= \frac{1}{\sqrt{n}\sigma_\varepsilon}(u_j - \hat{\beta}_1 - \hat{\beta}_2 j) \\ &\Leftrightarrow W(t) + (6t-4)\int_0^1 W(s)\,ds - (12t-6)\int_0^1 sW(s)\,ds.\end{aligned}$$

The limiting distributions of $n(\hat{\rho}-1)$ under H_0 can be expressed as

$$F(x) = \lim_{n\to\infty} P\Big(n(\hat{\rho}-1) \le x\Big) = P(H(x) \ge 0),$$

where

$$H(x) = x\int_0^1 G^2(W)\,dt - \int_0^1 G(t)\,dW(t).$$

The distribution function $F(x)$ can be computed by deriving the FD associated with $H(x)$. The limiting local powers of the unit root tests can also be computed by deriving the associated FD, which was discussed in Nabeya and Tanaka (1990a, 1990b). This topic will be discussed in Chapter 7.

In subsequent discussions, as a continuation of the last two sections, we consider the LBI or LBIU type tests, assuming that $\{\varepsilon_j\} \sim \text{NID}(0, \sigma_\varepsilon^2)$. The model (1.125) can be expressed as

$$\boldsymbol{y} = X\boldsymbol{\beta} + \boldsymbol{u}, \qquad \boldsymbol{u} = C(\rho)\boldsymbol{\varepsilon}, \tag{1.130}$$

where

$$\boldsymbol{y} = \begin{pmatrix} y_1 \\ y_2 \\ \vdots \\ y_n \end{pmatrix}, \quad X = \begin{pmatrix} \boldsymbol{x}_1' \\ \boldsymbol{x}_2' \\ \vdots \\ \boldsymbol{x}_n' \end{pmatrix}, \quad \boldsymbol{u} = \begin{pmatrix} u_1 \\ u_2 \\ \vdots \\ u_n \end{pmatrix}, \quad \boldsymbol{\varepsilon} = \begin{pmatrix} \varepsilon_1 \\ \varepsilon_2 \\ \vdots \\ \varepsilon_n \end{pmatrix},$$

whereas

$$C(\rho) = \begin{pmatrix} 1 & & & & 0 \\ \rho & 1 & & & \\ \cdot & \cdot & \cdot & & \\ \cdot & & \cdot & \cdot & \\ \cdot & & & \cdot & \cdot \\ \rho^{n-1} & \cdot & \cdot & \cdot & \rho \quad 1 \end{pmatrix}. \tag{1.131}$$

Here $C(1) = C$ is the random walk-generating matrix defined in (1.93).

Putting $\Phi(\rho) = C(\rho)C'(\rho)$, $\Phi = \Phi(1) = CC'$, and noting (Exercise 1.5.9) that

$$\left.\frac{d\Phi(\rho)}{d\rho}\right|_{\rho=1} = \boldsymbol{d}\boldsymbol{d}' - \Phi, \quad \boldsymbol{d} = \begin{pmatrix} 1 \\ 2 \\ \vdots \\ n \end{pmatrix}, \tag{1.132}$$

it follows from (1.97) that the LBI test rejects H_0 when

$$R_{n3} = \frac{\boldsymbol{y}'H\left(H'\Phi H\right)^{-1}H'\boldsymbol{d}\boldsymbol{d}'H\left(H'\Phi H\right)^{-1}H'\boldsymbol{y}}{\boldsymbol{y}'H\left(H'\Phi H\right)^{-1}H'\boldsymbol{y}} \tag{1.133}$$

takes small values, where H is an $n \times (n-p)$ matrix such that $H'H = I_{n-p}$, and $HH' = I_n - X(X'X)^{-1}X' = M$. It follows from Lemma 1.12 that

$$H(H'\Phi H)^{-1}H' = \Phi^{-1} - \Phi^{-1}X(X'\Phi^{-1}X)^{-1}X'\Phi^{-1} = \tilde{M},$$

which gives

$$R_{n3} = \frac{\boldsymbol{y}'\tilde{M}\boldsymbol{d}\boldsymbol{d}'\tilde{M}\boldsymbol{y}}{\boldsymbol{y}'\tilde{M}\boldsymbol{y}}. \tag{1.134}$$

It is recognized that R_{n3} in (1.134) reduces to 0 if $\tilde{M}\boldsymbol{d} = \boldsymbol{0}$. This occurs if $\boldsymbol{d}$ belongs to the column space of X because it holds that $\tilde{M}X = 0$. Avoiding this case, let us consider the case where $X = (1, \ldots, 1)' = \boldsymbol{e}$. Then we have

$$\begin{aligned}
\boldsymbol{y}'\tilde{M}\boldsymbol{y} &= \sum_{j=2}^{n}(y_j - y_{j-1})^2 = \sum_{j=2}^{n}(u_j - u_{j-1})^2, \\
\boldsymbol{y}'\tilde{M}\boldsymbol{d} &= y_n - y_1 = u_n - u_1,
\end{aligned}$$

so that

$$R_{n3} = \frac{\boldsymbol{y}'\tilde{M}\boldsymbol{d}\boldsymbol{d}'\tilde{M}\boldsymbol{y}}{\boldsymbol{y}'\tilde{M}\boldsymbol{y}} = \frac{(u_n - u_1)^2}{\sum_{j=2}^{n}(u_j - u_{j-1})^2} = \frac{U_n}{V_n},$$

where $U_n = (u_n - u_1)^2$ and $V_n = \sum_{j=2}^{n}(u_j - u_{j-1})^2$. It follows that, as $n \to \infty$ under H_0, $R_{n3} \Rightarrow \chi^2(1)$. The power performance of the present LBI test is found to be quite poor and is not recommended; this will be clarified in Chapter 7.

We next consider the LBIU test for the case where $\boldsymbol{d} = (1, 2, \ldots, n)'$ belongs to the column space of X. To derive the test, it is more convenient to use the log-likelihood for $\boldsymbol{y} \sim \mathrm{N}(X\boldsymbol{\beta}, \sigma_\varepsilon^2 \Phi(\rho))$. Noting that $|\Phi(\rho)| = 1$, the log-likelihood is given by

$$L(\boldsymbol{\delta}) = -\frac{n}{2} \log 2\pi\sigma_\varepsilon^2 - \frac{1}{2\sigma_\varepsilon^2}(\boldsymbol{y} - X\boldsymbol{\beta})'\Phi^{-1}(\rho)(\boldsymbol{y} - X\boldsymbol{\beta}), \tag{1.135}$$

where $\boldsymbol{\delta} = (\rho, \boldsymbol{\beta}', \sigma_\varepsilon^2)'$. Then the LBIU test rejects H_0 when

$$\begin{aligned} \left.\frac{\partial^2 L(\boldsymbol{\delta})}{\partial \rho^2}\right|_{H_0} &= -\frac{1}{2\sigma_\varepsilon^2}(\boldsymbol{y} - X\boldsymbol{\beta})'\frac{d^2\Phi^{-1}(\rho)}{d^2\rho}(\boldsymbol{y} - X\boldsymbol{\beta})\Bigg|_{H_0} \\ &= -\frac{1}{2\tilde{\sigma}_\varepsilon^2}\boldsymbol{y}'N'\left.\frac{d^2\Phi^{-1}(\rho)}{d^2\rho}\right|_{\rho=1} N\boldsymbol{y} \end{aligned}$$

becomes large, where $N = I_n - X(X'\Phi^{-1}X)^{-1}X'\Phi^{-1}$ and $\tilde{\sigma}_\varepsilon^2 = \boldsymbol{y}'N'\Phi^{-1} N\boldsymbol{y}/n$. We also obtain (Exercise 1.5.10)

$$\left.\frac{d^2\Phi^{-1}(\rho)}{d\rho^2}\right|_{\rho=1} = 2(I_n - \boldsymbol{e}_n\boldsymbol{e}_n'), \quad \boldsymbol{e}_n = (0, \ldots, 0, 1)' : n \times 1,$$

which yields

$$\left.\frac{\partial^2 L(\boldsymbol{\delta})}{\partial \rho^2}\right|_{H_0} = -n\frac{\boldsymbol{y}'N'(I_n - \boldsymbol{e}_n\boldsymbol{e}_n')N\boldsymbol{y}}{\boldsymbol{y}'N'\Phi^{-1}N\boldsymbol{y}}. \tag{1.136}$$

Since it holds that $N'\Phi^{-1}X = 0$, $\boldsymbol{y}'N'\boldsymbol{e}_n$ reduces to 0 if $\boldsymbol{e}_n$ belongs to the column space of $\Phi^{-1}X$ or if $\Phi\boldsymbol{e}_n = CC'\boldsymbol{e}_n = \boldsymbol{d}$ belongs to the column space of X. In that case we have the LBIU statistic

$$R_{n4} = \frac{\boldsymbol{y}'N'N\boldsymbol{y}}{\boldsymbol{y}'N'\Phi^{-1}N\boldsymbol{y}} = \frac{\boldsymbol{y}'N'N\boldsymbol{y}}{\boldsymbol{y}'\tilde{M}\boldsymbol{y}}, \tag{1.137}$$

and H_0 is rejected when R_{n4} takes small values.

Consider $X = (\boldsymbol{e}, \boldsymbol{d})$, which yields (Exercise 1.5.11)

$$\begin{aligned} \boldsymbol{y}'N'N\boldsymbol{y} &= \sum_{j=2}^{n}\left(y_j - y_1 - \frac{j-1}{n-1}(y_n - y_1)\right)^2 \\ &= \sum_{j=2}^{n}\left(u_j - u_1 - \frac{j-1}{n-1}(u_n - u_1)\right)^2, \\ \boldsymbol{y}'\tilde{M}\boldsymbol{y} &= \sum_{j=2}^{n}\left(y_j - y_{j-1} - \frac{y_n - y_1}{n-1}\right)^2 = \sum_{j=2}^{n}\left(u_j - u_{j-1} - \frac{u_n - u_1}{n-1}\right)^2, \end{aligned}$$

where it holds (Exercise 1.5.12) that, as $n \to \infty$ under H_0, $\mathbf{y}'\tilde{M}\mathbf{y}/(n\sigma_\varepsilon^2) \to 1$ in probability and

$$S_{n4} = \frac{1}{n} R_{n4} = \frac{\mathbf{y}'N'N\mathbf{y}/\left(n^2\sigma_\varepsilon^2\right)}{\mathbf{y}'\tilde{M}\mathbf{y}/\left(n\sigma_\varepsilon^2\right)} \Rightarrow S_4,$$

$$S_4 = \int_0^1 (W(t) - tW(1))^2 \, dt \overset{\mathcal{D}}{=} \int_0^1 \int_0^1 [\min(s,t) - st] \, dW(s) \, dW(t).$$

The c.f. of S_4 is $(D(2i\theta))^{-1/2}$, where $D(\lambda) = \sin\sqrt{\lambda}/\sqrt{\lambda}$.

The performance of the LBIU test is found to be quite good, unlike the LBI test; this will be clarified in Chapter 7.

Exercises for Section 1.5

Exercise 1.5.1 Show that the modified Cramér–von Mises–Smirnov statistic $W_n^2(\hat{\xi})$ defined in (1.75) can be rewritten as

$$W_n^2(\hat{\xi}) = \sum_{j=1}^n \left(F(X_j'; \hat{\xi}) - \frac{2j-1}{2n} \right)^2 + \frac{1}{12n},$$

where $X_1' \le X_2' \le \cdots \le X_n'$ are ordered observations of $X_1, X_2, \ldots, X_n$.

Exercise 1.5.2 Let $X_1, \ldots, X_n$ be a random sample from the continuous distribution $F(x)$. Prove that

$$\mathrm{E}(H_n(t)) = t, \qquad \mathrm{Cov}(H_n(s), H_n(t)) = \frac{1}{n} (\min(s,t) - st),$$

where $H_n(t)$ is the empirical distribution function defined by

$$H_n(t) = \frac{1}{n} \sum_{j=1}^n \epsilon\big(t - F(X_j)\big), \quad \epsilon(x) = \begin{cases} 1 & (x \ge 0), \\ 0 & (x < 0). \end{cases}$$

Exercise 1.5.3 Let $F(x;\xi)$ be the continuous distribution function of X with a parameter ξ, and $f(x;\xi)$ be its probability density. Put

$$\sigma^{-2}(\xi) = \int_{-\infty}^{\infty} \left(\frac{\partial \log f(x;\xi)}{\partial \xi} \right)^2 f(x;\xi) \, dx,$$

$$g_\xi(t) = \left. \frac{\partial F(x;\xi)}{\partial \xi} \right|_{x = x(t;\xi)}, \quad x(t;\xi) = \inf\{x : F(x;\xi) = t\}.$$

Then prove that, when $F(x;\xi) = R(x - \xi)$ and $r(x) = R'(x)$,

$$\sigma(\xi) \, g_\xi(t) = -\left(\int_{-\infty}^{\infty} \frac{(r'(x))^2}{r(x)} \, dx \right)^{-1/2} r(R^{-1}(t)).$$

Prove also that, when $F(x;\xi) = R(x/\xi)$,

$$\sigma(\xi)\, g_\xi(t) = -\left(\int_{-\infty}^{\infty} \frac{(x\, r'(x))^2}{r(x)}\, dx - 1\right)^{-1/2} R^{-1}(t)\, r(R^{-1}(t)).$$

Exercise 1.5.4 For the distribution function $F(x;\xi) = (R(x))^\xi$ $(\xi > 0)$, prove that

$$\sigma(\xi)\, g_\xi(t) = t\, \log t,$$

where $\sigma(\xi)$ and $g_\xi(t)$ are as defined in Exercise 1.5.3.

Exercise 1.5.5 Consider the model

$$y_j = \beta_j + \varepsilon_j, \quad \beta_j = \beta_{j-1} + \eta_j \quad (j = 1, \ldots, n),$$

where $\beta_0 = 0$, $\{\varepsilon_j\} \sim \mathrm{NID}(0, \sigma_\varepsilon^2)$ and $\{\eta_j\} \sim \mathrm{NID}(0, \sigma_\eta^2)$, with $\{\varepsilon_j\}$ and $\{\eta_j\}$ being independent of each other. Putting $\rho = \sigma_\eta^2/\sigma_\varepsilon^2$, show that $\partial L(\rho, \sigma_\varepsilon^2)/\partial\rho$ does not tend to normality when $\rho = 0$, where $L(\rho, \sigma_\varepsilon^2)$ is the log-likelihood for $\boldsymbol{y} = (y_1, \ldots, y_n)'$.

Exercise 1.5.6 Prove that

$$H(H'\Omega H)^{-1} H' = N'\Omega^{-1} N, \quad H(H'\Omega H)^{-2} H' = N'\Omega^{-2} N,$$

where Ω is any $n \times n$ nonsingular symmetric matrix, H is an $n \times (n-p)$ matrix with rank $n - p$, $N = I_n - X(X'\Omega^{-1}X)^{-1}X'\Omega^{-1}$, and X is an $n \times p$ matrix with rank p such that $H'X = 0$.

Exercise 1.5.7 Consider the testing problem $H_0 : \alpha = 1$ for the MA(1) model

$$y_j = \varepsilon_j - \alpha\varepsilon_{j-1} \quad (j = 1, \ldots, n),$$

where $\{\varepsilon_j\} \sim \mathrm{NID}(0, \sigma_\varepsilon^2)$. Derive the LBIU test statistic and its asymptotic distribution under H_0.

Exercise 1.5.8 Consider the model

$$y_j = \beta_1 + \beta_2\, j + u_j, \quad u_j = u_{j-1} + \varepsilon_j \qquad (j = 1, \ldots, n),$$

where $\{\varepsilon_j\} \sim \mathrm{i.i.d.}(0, \sigma_\varepsilon^2)$. Denote by $\hat{u}_j$ the residual obtained from the least squares regression of $\{y_j\}$ on $(1, j)$ $(j = 1, \ldots, n)$. Prove that

$$\frac{1}{n^2\sigma_\varepsilon^2} \sum_{j=1}^{n} \hat{u}_j^2 \Rightarrow \int_0^1 \left(W(t) + (6t - 4)\int_0^1 W(s)\, ds \right.$$
$$\left. -(12t - 6)\int_0^1 sW(s)\, ds\right)^2 dt.$$

Exercise 1.5.9 Define the $n \times n$ matrix $C(\rho)$ by (1.131), where its (j,k)th element is ρ^{j-k} for $j \geq k$ and 0 for $j < k$. Show that

$$\left.\frac{dC(\rho)C'(\rho)}{d\rho}\right|_{\rho=1} = \begin{pmatrix} 1 \\ 2 \\ \vdots \\ n \end{pmatrix} (1,\, 2,\, \ldots,\, n) - C(1)C'(1).$$

Exercise 1.5.10 For the matrix $C(\rho)$ given in Exercise 1.5.9, show that

$$\left.\frac{d^2\left\{\left(C(\rho)C'(\rho)\right)^{-1}\right\}}{d\rho^2}\right|_{\rho=1} = 2\left(I_n - e_n e_n'\right), \quad e_n = (0,\, \ldots,\, 0, 1)' : n \times 1.$$

Exercise 1.5.11 Consider the model

$$\boldsymbol{y} = X\boldsymbol{\beta} + \boldsymbol{u}, \quad \boldsymbol{u} = C(1)\boldsymbol{\varepsilon}, \quad \boldsymbol{\varepsilon} \sim \mathrm{N}\left(\boldsymbol{0}, \sigma_\varepsilon^2 I_n\right),$$

where $X = (\boldsymbol{e}, \boldsymbol{d})$, $\boldsymbol{e} = (1,\, \ldots,\, 1)' : n \times 1$ and $\boldsymbol{d} = (1, 2,\, \ldots,\, n)'$, whereas $C(\rho)$ is as given in Exercise 1.5.9 with $C = C(1)$. Then show that

$$\boldsymbol{y}'N'N\boldsymbol{y} = \sum_{j=2}^{n} \left(u_j - u_1 - \frac{j-1}{n-1}(u_n - u_1)\right)^2,$$

$$\boldsymbol{y}'\tilde{M}\boldsymbol{y} = \sum_{j=2}^{n} \left(u_j - u_{j-1} - \frac{u_n - u_1}{n-1}\right)^2,$$

where $N = I_n - X(X'(CC')^{-1}X)^{-1}X'(CC')^{-1}$ and $\tilde{M} = N'(CC')^{-1}N$.

Exercise 1.5.12 Prove that

$$\frac{1}{n^2}\sum_{j=2}^{n} \left(u_j - u_1 - \frac{j-1}{n-1}(u_n - u_1)\right)^2 \Rightarrow \int_0^1 (W(t) - tW(1))^2\, dt,$$

where $u_j = u_{j-1} + \varepsilon_j$, $u_0 = 0$ and $\{\varepsilon_j\} \sim \mathrm{NID}(0, 1)$.

2

Integral Equations and the Fredholm Determinant

In this chapter we present a general method for computing the c.f.s of purely quadratic functionals of Bm expressed by the Riemann–Stieltjes double integral. This method originates from the Anderson and Darling (1952) theorem that the c.f.s of such statistics can be derived via the Fredholm determinant (FD), which is an important quantity in the theory of homogeneous integral equations of Fredholm type. For the actual computation of the FD we find it easier to convert the integral equation into the equivalent differential equation with some boundary conditions. We present by many examples how to compute the FD, classifying the associated differential equations into five categories. We also suggest, for special types of the functionals of Bm, computationally efficient methods for obtaining the FD without dealing with differential equations. It is recognized that computerized algebra is quite useful for the derivation of the FD. If linear or bilinear functionals are added to the quadratic functionals of Bm, we need to deal with the resolvent in the theory of nonhomogeneous integral equations, which is the main theme of the next chapter.

2.1 Fredholm Determinant

We already introduced the FD in Chapter 1 without mentioning its definition and properties. Here we first give the definition of the FD. For this purpose we introduce the homogeneous integral equation of Fredholm type. Given $K(s,t)$, consider the following equation for λ and $f(t)$:

$$f(t) = \lambda \int_0^1 K(s,t) f(s)\, ds, \tag{2.1}$$

where $K(s,t)$ is called a *kernel* and is assumed to be a symmetric and continuous function in $[0,1] \times [0,1]$. A value λ for which this integral equation possesses a nonvanishing continuous solution $f(t)$ is called an *eigenvalue* of $K(s,t)$; the corresponding solution $f(t)$ is called an *eigenfunction* for the eigenvalue λ. It is known that there exists at least one eigenvalue insofar as $K(s,t)$ is not identically equal to zero. Note also that every eigenvalue is real, whereas $\lambda = 0$ is certainly not an eigenvalue. The maximum number of linearly independent eigenfunctions corresponding to the eigenvalue λ is called the *multiplicity* of λ.

To define the FD of $K(s,t)$, we approximate the integral equation (2.1) by the algebraic system

$$f\left(\frac{k}{N}\right) = \frac{\lambda}{N}\sum_{j=1}^{N} K\left(\frac{j}{N}, \frac{k}{N}\right) f\left(\frac{j}{N}\right) \qquad (k = 1, \ldots, N),$$

or, in matrix notation,

$$\boldsymbol{f}_N = \frac{\lambda}{N} K_N \boldsymbol{f}_N \quad \Leftrightarrow \quad \left(I_N - \frac{\lambda}{N} K_N\right) \boldsymbol{f}_N = \boldsymbol{0}, \tag{2.2}$$

where $\boldsymbol{f}_N = [(f(k/N))]$ is an $N \times 1$ vector and $K_N = [(K(j/N, k/N))]$ is an $N \times N$ symmetric matrix. To obtain λ that satisfies (2.2), we consider

$$D_N(\lambda) = \left| I_N - \frac{\lambda}{N} K_N \right|. \tag{2.3}$$

The zero of $D_N(\lambda)$ gives the reciprocal of the eigenvalues of K_N/N in matrix theory. The definition of $D_N(\lambda)$ in (2.3) gives a normalization $D_N(0) = 1$, which is independent of the matrix K_N. Then the FD of $K(s,t)$ is defined (see Hochstadt (1973) and Exercise 2.1.1) by

$$D(\lambda) = \lim_{N\to\infty} D_N(\lambda) = \lim_{N\to\infty} \left| I_N - \frac{\lambda}{N} K_N \right| \tag{2.4}$$

$$= 1 + \sum_{n=1}^{\infty} \frac{(-1)^n d_n}{n!} \lambda^n, \tag{2.5}$$

where

$$d_n = \int_0^1 \cdots \int_0^1 K\begin{pmatrix} t_1 & \cdot & \cdot & \cdot & t_n \\ t_1 & \cdot & \cdot & \cdot & t_n \end{pmatrix} dt_1 \cdots dt_n,$$

$$K\begin{pmatrix} t_1 & \cdot & \cdot & \cdot & t_n \\ t_1 & \cdot & \cdot & \cdot & t_n \end{pmatrix} = \begin{vmatrix} K(t_1,t_1) & K(t_1,t_2) & \cdots & K(t_1,t_n) \\ \vdots & \vdots & & \vdots \\ K(t_n,t_1) & K(t_n,t_2) & \cdots & K(t_n,t_n) \end{vmatrix}.$$

Note here that

$$d_1 = \int_0^1 K(s,s)\,ds, \tag{2.6}$$

$$\begin{aligned} d_2 &= \int_0^1 \int_0^1 \begin{vmatrix} K(s,s) & K(s,t) \\ K(s,t) & K(t,t) \end{vmatrix} ds\,dt \\ &= d_1^2 - \int_0^1 \int_0^1 K^2(s,t)\,ds\,dt, \end{aligned} \tag{2.7}$$

which will be used in Theorem 2.4.

The series in (2.5) converges for all λ (see Hochstadt (1973) and Exercise 2.1.2), that is, $D(\lambda)$ is an *entire function* of λ with $D(0) = 1$. Then $D(\lambda)$ has an infinite product expansion (Hochstadt, 1973, p.249) expressed as

$$D(\lambda) = \exp\left\{-\lambda \int_0^1 K(t,t)\,dt\right\} \prod_{n=1}^{\infty} \left\{\left(1 - \frac{\lambda}{\lambda_n}\right) \exp\left(\frac{\lambda}{\lambda_n}\right)\right\}, \tag{2.8}$$

where $\lambda_1, \lambda_2, \ldots$ are eigenvalues of $K(s,t)$ repeated as many times as their multiplicities. A much simpler product expansion for $D(\lambda)$ will be obtained shortly by imposing a condition on $K(s,t)$ in addition to symmetry and continuity.

It is recognized in (2.8) that the zeros of $D(\lambda)$ are eigenvalues of $K(s,t)$. In fact, we have (Hochstadt, 1973, p.243)

Theorem 2.1 *Suppose that $K(s,t)$ is symmetric and continuous in* $[0,1] \times [0,1]$. *Then every zero of $D(\lambda)$ is an eigenvalue of $K(s,t)$, and in turn every eigenvalue of $K(s,t)$ is a zero of $D(\lambda)$.*

It follows from Theorem 2.1 that $D(\lambda)$ has necessary and sufficient information about eigenvalues of the kernel $K(s,t)$. If $K(s,t)$ has an infinite number of eigenvalues, $K(s,t)$ is said to be *nondegenerate*; otherwise it is *degenerate*. These notions were already introduced in Section 1.3 without mentioning the eigenvalues. If all of the eigenvalues are positive (negative), then $K(s,t)$ is said to be *positive (negative) definite*. This is equivalent to

$$\int_0^1 \int_0^1 K(s,t)\,g(s)\,g(t)\,ds\,dt \geq 0 \quad (\leq 0) \tag{2.9}$$

for any continuous function $g(t)$ defined on $[0,1]$. The kernel $K(s,t)$ is said to be *nearly definite* if all but a finite number of eigenvalues have the same sign. Any degenerate kernel is always nearly definite; so is the sum of a definite kernel and degenerate kernels.

The following theorem called *Mercer's theorem* is quite useful, not only for obtaining a simpler expression for $D(\lambda)$ in (2.8) but also for deriving distributions used in subsequent discussions. For the proof, see Courant and Hilbert (1953, p.138) and Hochstadt (1973, p.91).

Theorem 2.2 (Mercer's theorem) *Let $K(s,t)$ be symmetric, continuous, and nearly definite in $[0,1] \times [0,1]$. Then $K(s,t)$ has the series expansion given by*

$$K(s,t) = \sum_{n=1}^{\infty} \frac{f_n(s) f_n(t)}{\lambda_n}, \tag{2.10}$$

where $\{\lambda_n\}$ is a sequence of eigenvalues of $K(s,t)$ repeated as many times as their multiplicities, $\{f_n(t)\}$ is an orthonormal sequence of eigenfunctions corresponding to λ_n, and the series on the right-hand side converges absolutely and uniformly to $K(s,t)$.

It can be seen from (2.10) that $K(s,t)$ and $K(1-s, 1-t)$ have the same set of eigenvalues, which implies that the two kernels have the same FD because of Theorem 2.3 described below. This can also be verified from (2.5) by checking the definition of d_n. It follows from Mercer's theorem that, if $K(s,t)$ is symmetric, continuous, and nearly definite,

$$\int_0^1 K(t,t)\, dt = \sum_{n=1}^{\infty} \frac{1}{\lambda_n}, \tag{2.11}$$

where each eigenvalue is repeated as many times as its multiplicity, and the sum converges absolutely. Thus we have, from (2.8),

$$\begin{aligned} D(\lambda) &= \exp\left\{ -\lambda \int_0^1 K(t,t)\, dt + \sum_{n=1}^{\infty} \frac{\lambda}{\lambda_n} \right\} \prod_{n=1}^{\infty} \left(1 - \frac{\lambda}{\lambda_n}\right) \\ &= \prod_{n=1}^{\infty} \left(1 - \frac{\lambda}{\lambda_n}\right). \end{aligned}$$

We can now prove the theorem by Anderson and Darling (1952) presented in (1.21), which relates the distribution of a quadratic functional of Bm to the FD. Here we update the *Anderson–Darling theorem* in the following way:

Theorem 2.3 *Consider the statistic*

$$S = \int_0^1 \int_0^1 K(s,t)\, dW(s)\, dW(t) \overset{\mathcal{D}}{=} \int_0^1 \int_0^1 K(1-s, 1-t)\, dW(s)\, dW(t), \tag{2.12}$$

where $K(s,t)$ is symmetric, continuous, and nearly definite. Then the c.f. of S is given by

$$E\left(e^{i\theta S}\right) = E\left[\exp\left\{i\theta \int_0^1 \int_0^1 K(s,t)\, dW(s)\, dW(t)\right\}\right]$$
$$= (D(2i\theta))^{-1/2} \tag{2.13}$$
$$= \prod_{n=1}^{\infty}\left(1 - \frac{2i\theta}{\lambda_n}\right)^{-1/2}, \tag{2.14}$$

where $D(\lambda)$ is the FD of $K(s,t)$, and $\{\lambda_n\}$ is a sequence of eigenvalues of $K(s,t)$ repeated as many times as their multiplicities.

Proof We have, from Mercer's theorem,

$$S = \int_0^1 \int_0^1 K(s,t)\, dW(s)\, dW(t) = \int_0^1 \int_0^1 \sum_{n=1}^{\infty} \frac{f_n(s) f_n(t)}{\lambda_n}\, dW(s)\, dW(t)$$
$$= \sum_{n=1}^{\infty} \frac{1}{\lambda_n}\left(\int_0^1 f_n(t)\, dW(t)\right)^2 \stackrel{\mathcal{D}}{=} \sum_{n=1}^{\infty} \frac{1}{\lambda_n} Z_n^2, \quad \{Z_n\} \sim \mathrm{NID}(0,1).$$

Then we can compute the c.f. of S by using this last expression to obtain (2.13) and (2.14), which establishes the theorem. □

Once the FD associated with the statistic S in (2.12) is derived, the computation of moments of S becomes easier, although there are various approaches. In particular, $\mathrm{Var}(S)$ can be computed quite easily, as the following theorem indicates.

Theorem 2.4 *For the statistic S in (2.12), its kth moment is given by*

$$E\left(S^k\right) = E\left[\left(\sum_{n=1}^{\infty} \frac{1}{\lambda_n} Z_n^2\right)^k\right]$$
$$= E\left[\left(\int_0^1 \int_0^1 K(s,t)\, dW(s)\, dW(t)\right)^k\right]$$
$$= \left.\frac{d^k\,(D(2i\theta))^{-1/2}}{i^k\, d\theta^k}\right|_{\theta=0}.$$

In particular, we have

$$E(S) = \sum_{n=1}^{\infty} \frac{1}{\lambda_n} = \int_0^1 K(t,t)\,dt = d_1, \tag{2.15}$$

$$E(S^2) = 3d_1^2 - 2d_2, \tag{2.16}$$

$$\text{Var}(S) = 2\sum_{n=1}^{\infty} \frac{1}{\lambda_n^2} = 2\int_0^1 \int_0^1 K^2(s,t)\,ds\,dt \tag{2.17}$$

$$= 2\left(d_1^2 - d_2\right), \tag{2.18}$$

$$E(S^3) = 15d_1^3 - 18d_1d_2 + 4d_3,$$

$$E(S^4) = 105d_1^4 - 180d_1^2d_2 + 48d_1d_3 + 36d_2^2 - 8d_4,$$

where d_n $(n = 1,2,3,4)$ *are coefficients in the series expansion for* $D(\lambda)$ *given in (2.5) with* d_1 *and* d_2 *defined in (2.6) and (2.7), respectively.*

Proof Since it follows from Theorem 2.3 that

$$S = \int_0^1 \int_0^1 K(s,t)\,dW(s)\,dW(t) \overset{\mathcal{D}}{=} \sum_{n=1}^{\infty} \frac{1}{\lambda_n} Z_n^2,$$

the first and second equalities hold. The third equality comes from the property of the $D(2i\theta)$. The first and second equalities in (2.15) are obvious, and the third equality comes from

$$\mathrm{E}(S) = \left.\frac{d\,(D(2\theta))^{-1/2}}{d\theta}\right|_{\theta=0} = \left.\frac{d}{d\theta}\left(1 + \sum_{n=1}^{\infty} \frac{(-1)^n(2\theta)^n}{n!} d_n\right)^{-1/2}\right|_{\theta=0}$$

$$= \left.\frac{d}{d\theta}\left(1 - 2\theta d_1 + \frac{4\theta^2}{2}d_2 - \frac{8\theta^3}{6}d_3 + \cdots\right)^{-1/2}\right|_{\theta=0} = d_1.$$

We obtain (2.16) by computing $d^2\,(D(2\theta))^{-1/2}\,/d\theta^2|_{\theta=0}$. The first relation in (2.17) is obvious, whereas the second relation comes from (1.20). The relation in (2.18) is obtained from (2.15) and (2.16). Similarly we can obtain the expressions for $\mathrm{E}(S^3)$ and $\mathrm{E}(S^4)$ by computing $d^k\,(D(2\theta))^{-1/2}\,/d\theta^k|_{\theta=0}$ for $k = 3,4$, which establishes the theorem. □

How to obtain the FD is an important statistical problem. Here we present three examples for which we compute the FD using its definition given in (2.5).

Example 2.5 Let us consider the degenerate kernel $K(s,t) = 1 + st$, which was earlier dealt with in (1.25), and the FD was derived by a different method.

Here we have, from (2.5),

$$K\begin{pmatrix} t_1 & \cdot & \cdot & \cdot & t_n \\ t_1 & \cdot & \cdot & \cdot & t_n \end{pmatrix} = \begin{cases} 1+t_1^2 & (n=1), \\ (t_2-t_1)^2 & (n=2), \\ 0 & (n \geq 3). \end{cases}$$

Thus it holds that

$$d_1 = \frac{4}{3}, \quad d_2 = \frac{1}{6}, \quad d_n = 0 \quad (n \geq 3),$$

which gives $D(\lambda) = 1 - 4\lambda/3 + \lambda^2/12$, and

$$\begin{aligned} \mathrm{E}\left(e^{i\theta S}\right) &= \mathrm{E}\left[\exp\left\{i\theta \int_0^1 \int_0^1 (1+st)\, dW(s)\, dW(t)\right\}\right] \\ &= (D(2i\theta))^{-1/2} = \left(1 - \frac{8i\theta}{3} - \frac{\theta^2}{3}\right)^{-1/2}. \end{aligned}$$

It follows from (2.15) and (2.18) that

$$\mathrm{E}(S) = d_1 = \frac{4}{3}, \quad \mathrm{Var}(S) = 2\left(d_1^2 - d_2\right) = 2\left(\frac{16}{9} - \frac{1}{6}\right) = \frac{29}{9}.$$

The eigenvalues are zeros of $D(\lambda)$, which gives $\lambda_1, \lambda_2 = 8 \pm 2\sqrt{13}$, so that

$$S \stackrel{\mathcal{D}}{=} \frac{1}{\lambda_1} Z_1^2 + \frac{1}{\lambda_2} Z_2^2, \quad (Z_1, Z_2)' \sim \mathrm{N}(\mathbf{0}, I_2).$$

Similarly we can compute (Exercise 2.1.3) the FDs of the following kernels:

$$K(s,t) = 1+s+t, \quad 1+s+t+st, \quad 1+s+t+s^2t^2.$$

For the computation of the FD of a degenerate kernel, it is sometimes much simpler to use the method presented in Chapter 1 (Exercise 2.1.4). Let us consider

$$K(s,t) = \sum_{j=1}^{n} g_j(s) g_j(t), \tag{2.19}$$

where $g_j(t)$ $(j = 1, \ldots, n)$ are continuous and linearly independent. Note that $K(s,t)$ is positive definite and was earlier dealt with in Section 1.3. For this kernel it holds that

$$S = \int_0^1 \int_0^1 \sum_{j=1}^{n} g_j(s) g_j(t)\, dW(s)\, dW(t) = \sum_{j=1}^{n} \left(\int_0^1 g_j(t)\, dW(t)\right)^2 = \eta'\eta,$$

where

$$\eta = \left(\int_0^1 g_1(t)\, dW(t), \ \ldots, \ \int_0^1 g_n(t)\, dW(t) \right)' \sim \mathrm{N}(\mathbf{0}, \ \Sigma).$$

Thus we obtain

$$\mathrm{E}\left(e^{i\theta S}\right) = \left| I_n - 2i\theta\Sigma \right|^{-1/2},$$

so that the FD of $K(s,t)$ in (2.19) is given by

$$D(\lambda) = \left| I_n - \lambda\Sigma \right| = \prod_{j=1}^{n} \left(1 - \frac{\lambda}{\lambda_j} \right), \tag{2.20}$$

where $\lambda_1, \ldots, \lambda_n$ are eigenvalues of $K(s,t)$, and $1/\lambda_1, \ldots, 1/\lambda_n$ are eigenvalues of the covariance matrix Σ.

Example 2.6 Let us consider $K(s,t) = \min(s,t)$, which was dealt with in (1.31). It turns out that $K(s,t)$ is positive definite and nondegenerate. We have, for $0 \le t_1 \le \cdots \le t_n \le 1$,

$$K\begin{pmatrix} t_1 & \cdot & \cdot & \cdot & t_n \\ t_1 & \cdot & \cdot & \cdot & t_n \end{pmatrix} = \begin{vmatrix} t_1 & t_1 & \cdot & \cdot & \cdot & t_1 & t_1 \\ t_1 & t_2 & \cdot & \cdot & \cdot & t_2 & t_2 \\ \vdots & \vdots & & & & \vdots & \vdots \\ t_1 & t_2 & \cdot & \cdot & \cdot & t_{n-1} & t_n \end{vmatrix}$$

$$= t_1(t_2 - t_1) \cdots (t_{n-1} - t_{n-2})(t_n - t_{n-1}),$$

and

$$\begin{aligned} d_n &= \int_0^1 \cdots \int_0^1 K\begin{pmatrix} t_1 & \cdot & \cdot & \cdot & t_n \\ t_1 & \cdot & \cdot & \cdot & t_n \end{pmatrix} dt_1 \cdots dt_n \\ &= n! \int \cdots \int_A t_1(t_2 - t_1) \cdots (t_{n-1} - t_{n-2})(t_n - t_{n-1})\, dt_1 \cdots dt_n = \frac{n!}{(2n)!}, \end{aligned}$$

where $A = \{0 \le t_1 \le \cdots \le t_n \le 1\}$. Then the FD of $K(s,t) = \min(s,t)$ is given by

$$D(\lambda) = \sum_{n=0}^{\infty} \frac{(-\lambda)^n}{n!} \frac{n!}{(2n)!} = \sum_{n=0}^{\infty} \frac{(-\lambda)^n}{(2n)!} = \cos\sqrt{\lambda} = \prod_{n=1}^{\infty} \left(1 - \frac{\lambda}{\left(n - \frac{1}{2}\right)^2 \pi^2} \right),$$

where the final and penultimate equalities come from the infinite product and series expansions, respectively, for the cos function. Thus we obtain

$$\begin{aligned}
\mathrm{E}\left(e^{i\theta S}\right) &= \mathrm{E}\left[\exp\left\{i\theta\int_0^1\int_0^1 \min(s,t)\,dW(s)\,dW(t)\right\}\right] \\
&= (D(2i\theta))^{-1/2} = \left(\cos\sqrt{2i\theta}\right)^{-1/2} \\
&= \prod_{n=1}^{\infty}\left(1 - \frac{2i\theta}{\left(n-\frac{1}{2}\right)^2\pi^2}\right)^{-1/2},
\end{aligned}$$

and eigenvalues are given by $\lambda_n = (n-1/2)^2\pi^2$ $(n = 1, 2, \ldots)$. Since $D(\lambda) = \cos\sqrt{\lambda} = 1 - d_1\lambda + d_2\lambda^2/2 - \cdots$ with $d_1 = 1/2$ and $d_2 = 1/12$, we have

$$\mathrm{E}(S) = \frac{1}{2}, \quad \mathrm{Var}(S) = 2\left(\frac{1}{4} - \frac{1}{12}\right) = \frac{1}{3}.$$

Example 2.7 Similarly, if $K(s,t) = \min(s,t) - st$, it can be shown (Exercise 2.1.5) that, for $0 \le t_1 \le \cdots \le t_n \le 1$,

$$\begin{aligned}
K\begin{pmatrix} t_1 & \cdot & \cdot & \cdot & t_n \\ t_1 & \cdot & \cdot & \cdot & t_n \end{pmatrix} &= \begin{vmatrix} t_1 - t_1^2 & t_1 - t_1t_2 & \cdot & \cdot & \cdot & t_1 - t_1t_{n-1} & t_1 - t_1t_n \\ t_1 - t_2t_1 & t_2 - t_2^2 & \cdot & \cdot & \cdot & t_2 - t_2t_{n-1} & t_2 - t_2t_n \\ \vdots & \vdots & & & & \vdots & \vdots \\ t_1 - t_nt_1 & t_2 - t_nt_2 & \cdot & \cdot & \cdot & t_{n-1} - t_{n-1}t_n & t_n - t_n^2 \end{vmatrix} \\
&= t_1(t_2 - t_1)\cdots(t_n - t_{n-1})(1 - t_n),
\end{aligned}$$

and

$$D(\lambda) = \sum_{n=0}^{\infty}\frac{(-\lambda)^n}{n!}\frac{n!}{(2n+1)!} = \sum_{n=0}^{\infty}\frac{(-\lambda)^n}{(2n+1)!} = \frac{\sin\sqrt{\lambda}}{\sqrt{\lambda}} = \prod_{n=1}^{\infty}\left(1 - \frac{\lambda}{n^2\pi^2}\right),$$

where the final and penultimate equalities come from the infinite product and series expansions, respectively, for the sin function. Thus we obtain

$$\begin{aligned}
\mathrm{E}\left(e^{i\theta S}\right) &= \mathrm{E}\left[\exp\left\{i\theta\int_0^1\int_0^1 [\min(s,t) - st]\,dW(s)\,dW(t)\right\}\right] \\
&= (D(2i\theta))^{-1/2} = \left(\frac{\sin\sqrt{2i\theta}}{\sqrt{2i\theta}}\right)^{-1/2} \\
&= \prod_{n=1}^{\infty}\left(1 - \frac{2i\theta}{n^2\pi^2}\right)^{-1/2},
\end{aligned}$$

and eigenvalues are given by $\lambda_n = n^2\pi^2$ $(n = 1, 2, \ldots)$. Since $D(\lambda) = \sin\sqrt{\lambda}/\sqrt{\lambda} = 1 - d_1\lambda + d_2\lambda^2/2 - \cdots$ with $d_1 = 1/6$ and $d_2 = 1/60$, we have

$$\mathrm{E}(S) = \frac{1}{6}, \quad \mathrm{Var}(S) = 2\left(\frac{1}{36} - \frac{1}{60}\right) = \frac{1}{45}.$$

The above examples indicate that the computation of the FD is cumbersome if we follow the series expansion formula given in (2.5). Instead we shall present an alternative method for computing the FD in Section 2.3. For this purpose the next section gives more details of the FD in connection with eigenvalues and eigenfunctions of the integral equation.

Exercises for Section 2.1

Exercise 2.1.1 Show that the following relation holds:

$$D_N(\lambda) = \left| I_N - \frac{\lambda}{N} K_N \right| = \sum_{n=0}^{N} \frac{a_n(N)}{n!} \lambda^n,$$

where

$$a_n(N) = \frac{(-1)^n}{N^n} \sum_{1 \le j_1, \ldots, j_n \le N} K\begin{pmatrix} j_1/N & \cdot & \cdot & \cdot & j_n/N \\ j_1/N & \cdot & \cdot & \cdot & j_n/N \end{pmatrix}.$$

Exercise 2.1.2 Prove that the infinite series of $D(\lambda)$ defined in (2.5) converges for all λ, using the Hadamard inequality

$$|A|^2 \le \prod_{j=1}^{n} \left(\sum_{k=1}^{n} A_{jk}^2 \right),$$

where A is an arbitrary $n \times n$ matrix with A_{jk} in the (j,k)th place.

Exercise 2.1.3 Derive the FDs of the following three kernels from its definition given in (2.5).

$$K(s,t) = 1 + s + t, \quad 1 + s + t + st, \quad 1 + s + t + s^2t^2.$$

Exercise 2.1.4 Derive the FDs of the following four kernels:

$$K_1(s,t) = \sum_{j=1}^{n} \cos j\pi s \cos j\pi t, \quad K_2(s,t) = \sum_{j=1}^{n} \sin j\pi s \sin j\pi t,$$

$$K_3(s,t) = \sum_{j=1}^{n} \cos j\pi(s - t), \quad K_4(s,t) = \sum_{j=1}^{n} \sin j\pi(s + t).$$

Exercise 2.1.5 Derive the FD of $K(s,t) = \min(s,t) - st$ from its definition given in (2.5).

2.2 Eigenvalues and Eigenfunctions

In the last section we derived the FDs of various kernels, for which eigenvalues were obtained explicitly. For

$$K(s,t) = 1 + st, \quad D(\lambda) = 1 - \frac{4}{3}\lambda + \frac{\lambda^2}{12},$$

eigenvalues are given by $8 \pm 2\sqrt{13}$. For

$$K(s,t) = \min(s,t), \quad D(\lambda) = \cos\sqrt{\lambda},$$

eigenvalues are given by $(n - 1/2)^2\pi^2$ $(n = 1, 2, \ldots)$. For

$$K(s,t) = \min(s,t) - st, \quad D(\lambda) = \frac{\sin\sqrt{\lambda}}{\sqrt{\lambda}},$$

eigenvalues are given by $n^2\pi^2$ $(n = 1, 2, \ldots)$.

However, it is often the case that the solutions to $D(\lambda) = 0$ cannot be computed explicitly, so that eigenvalues cannot be obtained explicitly. As a simple example, consider

$$K(s,t) = \min(s,t) - st - 3st(1-s)(1-t),$$

which was introduced in (1.78), and its FD was presented in (1.79) as

$$D(\lambda) = \frac{12}{\lambda}\left(\frac{2(1-\cos\sqrt{\lambda})}{\lambda} - \frac{\sin\sqrt{\lambda}}{\sqrt{\lambda}}\right).$$

It is impossible to obtain eigenvalues explicitly in the present case, but there is no problem in computing the c.f. of S in (2.12) with the present $K(s,t)$. Noting also that

$$D(\lambda) = 1 - d_1\lambda + \frac{d_2}{2}\lambda^2 - \cdots, \quad d_1 = \frac{1}{15}, \quad d_2 = \frac{1}{280},$$

we have

$$\mathrm{E}(S) = d_1 = \frac{1}{15}, \quad \mathrm{Var}(S) = 2\left(d_1^2 - d_2\right) = \frac{11}{6300}.$$

Some important properties (a) through (e) of the eigenvalues and eigenfunctions associated with a symmetric, continuous, and nearly definite kernel $K(s,t)$ are summarized here (Courant and Hilbert, 1953; Whittaker and Watson, 1958; Hochstadt, 1973).

(a) Every eigenvalue is nonzero and real.
(b) The multiplicity of each eigenvalue is finite.
(c) $K(s,t)$ and $K(1-s,1-t)$ have the same set of eigenvalues.
(d) The set of eigenvalues is at most denumerable, and they accumulate at infinity in the case of an infinite number of eigenvalues $\{\lambda_n\}$ $(n=1,2,\dots)$, where $|\lambda_n| \to \infty$ as $n \to \infty$.
(e) Eigenfunctions corresponding to different eigenvalues are orthogonal and can be taken to be orthonormal. If the multiplicity is greater than unity, the set of linearly independent eigenfunctions can be taken to be orthonormal, and the whole set of eigenfunctions is orthonormal.

Determination of the multiplicity is an important problem, but we defer the discussion until the next section because some more needs to be prepared. Here we just give an example where the multiplicity is two for each eigenvalue. Let us consider

$$K(s,t) = \min(s,t) - st + \frac{1}{2}(s^2+t^2-s-t) + \frac{1}{12}, \quad K(t,t) = \frac{1}{12}, \tag{2.21}$$

whose FD is given (see (2.122) in Section 2.4) by

$$D(\lambda) = \left(\frac{\sin\sqrt{\lambda}/2}{\sqrt{\lambda}/2}\right)^2 = \prod_{n=1}^{\infty}\left(1-\frac{\lambda}{4n^2\pi^2}\right)^2,$$

and the multiplicity will be shown to be two for each eigenvalue $\lambda_n = 4n^2\pi^2$ $(n=1,2,\dots)$. In fact, we have $\mathrm{E}(S) = \int_0^1 K(t,t)\,dt = 1/12$, whereas it holds that $\mathrm{E}(S) = \sum_{n=1}^{\infty} 1/\lambda_n$, which reduces to $1/24$ if eigenvalues are not repeated in the sum since $\sum_{n=1}^{\infty} 1/n^2 = \pi^2/6$.

It will also be shown that two linearly independent orthonormal eigenfunctions corresponding to $\lambda_n = 4n^2\pi^2$ are given by

$$f_{n1}(t) = \sqrt{2}\cos 2n\pi t, \qquad f_{n2}(t) = \sqrt{2}\sin 2n\pi t.$$

We note in passing that

$$D(\lambda) = \left(\frac{\sin\sqrt{\lambda}/2}{\sqrt{\lambda}/2}\right)^2 = 1 - d_1\lambda + \frac{d_2}{2}\lambda^2 - \cdots, \quad d_1 = \frac{1}{12}, \quad d_2 = \frac{1}{180},$$

which gives $\mathrm{Var}(S) = 2(d_1^2 - d_2) = 1/360$. Similarly, it holds that $\mathrm{Var}(S) = 2\sum_{n=1}^{\infty} 1/\lambda_n^2$, which reduces to $1/720$ if eigenvalues are not repeated in the sum since $\sum_{n=1}^{\infty} 1/n^4 = \pi^4/90$.

It is usually the case that the kernels associated with quadratic functionals of Bm are positive definite. Suppose that $K(s,t)$ is positive definite with eigenvalues $0 < \lambda_1 \le \lambda_2 \le \cdots$. Since it holds that

$$S = \int_0^1 \int_0^1 K(s,t)\,dW(s)\,dW(t) \stackrel{\mathcal{D}}{=} \sum_{n=1}^{\infty} \frac{1}{\lambda_n} Z_n^2, \quad \{Z_n\} \sim \mathrm{NID}(0,1), \tag{2.22}$$

where λ_n is repeated as many times as its multiplicity in the above sum, it is recognized that the smallest eigenvalue λ_1 is the most important. When eigenvalues are unknown, it may be interesting to estimate the magnitude of $1/\lambda_1$ relative to the total sum $\sum_{n=1}^{\infty} 1/\lambda_n$. Mercer's theorem (Theorem 2.2) can also be used for this purpose, which we now discuss.

Let us assume that the multiplicity of every eigenvalue is the same and equal to ℓ. It follows from (2.22) that

$$\mathrm{E}(S) = \int_0^1 K(t,t)\,dt = \frac{\ell}{\lambda_1} + \sum_{n=2}^{\infty} \frac{1}{\lambda_n}, \tag{2.23}$$

where λ_n is repeated ℓ times in the above sum. Then it holds that

$$\frac{1}{\lambda_1} < \frac{1}{\ell} \int_0^1 K(t,t)\,dt = \frac{1}{\ell} \sum_{n=1}^{\infty} \frac{1}{\lambda_n}.$$

It also follows from (2.17) that

$$\int_0^1 \int_0^1 K^2(s,t)\,ds\,dt = \frac{\ell}{\lambda_1^2} + \sum_{n=2}^{\infty} \frac{1}{\lambda_n^2}, \tag{2.24}$$

which yields

$$\frac{1}{\lambda_1} < \sqrt{\frac{1}{\ell} \int_0^1 \int_0^1 K^2(s,t)\,ds\,dt} = \sqrt{\frac{1}{\ell} \sum_{n=1}^{\infty} \frac{1}{\lambda_n^2}}.$$

Then we have the following relation between the smallest eigenvalue and the integrals of the kernel (Hochstadt, 1973).

Theorem 2.8 *Suppose that $K(s,t)$ is positive definite with eigenvalues $0 < \lambda_1 \le \lambda_2 \le \cdots$, where the multiplicity of every eigenvalue is ℓ. Then we have the following inequalities:*

$$\frac{1}{\lambda_1} < B_2 = \sqrt{\frac{1}{\ell} \int_0^1 \int_0^1 K^2(s,t)\,ds\,dt} < B_1 = \frac{1}{\ell} \int_0^1 K(t,t)\,dt. \tag{2.25}$$

It is seen that the upper bound B_2 for $1/\lambda_1$ is sharper than B_1. To find a more general upper bound, let us define

$$K^{(2)}(s,t) = \int_0^1 K(s,u)\,K(u,t)\,du, \tag{2.26}$$

which is called the *iterated kernel* of $K(s,t)$. It holds that

$$\int_0^1 K^{(2)}(t,t)\,dt = \int_0^1\int_0^1 K^2(s,t)\,ds\,dt.$$

It also follows from Mercer's theorem (Theorem 2.2) that we can apply term-by-term integration to get

$$\begin{aligned}
K^{(2)}(s,t) &= \int_0^1 \sum_{m=1}^{\infty} \frac{1}{\lambda_m} f_m(s) f_m(u) \sum_{n=1}^{\infty} \frac{1}{\lambda_n} f_n(u) f_n(t)\,du \\
&= \sum_{m=1}^{\infty}\sum_{n=1}^{\infty} \frac{1}{\lambda_m\,\lambda_n} f_m(s) f_n(t) \int_0^1 f_m(u) f_n(u)\,du \\
&= \sum_{n=1}^{\infty} \frac{1}{\lambda_n^2} f_n(s) f_n(t).
\end{aligned} \tag{2.27}$$

Notice here that the eigenvalues and eigenfunctions of $K^{(2)}(s,t)$ are given by

$$\left\{(\lambda_n^2,\ f_n(t)):\ n = 1,2,\ldots\right\},$$

so that $K^{(2)}(s,t)$ is positive definite and (2.27) is an infinite-dimensional extension of the property of symmetric matrices on the relationship with eigenvalues and eigenvectors, that is, if $A = P\Lambda P'$, then $A^2 = P\Lambda^2 P'$, where Λ is a diagonal matrix with eigenvalues on diagonals and P is an orthogonal matrix composed of eigenvectors.

More generally, the subsequent iterated kernels admit the expansions (Exercise 2.2.1)

$$K^{(j)}(s,t) = \int_0^1 K^{(j-1)}(s,u) K(u,t)\,du = \sum_{n=1}^{\infty} \frac{1}{\lambda_n^j} f_n(s) f_n(t), \tag{2.28}$$

where $j \geq 2$ and $K^{(1)}(s,t) = K(s,t)$. Then it holds that

$$\int_0^1 K^{(j)}(t,t)\,dt = \sum_{n=1}^{\infty} \frac{1}{\lambda_n^j} = \frac{\ell}{\lambda_1^j} + \sum_{n=2}^{\infty} \frac{1}{\lambda_n^j} \qquad (j = 1,2,\ldots),$$

which yields

$$\frac{1}{\lambda_1} < B_j = \left(\frac{1}{\ell}\int_0^1 K^{(j)}(t,t)\,dt\right)^{1/j} = \left(\frac{1}{\ell}\sum_{n=1}^{\infty}\frac{1}{\lambda_n^j}\right)^{1/j} \qquad (j = 1,2,\ldots).$$

It is recognized that the upper bound B_j comes closer to $1/\lambda_1$ with j. For example, consider $K(s,t)$ given in (2.21), where $\lambda_n = 4n^2\pi^2$ $(n = 1,2,\ldots)$ and $\ell = 2$. Then we have $1/\lambda_1 = 1/(4\pi^2) = 0.025330$, and

$$B_1 = \frac{1}{4\times 6} = 0.041667, \quad B_2 = \left(\frac{1}{16\times 90}\right)^{1/2} = 0.026352,$$

$$B_3 = \left(\frac{1}{64\times 945}\right)^{1/3} = 0.025476, \quad B_4 = \left(\frac{1}{256\times 9450}\right)^{1/4} = 0.025356.$$

In general, however, it turns out to be quite burdensome to compute B_j for $j \ge 3$ since the computation of $\int_0^1 K^{(j)}(t,t)\,dt$ is very involved for $j \ge 3$.

Restricting to $j = 1$ and 2, we compare $1/\lambda_1$ with the upper bounds B_1 and B_2 for the following four kernels (Exercise 2.2.2):

$$K_1(s,t) = \min(s,t), \quad K_2(s,t) = \min(s,t) - st,$$

$$K_3(s,t) = \frac{1}{6}s^2(3t-s), \quad K_4(s,t) = \frac{1}{120}s^3\left(s^2 - 5st + 10t^2\right) \quad (s \le t).$$

All of the eigenvalues of the above kernels are known to have multiplicity one. The kernels K_3 and K_4 arise from $\int_0^1 F_g^2(t)\,dt$, where $F_g(t)$ is g-fold integrated Bm defined in (1.15), where $g = 1$ and 2 correspond to K_3 and K_4, respectively (see (2.74) in the next section).

Table 2.1 shows $1/\lambda_1$, B_1, and B_2 of each kernel together with $1/(B_1\lambda_1)$, $1/(B_2\lambda_1)$, and B_2/B_1, where the smallest eigenvalues of K_3 and K_4 were computed numerically from the corresponding FD (see Section 2.4.5). For kernels K_i $(i = 1,2,3,4)$, the multiplicity ℓ of λ_1 being unity can be ensured from the inequality

$$\frac{1}{\lambda_1} < B_1 = \frac{1}{\ell}\int_0^1 K_i(t,t)\,dt,$$

which is violated if $\ell \ge 2$. It is seen that the entries in B_2 are closer to $1/\lambda_1$ than those in B_1. This is especially true for the kernels K_3 and K_4, and this fact is reflected in entries for $1/(B_2\lambda_1)$ and B_2/B_1. For K_1 and K_2, the entries

Table 2.1 *Smallest eigenvalues and upper bounds in (2.25).*

	$1/\lambda_1$	B_1	B_2	$1/(B_1\lambda_1)$	$1/(B_2\lambda_1)$	B_2/B_1
K_1	0.4053	0.5	0.4082	0.8106	0.9927	0.8165
K_2	0.1013	0.1667	0.1054	0.6079	0.9612	0.6325
K_3	0.08089	0.08333	0.08092	0.97069	0.99967	0.97100
K_4	0.00825	0.00833	0.00825	0.98962	0.99995	0.98967

for $1/(B_1\lambda_1)$, which is the ratio of $1/\lambda_1$ to the total sum $B_1 = \sum_{n=1}^{\infty} 1/\lambda_n$, are not high enough, which means that the contributions of $1/\lambda_n$ $(n \geq 2)$ are still high. On the other hand, $1/\lambda_1$ is dominant for K_3 and K_4, implying that the distribution may be approximated by that of $B_2\, Z^2$ without computing eigenvalues, where $Z \sim \mathrm{N}(0,1)$. This last fact will be examined in detail in Section 2.4.

Equation (2.28) gives a relationship between the iterated kernel $K^{(j)}(s,t)$ and its eigenvalues and eigenfunctions. The following theorem extends that relationship further (Exercise 2.2.3).

Theorem 2.9 *Suppose that $K(s,t)$ is symmetric, continuous, and nearly definite, which admits the expansion*

$$K(s,t) = \sum_{n=1}^{\infty} \frac{f_n(s) f_n(t)}{\lambda_n},$$

where $\{(\lambda_n,\ f_n)\}$ is an orthonormal set of eigenvalues and eigenfunctions of $K(s,t)$. Then it holds that, for any positive integer m,

$$L_m(s,t) = \sum_{j=1}^{m} \alpha_j K^{(j)}(s,t) = \sum_{n=1}^{\infty} \left(\sum_{j=1}^{m} \frac{\alpha_j}{\lambda_n^j} \right) f_n(s) f_n(t), \tag{2.29}$$

$$\begin{aligned} S_m &= \int_0^1 \int_0^1 L_m(s,t)\, dW(s)\, dW(t) \\ &\overset{\mathcal{D}}{=} \sum_{n=1}^{\infty} \left(\sum_{j=1}^{m} \frac{\alpha_j}{\lambda_n^j} \right) Z_n^2, \end{aligned} \tag{2.30}$$

where $K^{(1)}(s,t) = K(s,t)$, α_j $(j = 1, \ldots, m)$ are constants, and $\{Z_n\} \sim$ NID(0, 1). Moreover, it holds that

$$\int_0^1 L_m(s,t)\, f_n(s)\, ds = \left(\sum_{j=1}^{m} \frac{\alpha_j}{\lambda_n^j} \right) f_n(t), \tag{2.31}$$

and, for $\sum_{j=1}^{m} \alpha_j/\lambda_n^j \neq 0$, the orthonormal set of eigenvalues and eigenfunctions of $L_m(s,t)$ is given by

$$\left\{ \left(\frac{1}{\sum_{j=1}^{m} \alpha_j/\lambda_n^j},\ f_n \right) : n = 1, 2, \ldots \right\}. \tag{2.32}$$

Note that almost all the eigenvalues in the set (2.32) have the same sign because almost all λ_ns have the same sign and $|\lambda_n| \to \infty$ as $n \to \infty$. Note also that there exists no eigenfunction except $\{f_n(t)\}$ that is orthogonal to every $f_n(t)$. In fact, if such an eigenfunction $f_0(t)$ existed, it follows from (2.29) that

$$\int_0^1 L_m(s,t) f_0(s)\, ds = 0,$$

which contradicts the assumption that $f_0(t)$ is an eigenfunction. The set of eigenvalues and eigenfunctions in (2.32) is complete in this sense. The c.f. of S_m in (2.30) is given by

$$\mathrm{E}\left(e^{i\theta S_m}\right) = \prod_{n=1}^{\infty} \left(1 - 2i\theta \sum_{j=1}^{m} \frac{\alpha_j}{\lambda_n^j}\right)^{-1/2},$$

so that the FD $D_m(\lambda)$ of the kernel $L_m(s,t)$ in (2.29) is

$$D_m(\lambda) = \prod_{n=1}^{\infty} \left(1 - \lambda \sum_{j=1}^{m} \frac{\alpha_j}{\lambda_n^j}\right). \tag{2.33}$$

Suppose that $m = 2$, which is the most important case in applications. Then we can show that the computation of the FD is possible without the knowledge of λ_n $(n = 1, 2, \ldots)$. In fact, the FD $D_2(\lambda)$ of $\alpha_1 K(s,t) + \alpha_2 K^{(2)}(s,t)$ can be expressed as

$$D_2(\lambda) = \prod_{n=1}^{\infty} \left[1 - \lambda\left(\frac{\alpha_1}{\lambda_n} + \frac{\alpha_2}{\lambda_n^2}\right)\right] = \prod_{n=1}^{\infty} \left(1 - \frac{\beta_1(\lambda)}{\lambda_n}\right)\left(1 - \frac{\beta_2(\lambda)}{\lambda_n}\right),$$

where

$$\beta_1(\lambda) = \frac{\alpha_1\lambda + \sqrt{\alpha_1^2\lambda^2 + 4\alpha_2\lambda}}{2}, \quad \beta_2(\lambda) = \frac{\alpha_1\lambda - \sqrt{\alpha_1^2\lambda^2 + 4\alpha_2\lambda}}{2}. \tag{2.34}$$

Then we have the following theorem (Exercise 2.2.4).

Theorem 2.10 *The c.f. of*

$$S_2 = \int_0^1 \int_0^1 \left[\alpha_1 K(s,t) + \alpha_2 K^{(2)}(s,t)\right] dW(s)\, dW(t)$$

is given by

$$\begin{aligned} E\left(e^{i\theta S_2}\right) &= E\left[\exp\left\{i\theta \int_0^1 \int_0^1 \left[\alpha_1 K(s,t) + \alpha_2 K^{(2)}(s,t)\right] dW(s)\, dW(t)\right\}\right] \\ &= \left[D\Big(\beta_1(2i\theta)\Big) D\Big(\beta_2(2i\theta)\Big)\right]^{-1/2}, \end{aligned} \tag{2.35}$$

where $D(\lambda)$ is the FD of $K(s,t)$, and $\beta_1(\lambda)$ and $\beta_2(\lambda)$ are defined in (2.34).

This theorem will be used in Chapters 5 and 6 when we discuss testing problems and examine limiting powers of various tests under a sequence of local alternatives.

So far we have dealt with quadratic functionals of Bm in the form of the Riemann–Stieltjes double integral. However, it is often the case that the statistics are expressed in the form of the Riemann integral with the integrand being the square of a zero-mean Gaussian process. Here we consider such a case and present a theorem that enables us to obtain its c.f. (see Loève (1978), Anderson and Darling (1952), and Exercise 2.2.5).

Theorem 2.11 *Let* $\{Y(t)\}$ *be a zero-mean Gaussian process defined on* $[0, 1]$. *Then* $Y(t)$ *has a series expansion given by*

$$Y(t) = \sum_{n=1}^{\infty} \frac{f_n(t)}{\sqrt{\lambda_n}} Z_n, \qquad \{Z_n\} \sim NID(0,1), \tag{2.36}$$

where $\{\lambda_n\}$ *is a sequence of eigenvalues of* $K(s,t) = Cov\,(Y(s), Y(t))$ *repeated as many times as their multiplicities, whereas* $\{f_n(t)\}$ *is an orthonormal sequence of eigenfunctions corresponding to* λ_n. *The series on the right-hand side converges in the m.s. sense and with probability one uniformly on* $[0, 1]$. *Moreover, it holds that*

$$\int_0^1 Y^2(t)\,dt = \sum_{n=1}^{\infty} \frac{1}{\lambda_n} Z_n^2 \overset{\mathcal{D}}{=} \int_0^1 \int_0^1 \mathrm{Cov}\,(Y(s), Y(t))\, dW(s)\, dW(t). \tag{2.37}$$

All of the eigenvalues are positive because the kernel $\mathrm{Cov}(Y(s), Y(t))$ is positive definite. The series expansion in (2.36) is called the *Karhunen–Loève expansion*, which gives, in conjunction with Mercer's theorem, the distributional equivalence described in (2.37). In Section 1.3 we employed this fact. For example,

$$\begin{aligned}
\int_0^1 W^2(t)\,dt &= \int_0^1 \int_0^1 [1 - \max(s,t)]\, dW(s)\, dW(t) \\
&\overset{\mathcal{D}}{=} \int_0^1 \int_0^1 \mathrm{Cov}(W(s), W(t))\, dW(s)\, dW(t) \\
&= \int_0^1 \int_0^1 \min(s,t)\, dW(s)\, dW(t) \\
&= \int_0^1 (W(1) - W(t))^2\, dt,
\end{aligned}$$

$$\begin{aligned}\int_0^1 \bar{W}^2(t)\,dt &= \int_0^1\int_0^1 [\min(s,t) - st]\,dW(s)\,dW(t)\\ &\stackrel{\mathcal{D}}{=} \int_0^1\int_0^1 \mathrm{Cov}(\bar{W}(s), \bar{W}(t))\,dW(s)\,dW(t)\\ &= \int_0^1\int_0^1 \left[\frac{1}{3} - \max(s,t) + \frac{s^2+t^2}{2}\right] dW(s)\,dW(t)\\ &= \int_0^1 (W(t) - tW(1))^2\,dt,\end{aligned}$$

where $\bar{W}(t) = W(t) - \int_0^1 W(s)\,ds$.

As another example, for g-fold integrated Bm $F_g(t)$ defined in (1.14), we have

$$\begin{aligned}\int_0^1 F_g^2(t)\,dt &= \int_0^1\int_0^1 \left(\int_{\max(s,t)}^1 \frac{((u-s)(u-t))^g}{(g!)^2}\,du\right) dW(s)\,dW(t)\\ &\stackrel{\mathcal{D}}{=} \int_0^1\int_0^1 \mathrm{Cov}(F_g(s), F_g(t))\,dW(s)\,dW(t)\\ &= \int_0^1\int_0^1 \left(\int_0^{\min(s,t)} \frac{((s-u)(t-u))^g}{(g!)^2}\,du\right) dW(s)\,dW(t).\end{aligned}$$

It will be recognized that the kernel in this last expression is simpler than the first in terms of the derivation of the FD, although it is not easy for large g, which we discuss in Section 2.4.

Exercises for Section 2.2

Exercise 2.2.1 Show that, when $K(s,t)$ is symmetric, continuous, and nearly definite,

$$K^{(j)}(s,t) = \sum_{n=1}^{\infty} \frac{1}{\lambda_n^j} f_n(s) f_n(t), \qquad \int_0^1 K^{(j)}(t,t)\,dt = \sum_{n=1}^{\infty} \frac{1}{\lambda_n^j},$$

where $j = 2, 3, \ldots$ and $K^{(j)}(s,t)$ is the iterated kernel defined by

$$K^{(j)}(s,t) = \int_0^1 K^{(j-1)}(s,u) K(u,t)\,du, \quad K^{(1)}(s,t) = K(s,t).$$

Exercise 2.2.2 Compute the upper bound B_2 for $1/\lambda_1$ of the symmetric kernel given by $K_3(s,t) = s^2(3t - s)/6$ for $s \le t$, where B_2 is defined in (2.25), whereas λ_1 is the smallest eigenvalue of $K_3(s,t)$.

Exercise 2.2.3 Suppose that the kernel is given by

$$L_m(s,t) = \sum_{j=1}^{m} \alpha_j K^{(j)}(s,t),$$

where $K^{(j)}(s,t)$ is the iterated kernel of $K(s,t)$ defined in (2.28). Prove that, if $\sum_{j=1}^m \alpha_j/\lambda_n^j \neq 0$, the set of eigenvalues and eigenfunctions of $L_m(s,t)$ is given by

$$\left\{ \left(\frac{1}{\sum_{j=1}^m \alpha_j/\lambda_n^j},\ f_n \right) : \ n = 1, 2, \ldots \right\},$$

where λ_n and $f_n(t)$ are the eigenvalue and eigenfunction of $K(s,t)$.

Exercise 2.2.4 For the MA(1) model

$$y_j - \mu = \varepsilon_j - \alpha\varepsilon_{j-1} \qquad (j = 1, \ldots, n),$$

where $\varepsilon_0, \varepsilon_1, \ldots, \varepsilon_n \sim \text{NID}(0, \sigma_\varepsilon^2)$, the LBIU statistic S_{n2} for testing $H_0 : \alpha = 1$ was derived in Section 1.5.3. It holds (see Chapter 6) that, under $\alpha = 1 - c/n$ as $n \to \infty$,

$$S_{n2} \Rightarrow S_2 = \int_0^1 \int_0^1 \left(K(s,t) + c^2 K^{(2)}(s,t) \right) dW(s)\, dW(t),$$

where $K(s,t) = \min(s,t) - st - 3st(1-s)(1-t)$. Derive the c.f. of S_2.

Exercise 2.2.5 Let $\{Y(t)\}$ be a zero-mean Gaussian process. Then prove that

$$\int_0^1 Y^2(t)\, dt = \sum_{n=1}^{\infty} \frac{1}{\lambda_n} Z_n^2 \stackrel{\mathcal{D}}{=} \int_0^1 \int_0^1 K(s,t)\, dW(s)\, dW(t),$$

where $\{Z_n\} \sim \text{NID}(0,1)$, $K(s,t) = \text{Cov}(Y(s), Y(t))$, and λ_n $(n = 1, 2, \ldots)$ are the eigenvalues of $K(s,t)$.

2.3 Derivation of the Fredholm Determinant

In this section we discuss how to compute the FD $D(\lambda)$ of a symmetric, continuous, and nearly definite kernel $K(s,t)$. The definition of $D(\lambda)$ can be given in several ways, among which are

$$D(\lambda) = \lim_{N\to\infty} \left| I_N - \frac{\lambda}{N} K_N \right| = \prod_{n=1}^{\infty} \left(1 - \frac{\lambda}{\lambda_n} \right) = \sum_{n=0}^{\infty} \frac{(-1)^n d_n}{n!} \lambda^n, \quad (2.38)$$

where K_N is an $N \times N$ matrix with $K(j/N, k/N)$ in the (j,k)th place, λ_n $(n = 1, 2, \dots)$ are eigenvalues of $K(s,t)$, and

$$d_n = \int_0^1 \cdots \int_0^1 \begin{vmatrix} K(s_1,t_1) & K(s_1,t_2) & \cdot & \cdot & \cdot & K(s_1,t_n) \\ \vdots & \vdots & & & & \vdots \\ K(s_n,t_1) & K(s_n,t_2) & \cdot & \cdot & \cdot & K(s_n,t_n) \end{vmatrix} dt_1 \cdots dt_n.$$

It was recognized in Section 2.1 that the computation of $D(\lambda)$ via the above formulas is burdensome or impossible in general.

In Section 2.3.1 we present a general method for obtaining the FD, whereas Section 2.3.2 discusses Darling's formula and its extensions, imposing some restrictions on the kernel.

2.3.1 General Method for Obtaining the Fredholm Determinant

We present a general method for obtaining the FD demonstrated in Nabeya and Tanaka (1988, 1990a). We first present a set of sufficient conditions for a function of λ to be the FD.

Theorem 2.12 *Let $K(s,t)$ be symmetric, continuous, and nearly definite in* $[0,1] \times [0,1]$ *and $\{\lambda_n\}$ be a sequence of eigenvalues of K. Suppose that $\tilde{D}(\lambda)$ is an entire function of λ with $\tilde{D}(0) = 1$. Then $\tilde{D}(\lambda)$ becomes the FD of K if*

(i) every zero of $\tilde{D}(\lambda)$ is an eigenvalue of K, and in turn every eigenvalue of K is a zero of $\tilde{D}(\lambda)$;
(ii) $\tilde{D}(\lambda)$ can be expanded as

$$\tilde{D}(\lambda) = \prod_{n=1}^{\infty} \left(1 - \frac{\lambda}{\lambda_n}\right)^{\ell_n}, \tag{2.39}$$

where ℓ_n is the multiplicity of λ_n.

We shall describe in Theorem 2.14 how to find a candidate for the FD together with the determination of the multiplicity. Condition (ii) in the above theorem is our main concern. The possibility of the infinite product expansion in (2.39) is important, together with the determination of the multiplicity. We first deal with how to ensure the possibility of the product expansion. The determination of the multiplicity will be discussed shortly after introducing a differential equation with some conditions equivalent to the integral equation.

The following theorem gives a sufficient condition for an entire function to admit an infinite product expansion (Whittaker and Watson, 1958, p.137).

Theorem 2.13 *Let $h(z)$ be an entire function that has simple zeros at the points $a_1, a_2, \ldots$, where $a_n \neq 0$ and $\lim_{n\to\infty} |a_n| = \infty$. Suppose that there is a sequence $\{C_m\}$ of simple closed curves such that $h'(z)/h(z)$ is bounded on C_m as $m \to \infty$. Then $h(z)$ can be expanded as*

$$h(z) = h(0)\exp\left\{\frac{h'(0)z}{h(0)}\right\}\prod_{n=1}^{\infty}\left\{\left(1-\frac{z}{a_n}\right)\exp\left(\frac{z}{a_n}\right)\right\}. \tag{2.40}$$

To demonstrate the usefulness of Theorem 2.13, let us consider

$$h(z) = \cos z - bz\sin z, \quad h'(z) = -bz\cos z - (b+1)\sin z, \tag{2.41}$$

where b is a constant. It will be shown in (2.107) that $h(\sqrt{\lambda})$ is the FD of $K(s,t) = 1 - \max(s,t) + b$.

Suppose first that $b = 0$ so that $h(z) = \cos z$, which is an even function with $h(0) = 1$ and $h'(0) = 0$. The zeros of $h(z)$ are all simple, and we denote them by $\pm a_n = \pm(n - 1/2)\pi$ $(n = 1, 2, \ldots)$, where $\lim_{n\to\infty} |a_n| = \infty$. Let us consider $h'(z)/h(z) = -\sin z/\cos z$, and denote by C_m the square in the complex plane with vertices $2m\pi(\pm 1 \pm i)$ $(m = 1, 2, \ldots)$. Put $z = 2m\pi + i\beta$ with $|\beta| \leq 2m\pi$. Then it holds that

$$\cos z = \frac{e^{iz} + e^{-iz}}{2} = \frac{1}{2}\left(e^{i2m\pi-\beta} + e^{-i2m\pi+\beta}\right) = \frac{1}{2}\left(e^{-\beta} + e^{\beta}\right),$$

$$\sin z = \frac{e^{iz} - e^{-iz}}{2i} = \frac{1}{2i}\left(e^{2im\pi-\beta} - e^{-i2m\pi+\beta}\right) = \frac{1}{2i}\left(e^{-\beta} - e^{\beta}\right).$$

Conversely, when $z = \alpha + i2m\pi$ with $|\alpha| \leq 2m\pi$, we have

$$\cos z = \frac{1}{2}\left(e^{i\alpha-2m\pi} + e^{-i\alpha+2m\pi}\right),$$

$$\sin z = \frac{1}{2i}\left(e^{i\alpha-2m\pi} - e^{-i\alpha+2m\pi}\right).$$

It can be checked that $h'(z)/h(z) = -\sin z/\cos z$ is bounded on each side of C_m as $m \to \infty$. Using Theorem 2.13, we can now expand the even function $h(z)$ with $h(0) = 1$ and $h'(0) = 0$ as

$$\begin{aligned} h(z) = \cos z &= \prod_{n=1}^{\infty}\left\{\left(1-\frac{z}{a_n}\right)\exp\left(\frac{z}{a_n}\right)\left(1+\frac{z}{a_n}\right)\exp\left(-\frac{z}{a_n}\right)\right\} \\ &= \prod_{n=1}^{\infty}\left(1-\frac{z^2}{a_n^2}\right) = \prod_{n=1}^{\infty}\left(1-\frac{z^2}{(n-1/2)^2\,\pi^2}\right). \end{aligned} \tag{2.42}$$

Suppose next that $b \neq 0$ in (2.41). Then $h(z)$ is an even function with $h(0) = 1$, $h'(0) = 0$, and the zeros of $h(z)$ are all simple. It can be shown (Exercise 2.3.1) that $\cos z/\sin z$ is bounded on each side of squares C_m in the complex plane with vertices $(2m + 1/2)\,\pi(\pm 1 \pm i)$ $(m = 1, 2, \ldots)$, which implies that

$$\frac{h'(z)}{h(z)} = \frac{-bz\cos z - (b+1)\sin z}{\cos z - bz\sin z} = \frac{bz(\cos z/\sin z) + b + 1}{bz - (\cos z/\sin z)}$$

is also bounded on each side of C_m as $m \to \infty$. Then $h(z) = \cos z - bz\sin z$ admits the infinite product expansion as in (2.42), although a_n cannot be obtained explicitly in the present case.

As another example, consider

$$h(z) = \frac{3}{z^3}(\sin z - z\cos z), \quad h'(z) = \frac{3}{z^4}((z^2 - 3)\sin z + 3z\cos z), \quad (2.43)$$

where $h(\sqrt{\lambda})$ is the FD of $K(s,t) = 1 - \max(s,t) - 3(1-s^2)(1-t^2)/4$, which will be derived in (2.166). Noting that $\sin z/\cos z$ is bounded on squares C_m in the complex plane with vertices $2m\pi(\pm 1 \pm i)$ $(m = 1, 2, \ldots)$, we can show that $h'(z)/h(z)$ is bounded on C_m as $m \to \infty$. Since $h(z)$ is an even function with $h(0) = 1$ and $h'(0) = 0$, and the zeros of $h(z)$ are all simple, we have a product expansion similar to (2.42).

We now discuss how to obtain the FD by employing Theorems 2.12 and 2.13 together with the determination of the multiplicity. Suppose that the integral equation

$$f(t) = \lambda \int_0^1 K(s,t) f(s)\, ds \quad (2.44)$$

is equivalent to a differential equation with some boundary conditions. The differential equation is assumed to have a general solution

$$f(t) = c_1\phi_1(t) + \cdots + c_r\phi_r(t), \quad (2.45)$$

where $c_1, \ldots, c_r$ are arbitrary constants; $\phi_1(t), \ldots, \phi_r(t)$ are linearly independent, continuous functions; and the boundary conditions are given by

$$M(\lambda)\,\mathbf{c} = \mathbf{0}, \quad M(\lambda) : r \times r, \quad \mathbf{c} = (c_1, c_2, \ldots, c_r)'. \quad (2.46)$$

Then we have the following theorem leading to the derivation of the FD.

Theorem 2.14 *Suppose that the integral equation (2.44) is equivalent to the general solution (2.45) of a differential equation with the boundary conditions* $M(\lambda)\,\mathbf{c} = \mathbf{0}$*, where* $M(\lambda)$ *and* $\mathbf{c}$ *are defined in (2.46). Then* $|M(\lambda)| = 0$ $(\lambda \neq 0)$ *is a necessary and sufficient condition for* λ $(\neq 0)$ *to be an eigenvalue of* $K(s,t)$*, and the multiplicity* ℓ_n *of the eigenvalue* λ_n *is given by*

$$\ell_n = r - \mathit{rank}\,(M(\lambda_n)). \quad (2.47)$$

It is recognized that the multiplicity of λ_n is simply the dimension of a null space of $M(\lambda_n)$. For illustrative purposes of Theorem 2.14, let us consider $K(s,t) = 1 - \max(s,t)$. The integral equation (2.44) with this kernel is given by

$$\begin{aligned} f(t) &= \lambda \int_0^1 [1 - \max(s,t)]\, f(s)\, ds \\ &= \lambda \left[-t \int_0^t f(s)\, ds - \int_t^1 s f(s)\, ds + \int_0^1 f(s)\, ds \right], \end{aligned}$$

which yields

$$f'(t) = -\lambda \int_0^t f(s)\, ds$$

and the differential equation

$$f''(t) + \lambda f(t) = 0. \tag{2.48}$$

We also have the four boundary conditions

$$f(0) = \lambda \int_0^1 (1-s) f(s)\, ds, \quad f(1) = 0,$$

$$f'(0) = 0, \quad f'(1) = -\lambda \int_0^1 f(s)\, ds,$$

from which we choose the two conditions. The choice is somewhat arbitrary as far as they are linearly independent, but the simpler the better. In the present case we choose

$$f(1) = 0, \quad f'(0) = 0. \tag{2.49}$$

We next show that the differential equation (2.48) with the boudary conditions in (2.49) leads to the integral equation (2.44) with $K(s,t) = 1 - \max(s,t)$. Let us denote by $R(t)$ the right-hand side of the integral equation. Then we have

$$\begin{aligned} R(t) &= \lambda \int_0^1 [1 - \max(s,t)]\ f(s)\, ds \\ &= - \int_0^1 [1 - \max(s,t)]\ f''(s)\, ds \\ &= t \int_0^t f''(s)\, ds + \int_t^1 s f''(s)\, ds - \int_0^1 f''(s)\, ds \\ &= t(f'(t) - f'(0)) + \left[s f'(s) \right]_t^1 - \int_t^1 f'(s)\, ds - (f'(1) - f'(0)) \\ &= f(t), \end{aligned}$$

which is equal to the left-hand side of the integral equation.

The general solution to the differential equation (2.48) is given by

$$f(t) = c_1 \cos\sqrt{\lambda}t + c_2 \sin\sqrt{\lambda}t, \tag{2.50}$$

where c_1 and c_2 are arbitrary constants, from which the two boundary conditions in (2.49) are expressed as

$$f(1) = c_1 \cos\sqrt{\lambda} + c_2 \sin\sqrt{\lambda} = 0, \quad f'(0) = c_2\sqrt{\lambda} = 0.$$

Thus we have the following homogeneous equation in $\boldsymbol{c} = (c_1, c_2)'$:

$$M(\lambda)\,\boldsymbol{c} = \mathbf{0}, \quad M(\lambda) = \begin{pmatrix} \cos\sqrt{\lambda} & \sin\sqrt{\lambda} \\ 0 & \sqrt{\lambda} \end{pmatrix}. \tag{2.51}$$

The eigenfunction $f(t)$ in (2.50) must be nonvanishing, which occurs only when $\boldsymbol{c} \neq \mathbf{0}$. Then (2.51) implies that $|M(\lambda)| = \sqrt{\lambda}\cos\sqrt{\lambda} = 0$, and λ_n is an eigenvalue if and only if $\cos\sqrt{\lambda_n} = 0$. It now holds that $\text{rank}(M(\lambda_n)) = 1$ for each eigenvalue λ_n. Thus the multiplicity of λ_n is unity. We therefore obtain $\tilde{D}(\lambda) = \cos\sqrt{\lambda}$ with $\tilde{D}(0) = 1$ as a candidate for the FD of $K(s,t) = 1 - \max(s,t)$. Condition (i) in Theorem 2.12 has now been established. Condition (ii) with $\ell_n = 1$ is proved since Theorem 2.13 ensures that $h(z) = \cos z$ admits an infinite product expansion (see the discussions following (2.41)). Every zero of $\tilde{D}(\lambda) = h(\sqrt{\lambda})$ is of order unity, which is equal to the multiplicity of each eigenvalue. Therefore $\tilde{D}(\lambda) = \cos\sqrt{\lambda}$ is really the FD of $K(s,t) = 1 - \max(s,t)$.

The orthonormal eigenfunction corresponding to λ_n is given, from (2.50) with $c_2 = 0$, by

$$f_n(t) = \sqrt{2}\cos\left(n - \frac{1}{2}\right)\pi t,$$

so that we have

$$1 - \max(s,t) = \sum_{n=1}^{\infty} \frac{2\cos\left(n - \frac{1}{2}\right)\pi s\,\cos\left(n - \frac{1}{2}\right)\pi t}{\left(n - \frac{1}{2}\right)^2 \pi^2}$$

by Mercer's theorem. This last expression was introduced earlier in Section 1.3 without proof. Theorem 2.3 now yields

$$\mathrm{E}\left[\exp\left\{i\theta \int_0^1\!\!\int_0^1 [1 - \max(s,t)]\; dW(s)\, dW(t)\right\}\right] = \left(\cos\sqrt{2i\theta}\right)^{-1/2}, \tag{2.52}$$

which was also presented in Section 1.3, assuming that the infinite series expansion for $1 - \max(s,t)$ is known.

We give below various examples of the differential equations resulting from the integral equation (2.44). For later purposes we classify these into five cases, Case 2.1 through Case 2.5.

Case 2.1 This is a case of the second-order homogeneous differential equation given by

$$f''(t) + (\lambda - \delta) f(t) = 0, \tag{2.53}$$

where δ is a constant coming from $K(s,t)$. The kernels leading to (2.53) will be given in Section 2.4.1 as examples [1] through [10]. The example shown in (2.48) was the case where $\delta = 0$. The general solution to (2.53) is given by

$$f(t) = c_1 \cos\sqrt{\lambda - \delta}\, t + c_2 \sin\sqrt{\lambda - \delta}\, t, \tag{2.54}$$

where c_1 and c_2 are arbitrary constants. The two appropriate boundary conditions lead to $M(\lambda)\,\boldsymbol{c} = \boldsymbol{0}$, where $M(\lambda)$ is a 2×2 matrix and $\boldsymbol{c} = (c_1, c_2)'$. Then λ ($\neq 0$) is an eigenvalue of K if and only if $M(\lambda)\,\boldsymbol{c} = \boldsymbol{0}$ has a nontrivial solution $\boldsymbol{c} \neq \boldsymbol{0}$. Thus $|M(\lambda)| = 0$ is a necessary and sufficient condition for λ ($\neq 0$) to be an eigenvalue, from which we can find a candidate $\tilde{D}(\lambda)$ for $D(\lambda)$. The rank of $M(\lambda_n)$ for each eigenvalue λ_n is either 1 or 0 in the present case, and it follows from Theorem 2.14 that

$$\text{multiplicity of } \lambda_n = \begin{cases} 1 & (\text{rank}(M(\lambda_n)) = 1) \\ 2 & (\text{rank}(M(\lambda_n)) = 0) \end{cases} \qquad \text{for } 2 \times 2 \text{ matrix } M(\lambda_n). \tag{2.55}$$

Case 2.2 This is also a case of a second-order homogeneous differential equation, given by

$$f''(t) - \frac{2\alpha - 1}{t} f'(t) + \left(\beta^2\gamma^2 t^{2\gamma-2} + \frac{\alpha^2 - \nu^2\gamma^2}{t^2}\right) f(t) = 0. \tag{2.56}$$

This is called *Bessel's differential equation* (Bowman, 2010). The kernels leading to (2.56) will be given in Section 2.4.2 as examples [11] through [17]. When ν is a noninteger, the general solution to (2.56) is given by

$$f(t) = t^{\alpha}\left(c_1 J_\nu(\beta t^\gamma) + c_2 J_{-\nu}(\beta t^\gamma)\right), \tag{2.57}$$

where c_1 and c_2 are arbitrary constants, and $J_\nu(z)$ is the *Bessel function of the first kind* defined by

$$J_\nu(z) = \left(\frac{z}{2}\right)^{\nu} \sum_{k=0}^{\infty} \frac{(-z^2/4)^k}{k!\,\Gamma(\nu + k + 1)}. \tag{2.58}$$

Here $\Gamma(p)$ is the gamma function defined by

$$\Gamma(p) = \int_0^\infty x^{p-1} e^{-x}\, dx \quad (p > 0).$$

It is known (Watson, 1958, p.498) that $J_\nu(z)$ has an infinite product expansion given by

$$J_\nu(z) = \frac{(z/2)^\nu}{\Gamma(\nu+1)} \prod_{n=1}^{\infty} \left(1 - \frac{z^2}{a_n^2}\right), \tag{2.59}$$

where $a_1 < a_2 < \cdots$ are positive zeros of $J_\nu(z)$. The two appropriate boundary conditions lead to $M(\lambda)\,\boldsymbol{c} = \boldsymbol{0}$, and we can determine the multiplicity using (2.55).

Case 2.3 This is a special case of Case 2.4, discussed next, and is concerned with the nearly definite kernel

$$K(s,t) = K_0(s,t) - \psi(s)\psi(t), \tag{2.60}$$

where $K_0(s,t)$ is a positive definite kernel and $\psi(t)$ is a continuous function. To derive the FD of $K(s,t) = K_0(s,t) - \psi(s)\psi(t)$, we must be more specific about $K_0(s,t)$ and $\psi(t)$, where we assume that $\psi(t)$ is continuously differentiable. For example, let us consider the case where

$$K_0(s,t) = \min(s,t) - st, \quad \psi(0) = 0, \quad \psi(1) = 0.$$

If $\psi(t)$ is twice continuously differentiable, the integral equation with the kernel $K(s,t)$ yields the second-order nonhomogeneous differential equation

$$f''(t) + \lambda f(t) = -\lambda a \psi''(t), \tag{2.61}$$

where

$$a = \int_0^1 \psi(s) f(s)\, ds, \quad f(0) = 0, \quad f(1) = 0, \quad \psi(0) = 0, \quad \psi(1) = 0. \tag{2.62}$$

It can be shown (Exercise 2.3.2) that the differential equation (2.61) with the five conditions in (2.62) is equivalent to the integral equation (2.44) with the present kernel $K(s,t)$. We can proceed in the same way as in Case 2.4 discussed next, but Darling's formula (Darling, 1955, theorem 6.5) is more convenient in the present case, which will be dealt with in Section 2.3.2.

Case 2.4 This is an extension of Case 2.3. We consider the kernel

$$K(s,t) = K_0(s,t) + \sum_{k=1}^{q} \xi_k(s)\eta_k(t), \tag{2.63}$$

where $K_0(s,t) = \min(s,t) - st$ or $K_0(s,t) = 1 - \max(s,t)$. It is assumed that

(i) $\xi_k(s)$ and $\eta_k(t)$ $(k = 1, \ldots, q)$ are continuous, and each set is linearly independent in the space $C[0, 1]$.
(ii) $\eta_k''(t)$ $(k = 1, \ldots, q)$ are linearly independent.

For $K_0(s,t) = \min(s,t) - st$ or $1 - \max(s,t)$, the integral equation

$$f(t) = \lambda\left(\int_0^1 K_0(s,t)f(s)\,ds + \sum_{k=1}^{q} a_k\eta_k(t)\right) \tag{2.64}$$

leads to the nonhomogeneous differential equation

$$f''(t) + \lambda f(t) = \lambda\sum_{k=1}^{q} a_k\eta_k''(t), \tag{2.65}$$

where

$$a_k = \int_0^1 \xi_k(s)f(s)\,ds \qquad (k = 1, \ldots, q). \tag{2.66}$$

The derivation of the associated FD is suggested in Nabeya and Tanaka (1988) as follows, by modifying the technique of Kac, Kiefer, and Wolfowitz (1955). For given λ $(\neq 0)$, the general solution to (2.65) is

$$f(t) = c_1\cos\sqrt{\lambda}\,t + c_2\sin\sqrt{\lambda}\,t + \sum_{k=1}^{q} a_k g_k(t), \tag{2.67}$$

where $g_k(t)$ is a special solution of

$$g_k''(t) + \lambda g_k(t) = \lambda\eta_k''(t).$$

Then, by modifying Theorem 2.14, we have the following theorem leading to the derivation of the FD of $K(s,t)$ in (2.63).

Theorem 2.15 *Suppose that the integral equation (2.64) with* $K_0(s,t) = \min(s,t) - st$ *or* $K_0(s,t) = 1 - \max(s,t)$ *is equivalent to the general solution (2.67) of the differential equation (2.65) under the condition (2.66) on* a_k *for* $k = 1, \ldots, q$ *and two appropriate boundary conditions expressed as* $M(\lambda)\,\mathbf{c} = \mathbf{0}$, *where* $M(\lambda)$ *is a* $(q+2)\times(q+2)$ *matrix and* $\mathbf{c} = (a_1, \ldots, a_q, c_1, c_2)'$. *Then* $|M(\lambda)| = 0$ $(\lambda \neq 0)$ *is a necessary and sufficient condition for* λ $(\neq 0)$ *to be an eigenvalue of* $K(s,t)$, *and the multiplicity of the eigenvalue* λ_n *is given by*

$$\ell_n = q + 2 - \operatorname{rank}(M(\lambda_n)). \tag{2.68}$$

The kernels leading to (2.65) will be given in Section 2.4.4 as examples [24] through [27].

Case 2.5 This case arises from the quadratic functional S_g of g-fold integrated Bm $F_g(t)$ introduced in Section 1.2, where

$$S_g = \int_0^1 F_g^2(t)\,dt, \quad F_g(t) = \int_0^t \frac{(t-s)^g}{g!}\,dW(s) \quad (g = 0, 1, 2, \ldots). \tag{2.69}$$

The statistic S_g expressed by the Riemann integral can also be expressed by the Riemann–Stieltjes double integral as

$$S_g = \int_0^1 \int_0^1 K(s,t)\,dW(s)\,dW(t), \tag{2.70}$$

where

$$K(s,t) = \frac{1}{(g!)^2} \int_{\max(s,t)}^1 ((s-u)(t-u))^g\,du.$$

The expression (2.70) gives the integral equation

$$\begin{aligned} f(t) &= \lambda \int_0^1 K(s,t) f(s)\,ds \\ &= \lambda \left[\int_0^t K_1(s,t) f(s)\,ds + \int_t^1 K_2(s,t) f(s)\,ds \right], \end{aligned} \tag{2.71}$$

where

$$K_1(s,t) = \frac{1}{(g!)^2} \int_t^1 ((s-u)(t-u))^g\,du,$$
$$K_2(s,t) = \frac{1}{(g!)^2} \int_s^1 ((s-u)(t-u))^g\,du.$$

Let us define, for $j = 0, 1, \ldots, g$,

$$A_j(s,t) = \frac{\partial^j K_1(s,t)}{\partial t^j} = \frac{g(g-1)\cdots(g-j+1)}{(g!)^2} \int_t^1 (s-u)^g (t-u)^{g-j}\,du,$$
$$B_j(s,t) = \frac{\partial^j K_2(s,t)}{\partial t^j} = \frac{g(g-1)\cdots(g-j+1)}{(g!)^2} \int_s^1 (s-u)^g (t-u)^{g-j}\,du,$$

where $A_j(t,t) = B_j(t,t)$ and $A_j(s,1) = B_j(1,t) = 0$ for $j = 0, 1, \ldots, g$. Then it holds that, for $j = 0, 1, \ldots, g$,

$$f^{(j)}(t) = \lambda \left[\int_0^t A_j(s,t) f(s)\,ds + \int_t^1 B_j(s,t) f(s)\,ds \right], \quad f^{(j)}(1) = 0.$$

Since

$$\begin{aligned}
A_g(s,t) &= \frac{\partial^g K_1(s,t)}{\partial t^g} = \frac{1}{g!}\int_t^1 (s-u)^g\,du \\
&= \frac{1}{(g+1)!}\left((s-t)^{g+1} - (s-1)^{g+1}\right), \\
B_g(s,t) &= \frac{\partial^g K_2(s,t)}{\partial t^g} = \frac{1}{g!}\int_s^1 (s-u)^g\,du = \frac{-1}{(g+1)!}(s-1)^{g+1},
\end{aligned}$$

it holds that $B_j(s,t) = 0$ for $j > g$, and

$$\begin{aligned}
f^{(g)}(t) &= \frac{\lambda}{(g+1)!}\left[\int_0^t (s-t)^{g+1} f(s)\,ds - \int_0^1 (s-1)^{g+1} f(s)\,ds\right], \\
f^{(g+j)}(t) &= \frac{\lambda(-1)^j}{(g+1-j)!}\int_0^t (s-t)^{g+1-j} f(s)\,ds, \quad f^{(g+j)}(0) = 0, \\
f^{(2g+1)}(t) &= \lambda(-1)^{g+1}\int_0^t f(s)\,ds,
\end{aligned}$$

so that we obtain the following homogeneous differential equation of order $2g+2$:

$$f^{(2g+2)}(t) + (-1)^g \lambda f(t) = 0 \tag{2.72}$$

with $2g+2$ boundary conditions

$$f^{(j)}(1) = 0, \quad f^{(g+1+j)}(0) = 0 \qquad (j = 0, 1, \ldots, g). \tag{2.73}$$

Conversely, let us show that the differential equation (2.72) with the boundary conditions (2.73) implies the integral equation (2.71). Define

$$\begin{aligned}
C_j(s,t) &= \frac{\partial^j K_1(s,t)}{\partial s^j} = \frac{g(g-1)\cdots(g-j+1)}{(g!)^2}\int_t^1 (s-u)^{g-j}(t-u)^g\,du, \\
D_j(s,t) &= \frac{\partial^j K_2(s,t)}{\partial s^j} = \frac{g(g-1)\cdots(g-j+1)}{(g!)^2}\int_s^1 (s-u)^{g-j}(t-u)^g\,du,
\end{aligned}$$

so that

$$\begin{aligned}
C_g(s,t) &= \frac{1}{g!}\int_t^1 (t-u)^g\,du = \frac{-(t-1)^{g+1}}{(g+1)!}, \\
D_g(s,t) &= \frac{1}{g!}\int_s^1 (t-u)^g\,du = \frac{-1}{(g+1)!}\left((t-1)^{g+1} - (t-s)^{g+1}\right).
\end{aligned}$$

It holds that $C_j(t,t) = D_j(t,t)$ and $C_j(s,1) = D_j(1,t) = C_{g+1+j}(s,t) = 0$ for $j = 0, 1, \ldots, g$, whereas

$$D_{g+1+j}(t,t) = 0 \quad (j = 0, 1, \ldots, g-1), \quad D_{2g+1}(s,t) = (-1)^{g+1}.$$

Then we have, by integration by parts,

$$\begin{aligned}
\lambda \int_0^1 K(s,t) f(s)\, ds &= (-1)^{g+1} \int_0^1 K(s,t) f^{(2g+2)}(s)\, ds \\
&= (-1)^{g+1} \left[\left[K(s,t) f^{(2g+1)}(s) \right]_0^1 - \int_0^1 \frac{\partial K(s,t)}{\partial s} f^{(2g+1)}(s)\, ds \right] \\
&= (-1)^{g+1} \Big[K(1,t) f^{(2g+1)}(1) - K(0,t) f^{(2g+1)}(0) \\
&\qquad - \int_0^t C_1(s,t) f^{(2g+1)}(s)\, ds - \int_t^1 D_1(s,t) f^{(2g+1)}(s)\, ds \Big] \\
&= (-1)^{g+2} \left[\int_0^t C_1(s,t) f^{(2g+1)}(s)\, ds + \int_t^1 D_1(s,t) f^{(2g+1)}(s)\, ds \right],
\end{aligned}$$

where we have used $K(1,t) = 0$ and $f^{(2g+1)}(0) = 0$. Proceeding further by integration by parts and using the boundary conditions (2.73), we obtain

$$\begin{aligned}
&\lambda \int_0^1 K(s,t) f(s)\, ds \\
&\quad = -\left[\int_0^t C_g(s,t) f^{(g+2)}(s)\, ds + \int_t^1 D_g(s,t) f^{(g+2)}(s)\, ds \right] \\
&\quad = -\int_t^1 D_{g+2}(s,t) f^{(g)}(s)\, ds = (-1)^g \int_t^1 D_{2g+1}(s,t) f'(s)\, ds \\
&\quad = (-1)^{2g+1} \int_t^1 f'(s)\, ds = -f(1) + f(t) = f(t),
\end{aligned}$$

which gives the integral equation (2.71).

The statistic S_g defined in (2.69) can also be expressed as

$$S_g = \int_0^1 F_g^2(t)\, dt \stackrel{\mathcal{D}}{=} \int_0^1 \int_0^1 L(s,t)\, dW(s)\, dW(t), \tag{2.74}$$

where

$$L(s,t) = \mathrm{Cov}(F_g(s), F_g(t)) = \frac{1}{(g!)^2} \int_0^{\min(s,t)} ((s-u)(t-u))^g\, du.$$

The kernel $L(s,t)$ gives the integral equation

$$f(t) = \lambda \int_0^1 \frac{1}{(g!)^2} \left(\int_0^{\min(s,t)} ((s-u)(t-u))^g \, du \right) f(s)\, ds, \qquad (2.75)$$

which yields (Exercise 2.3.3) the differential equation (2.72) with $2g+2$ boundary conditions of

$$f^{(j)}(0) = 0, \quad f^{(g+1+j)}(1) = 0 \qquad (j = 0, 1, \ldots, g). \qquad (2.76)$$

Conversely, it can be shown (Exercise 2.3.4) that the differential equation (2.72) with the boundary conditions (2.76) leads to the integral equation (2.75).

The general solution to the differential equation (2.72) can be obtained as follows. Consider the characteristic equation corresponding to (2.72), which is given, for any integer n, by

$$x^{2g+2} = (-1)^{g+1}\lambda = \begin{cases} \lambda(\cos 2n\pi + i \sin 2n\pi) & (g : \text{odd}), \\ \lambda(\cos(2n+1)\pi + i\sin(2n+1)\pi) & (g : \text{even}). \end{cases}$$

The roots of this equation are given, for $j = 1, \ldots, 2g+2$, by

$$x_j = \begin{cases} \lambda^{1/(2g+2)} \left(\cos \dfrac{(j-1)\pi}{g+1} + i \sin \dfrac{(j-1)\pi}{g+1} \right) & (g : \text{odd}), \\ \lambda^{1/(2g+2)} \left(\cos \dfrac{(2j-1)\pi}{2g+2} + i \sin \dfrac{(2j-1)\pi}{2g+2} \right) & (g : \text{even}), \end{cases} \qquad (2.77)$$

where $x_{g+1+k} = -x_k$ for $k = 1, \ldots, g+1$. Then the general solution to (2.72) is obtained (Exercise 2.3.5) as

$$\begin{aligned} f(t) &= \sum_{j=1}^{2g+2} b_j e^{x_j t} = \sum_{k=1}^{g+1} \left(b_k e^{x_k t} + b_{g+1+k} e^{-x_k t} \right) \\ &= \sum_{k=1}^{g+1} \left(b_k e^{i(-i x_k t)} + b_{g+1+k} e^{i(i x_k t)} \right) \\ &= \sum_{k=1}^{g+1} \left(c_{2k-1} \cos(i y_k t) + c_{2k} \sin(i y_k t) \right), \end{aligned} \qquad (2.78)$$

where b_j and c_j $(j = 1, \ldots, 2g+2)$ are arbitrary constants, and $y_1, \ldots, y_{g+1}$ are distinct elements of $x_1, \ldots, x_{2g+2}$ for which $y_k \neq -y_\ell$.

The $2g+2$ boundary conditions in (2.73) or (2.76) will yield a homogeneous equation $M(\lambda)\boldsymbol{c} = \boldsymbol{0}$ in $\boldsymbol{c} : (2g+2) \times 1$. Then we can obtain the FD, although the computation becomes more burdensome as g gets large. The kernels leading to the general solution (2.78) will be given in Section 2.4.5 as examples [30] through [34].

2.3.2 Darling's Formula and Its Extensions

Here we deal with the kernel

$$K(s,t) = K_0(s,t) - \sum_{j=1}^{m} \psi_j(s)\psi_j(t), \tag{2.79}$$

where $K_0(s,t)$ is a symmetric, continuous, and positive definite kernel, and $\psi_j(t)$ $(j = 1, \ldots, m)$ are continuous and linearly independent functions. When $m = 1$, the derivation of the FD of $K(s,t)$ was suggested by Darling (1955, theorem 6.2) on the basis of

Theorem 2.16 *Consider the kernel $K(s,t)$ defined by (2.79) with $m = 1$, where $K_0(s,t)$ is a symmetric, continuous, and positive definite kernel whose FD is $D_0(\lambda)$, and $\psi_1(t) = \psi(t)$ is a continuous function. Let $\lambda_1, \lambda_2, \ldots$ be the eigenvalues of $K_0(s,t)$, where the multiplicity of each eigenvalue is unity, and let $f_1(t), f_2(t), \ldots$ be the corresponding orthonormal eigenfunctions. Then the FD of $K(s,t)$ is given by*

$$D(\lambda) = D_0(\lambda)P(\lambda), \tag{2.80}$$

where

$$P(\lambda) = 1 + \lambda \sum_{n=1}^{\infty} \frac{b_n^2}{1 - \lambda/\lambda_n}, \qquad b_n = \int_0^1 \psi(t) f_n(t)\, dt. \tag{2.81}$$

The expression for $D(\lambda)$ in (2.80) is simple, but its computation deals with $P(\lambda)$, which contains the infinite sum described in (2.81). To compute the infinite sum easily, we become more specific about $K_0(s,t)$ and $\psi(t)$, where we assume that $\psi(t)$ is continuously differentiable. We note in passing that the definition of b_n in (2.81) gives

$$\psi(t) = \sum_{n=1}^{\infty} b_n f_n(t), \qquad \int_0^1 \psi^2(t)\, dt = \sum_{n=1}^{\infty} b_n^2.$$

We first consider the case where

$$K_0(s,t) = \min(s,t) - st, \qquad \psi(0) = 0, \qquad \psi(1) = 0.$$

Note that the FD of K_0 is $D_0(\lambda) = \sin\sqrt{\lambda}/\sqrt{\lambda}$, and the eigenvalues of K_0 are given by $\lambda_n = n^2\pi^2$ $(n = 1, 2, \ldots)$ with the multiplicity of each eigenvalue equal to unity, whereas the orthonormal eigenfunction corresponding to λ_n is given by $f_n(t) = \sqrt{2}\sin n\pi t$.

Then we have Darling's formula (Darling, 1955, theorem 6.5) described as

Theorem 2.17 (Darling's formula) *Consider the kernel defined by*

$$K_1(s,t) = \min(s,t) - st - \psi(s)\psi(t), \quad \psi(0) = 0, \quad \psi(1) = 0, \tag{2.82}$$

where $\psi(t)$ is continuously differentiable. Then its FD is given by

$$D_1(\lambda) = \frac{\sin\sqrt{\lambda}}{\sqrt{\lambda}} - 2\int_0^1 \left(\int_0^t \psi'(s)\cos\sqrt{\lambda}s\,ds\right)\psi'(t)\cos\sqrt{\lambda}(1-t)\,dt. \tag{2.83}$$

Proof Note first that the FD of $\min(s,t) - st$ is given by $D_0(\lambda) = \sin\sqrt{\lambda}/\sqrt{\lambda}$ with eigenvalues $\lambda_n = n^2\pi^2$ and orthonormal eigenfunctions $f_n(t) = \sqrt{2}\sin n\pi t$. Since

$$b_n = \int_0^1 \psi(t) f_n(t)\,dt = \frac{\sqrt{2}}{n\pi}\int_0^1 \psi'(t)\cos n\pi t\,dt,$$

it follows from Theorem 2.16 that the FD of $K_1(s,t)$ is $D_1(\lambda) = D_0(\lambda)P(\lambda)$, where

$$\begin{aligned} P(\lambda) &= 1 + \frac{2\lambda}{\pi^2}\sum_{n=1}^{\infty}\frac{1}{n^2 - \lambda/\pi^2}\int_0^1\int_0^1 \psi'(s)\psi'(t)\cos n\pi s\,\cos n\pi t\,ds\,dt \\ &= 1 + \frac{\lambda}{\pi^2}\sum_{n=1}^{\infty}\frac{1}{n^2 - \lambda/\pi^2} \\ &\quad \times \int_0^1\int_0^1 \psi'(s)\psi'(t)\Big\{\cos n\pi(s+t) + \cos n\pi(s-t)\Big\}\,ds\,dt. \end{aligned}$$

Using the relation (Gradshteyn and Ryzhik, 1965, p.48)

$$\sum_{n=1}^{\infty}\frac{\cos n\pi x}{n^2 - \alpha^2} = \frac{1}{2\alpha^2} - \frac{\pi}{2}\frac{\cos\alpha\pi(1-x)}{\alpha\sin\alpha\pi} \qquad (0 \le x \le 2), \tag{2.84}$$

we have (Exercise 2.3.6)

$$P(\lambda) = 1 - \frac{\sqrt{\lambda}}{\sin\sqrt{\lambda}}\int_0^1\int_0^1 \psi'(s)\psi'(t)L_1(s,t)\,ds\,dt, \tag{2.85}$$

where

$$L_1(s,t) = \begin{cases} \cos\sqrt{\lambda}s\,\cos\sqrt{\lambda}(1-t) & (s \le t), \\ \cos\sqrt{\lambda}t\,\cos\sqrt{\lambda}(1-s) & (s \ge t), \end{cases}$$

which leads to (2.83) because $\psi'(s)\psi'(t)L_1(s,t)$ is symmetric. □

The kernel in (2.82) with $\psi(t) = \sqrt{3}t(1-t)$ was earlier introduced in Section 1.5 in connection with goodness-of-fit tests. The other kernels leading to (2.83) will be given in Section 2.4.3 as examples [18], [19], and [20]. For these examples, the computation of the integrals on the right-hand side of (2.83) can be done easily by computerized algebra.

We can also give a formula similar to Darling's formula for the FD of the kernel $K_2(s,t) = 1 - \max(s,t) - \psi(s)\psi(t)$, which is proved in Tanaka (2023).

Theorem 2.18 *Consider the kernel defined by*

$$K_2(s,t) = 1 - \max(s,t) - \psi(s)\psi(t), \quad \psi(1) = 0, \tag{2.86}$$

where $\psi(t)$ is continuously differentiable. Then its FD is given by

$$D_2(\lambda) = \cos\sqrt{\lambda} + 2\sqrt{\lambda}\int_0^1 \left(\int_0^t \psi'(s)\sin\sqrt{\lambda}s\,ds\right)\psi'(t)\cos\sqrt{\lambda}(1-t)\,dt. \tag{2.87}$$

Proof Note first that the FD of $1 - \max(s,t)$ is given by $D_0(\lambda) = \cos\sqrt{\lambda}$ with eigenvalues $\lambda_n = (n-1/2)^2\pi^2$ and orthonormal eigenfunctions $f_n(t) = \sqrt{2}\cos(n-1/2)\pi t$. Since

$$b_n = \int_0^1 \psi(t) f_n(t)\,dt = -\frac{\sqrt{2}}{(n-1/2)\pi}\int_0^1 \psi'(t)\sin\left(n-\frac{1}{2}\right)\pi t\,dt,$$

it follows from Theorem 2.16 that

$$\begin{aligned} P(\lambda) &= 1 + \frac{2\lambda}{\pi^2}\sum_{n=1}^{\infty}\frac{1}{(n-1/2)^2 - \frac{\lambda}{\pi^2}}\left(\int_0^1 \psi'(t)\sin\left(n-\frac{1}{2}\right)\pi t\,dt\right)^2 \\ &= 1 + \frac{\lambda}{\pi^2}\sum_{n=1}^{\infty}\frac{1}{(n-1/2)^2 - \frac{\lambda}{\pi^2}}\int_0^1\int_0^1 \psi'(s)\psi'(t) \\ &\quad \times \left\{\cos\left(n-\frac{1}{2}\right)\pi(s-t) - \cos\left(n-\frac{1}{2}\right)\pi(s+t)\right\}\,ds\,dt. \end{aligned}$$

Using the following formula derived from Gradshteyn and Ryzhik (1965, p.48) as

$$\sum_{n=1}^{\infty}\frac{\cos(n-1/2)\pi x}{(n-1/2)^2 - \alpha^2} = \frac{\pi}{2\alpha}\frac{\sin\alpha\pi(1-x)}{\cos\alpha\pi} \qquad (0 \le x \le 2), \tag{2.88}$$

we have (Exercise 2.3.7)

$$P(\lambda) = 1 + \frac{\sqrt{\lambda}}{\cos\sqrt{\lambda}} \int_0^1 \int_0^1 \psi'(s)\psi'(t) L_2(s,t)\, ds\, dt, \tag{2.89}$$

where

$$L_2(s,t) = \begin{cases} \sin\sqrt{\lambda}s \, \cos\sqrt{\lambda}(1-t) & (s \le t), \\ \sin\sqrt{\lambda}t \, \cos\sqrt{\lambda}(1-s) & (s \ge t). \end{cases}$$

Thus we obtain (2.87) because $\psi'(s)\psi'(t)L_2(s,t)$ is symmetric. □

It may be noted that Theorem 2.18 yields the FD of $K_0(s,t) = \min(s,t) - st$. In fact, the FD of $K_0(s,t)$ is the same as that of

$$K_0(1-s, 1-t) = \min(1-s, 1-t) - (1-s)(1-t) = 1 - \max(s,t) - \psi(s)\psi(t),$$

where $\psi(t) = 1 - t$ with $\psi(1) = 0$. Then the kernel $K_2(s,t) = K_0(1-s, 1-t)$ satisfies the condition described in Theorem 2.18 with $\psi'(t) = -1$. Thus the FD of $K_2(s,t)$ is given by

$$\begin{aligned} D_2(\lambda) &= \cos\sqrt{\lambda} + 2\sqrt{\lambda} \int_0^1 \left(\int_0^t \sin\sqrt{\lambda}s\, ds \right) \cos\sqrt{\lambda}(1-t)\, dt \\ &= \cos\sqrt{\lambda} + 2 \int_0^1 (1 - \cos\sqrt{\lambda}t) \cos\sqrt{\lambda}(1-t)\, dt \\ &= \cos\sqrt{\lambda} + \frac{\sin\sqrt{\lambda} - \sqrt{\lambda}\cos\sqrt{\lambda}}{\sqrt{\lambda}} = \frac{\sin\sqrt{\lambda}}{\sqrt{\lambda}}. \end{aligned}$$

The kernels leading to (2.87) will be given in Section 2.4.3 as examples [21], [22], and [23]. For these examples, the computation of the integrals on the right-hand side of (2.87) can be done easily by computerized algebra. The methods suggested in Theorems 2.17 and 2.18 may not be of wide applicability because of the restrictive assumptions imposed on the kernel $K(s,t)$, but they are quite useful in some cases for computing its FD with the aid of computerized algebra. Some applications will be given in Chapters 5 and 6.

Darling's result was extended by Sukhatme (1972) to deal with a general kernel given in (2.79) as

$$K(s,t) = K_0(s,t) - \sum_{j=1}^{m} \psi_j(s)\psi_j(t), \tag{2.90}$$

where $K_0(s,t)$ is positive definite and its FD is $D_0(\lambda)$, and $\psi_j(t)$ $(j = 1, \ldots, m)$ are continuous and linearly independent. Then we have (Sukhatme, 1972)

Theorem 2.19 *Consider the kernel $K(s,t)$ defined by (2.90). Let $\lambda_1, \lambda_2, \ldots$ be the eigenvalues of $K_0(s,t)$, where the multiplicity of each eigenvalue is unity, and let $f_1(t), f_2(t), \ldots$ be the corresponding orthonormal eigenfunctions. Then the FD of $K(s,t)$ is given by*

$$D(\lambda) = D_0(\lambda)\,|P(\lambda)|, \tag{2.91}$$

where $P(\lambda)$ is the $m \times m$ symmetric matrix whose (j,k)th elements are

$$P_{jj}(\lambda) = 1 + \lambda \sum_{n=1}^{\infty} \frac{b_{jn}^2}{1 - \lambda/\lambda_n}, \quad P_{jk}(\lambda) = \lambda \sum_{n=1}^{\infty} \frac{b_{jn} b_{kn}}{1 - \lambda/\lambda_n} \quad (j \neq k),$$

$$b_{jn} = \int_0^1 \psi_j(t) f_n(t)\, dt.$$

Specifying the kernel $K_0(s,t)$ in (2.90) as $\min(s,t) - st$, we can establish the following theorem (Tanaka, 2023).

Theorem 2.20 *Consider the kernel defined by*

$$K_{1m}(s,t) = \min(s,t) - st - \sum_{j=1}^{m} \psi_j(s)\psi_j(t), \quad \psi_j(0) = \psi_j(1) = 0, \tag{2.92}$$

where $\psi_j(t)$ $(j = 1, \ldots, m)$ are continuously differentiable. Then its FD is given by

$$D_{1m}(\lambda) = \frac{\sin\sqrt{\lambda}}{\sqrt{\lambda}}\,|P(\lambda)|, \tag{2.93}$$

where the (j,k)th elements $P_{jk}(\lambda)$ of the $m \times m$ symmetric matrix $P(\lambda)$ are

$$P_{jj}(\lambda) = 1 - \frac{2\sqrt{\lambda}}{\sin\sqrt{\lambda}} \int_0^1 \left(\int_0^t \psi_j'(s) \cos\sqrt{\lambda}s\, ds \right) \psi_j'(t) \cos\sqrt{\lambda}(1-t)\, dt,$$

$$P_{jk}(\lambda) = -\frac{\sqrt{\lambda}}{\sin\sqrt{\lambda}} \int_0^1 \int_0^1 \psi_j'(s)\psi_k'(t) L_1(s,t)\, ds\, dt \qquad (j \neq k),$$

$$L_1(s,t) = \begin{cases} \cos\sqrt{\lambda}s\, \cos\sqrt{\lambda}(1-t) & (s \le t), \\[2ex] \cos\sqrt{\lambda}t\, \cos\sqrt{\lambda}(1-s) & (s \ge t). \end{cases}$$

Proof Since $f_n(t) = \sqrt{2}\sin n\pi t$ and

$$b_{jn} = \int_0^1 \psi_j(t) f_n(t)\, dt = \frac{\sqrt{2}}{n\pi} \int_0^1 \psi_j'(t) \cos n\pi t\, dt,$$

we obtain $P_{jj}(\lambda)$ from Theorem 2.17. On the other hand, putting $\lambda_n = n^2\pi^2$, we have

$$\begin{aligned}
P_{jk}(\lambda) &= \lambda \sum_{n=1}^{\infty} \frac{b_{jn}b_{kn}}{1-\lambda/\lambda_n} \\
&= \frac{2\lambda}{\pi^2}\sum_{n=1}^{\infty}\frac{1}{n^2-\lambda/\pi^2}\int_0^1\int_0^1 \psi_j'(s)\psi_k'(t)\cos n\pi s\,\cos n\pi t\,ds\,dt \\
&= \frac{\lambda}{\pi^2}\sum_{n=1}^{\infty}\frac{1}{n^2-\lambda/\pi^2}\int_0^1\int_0^1 \psi_j'(s)\psi_k'(t) \\
&\qquad\qquad \times \{\cos n\pi(s+t)+\cos n\pi(s-t)\}\,ds\,dt,
\end{aligned}$$

which leads to $P_{jk}(\lambda)$ in the theorem because of Theorem 2.17. Thus the theorem is established. □

Specifying the kernel $K_0(s,t)$ in (2.90) as $1-\max(s,t)$, we can establish the following theorem (Tanaka, 2023), whose proof is left as Exercise 2.3.8.

Theorem 2.21 *Consider the kernel defined by*

$$K_{2m}(s,t) = 1-\max(s,t) - \sum_{j=1}^{m}\psi_j(s)\psi_j(t), \quad \psi_j(1)=0, \tag{2.94}$$

where $\psi_j(t)$ $(j = 1, \ldots, m)$ are continuously differentiable. Then its FD is given by

$$D_{2m}(\lambda) = \cos\sqrt{\lambda}\,|P(\lambda)|, \tag{2.95}$$

where the (j,k)th elements $P_{jk}(\lambda)$ of the $m \times m$ symmetric matrix $P(\lambda)$ are

$$\begin{aligned}
P_{jj}(\lambda) &= 1+\frac{2\sqrt{\lambda}}{\cos\sqrt{\lambda}}\int_0^1\left(\int_0^t \psi_j'(s)\sin\sqrt{\lambda}s\,ds\right)\psi_j'(t)\cos\sqrt{\lambda}(1-t)\,dt, \\
P_{jk}(\lambda) &= \frac{\sqrt{\lambda}}{\cos\sqrt{\lambda}}\int_0^1\int_0^1 \psi_j'(s)\psi_k'(t)L_2(s,t)\,ds\,dt \qquad (j\neq k), \\
L_2(s,t) &= \begin{cases} \sin\sqrt{\lambda}s\,\cos\sqrt{\lambda}(1-t) & (s\le t), \\ \sin\sqrt{\lambda}t\,\cos\sqrt{\lambda}(1-s) & (s\ge t). \end{cases}
\end{aligned}$$

The formulas described in Theorems 2.20 and 2.21 compute the FD simply and efficiently, which will be demonstrated in examples [25] through [29] of Section 2.4.4 and in Chapters 5 and 6.

Exercises for Section 2.3

Exercise 2.3.1 For $h(z) = \cos z - bz \sin z$ $(b \neq 0)$, show that $h'(z)/h(z)$ is bounded on each side of squares C_m with vertices $(2m + 1/2)\pi(\pm 1 \pm i)$ $(m = 1, 2, \ldots)$.

Exercise 2.3.2 Show that the differential equation

$$f''(t) + \lambda f(t) = -\lambda a \psi''(t)$$

with the conditions

$$a = \int_0^1 \psi(s) f(s)\, ds, \quad f(0) = f(1) = 0, \quad \psi(0) = \psi(1) = 0$$

implies the integral equation

$$f(t) = \lambda \left(\int_0^1 [\min(s,t) - st] f(s)\, ds - a\, \psi(t) \right).$$

Exercise 2.3.3 Show that the integral equation with the kernel

$$L(s,t) = \frac{1}{(g!)^2} \int_0^{\min(s,t)} ((s-u)(t-u))^g\, du$$

leads to the differential equation (2.72) with $2g + 2$ boundary conditions

$$f^{(j)}(0) = 0, \quad f^{(g+1+j)}(1) = 0 \quad (j = 0, 1, \ldots, g).$$

Exercise 2.3.4 Show that the differential equation (2.72) with $2g+2$ boundary conditions given in Exercise 2.3.3 leads to the integral equation (2.75).

Exercise 2.3.5 Show that the general solution to the differential equation (2.72) is given by (2.78).

Exercise 2.3.6 Using (2.84), prove that $P(\lambda)$ in (2.85) is given by

$$1 - \frac{\sqrt{\lambda}}{\sin \sqrt{\lambda}} \int_0^1 \int_0^1 \psi'(s)\psi'(t) L_1(s,t)\, ds\, dt.$$

Exercise 2.3.7 Using (2.88), prove that $P(\lambda)$ in (2.89) is given by

$$1 + \frac{\sqrt{\lambda}}{\cos \sqrt{\lambda}} \int_0^1 \int_0^1 \psi'(s)\psi'(t) L_2(s,t)\, ds\, dt.$$

Exercise 2.3.8 Prove Theorem 2.21.

2.4 Various Examples of the Fredholm Determinant

In Section 2.3.1 we indicated that the homogeneous integral equation of Fredholm type with kernel $K(s,t)$ is equivalent to a differential equation with some conditions including the boundary conditions. Then we presented a suitable procedure that enables us to obtain the FD $D(\lambda)$ of $K(s,t)$. In Section 2.3.2 we established formulas that make it easy to compute the FD more efficiently for kernels with specified assumptions.

In this section we consider 34 examples ([1] through [34]) of $K(s,t)$ and derive the corresponding $D(\lambda)$. For each $K(s,t)$ we also consider the statistic

$$S = \int_0^1 \int_0^1 K(s,t)\, dW(s)\, dW(t), \tag{2.96}$$

for which we report

$$\mathrm{E}\left(e^{i\theta S}\right) = (D(2i\theta))^{-1/2}, \tag{2.97}$$

$$\mathrm{E}(S) = \int_0^1 K(t,t)\, dt = \sum_{n=1}^{\infty} \frac{1}{\lambda_n} = d_1, \tag{2.98}$$

$$\begin{aligned} \mathrm{Var}(S) &= 4\int_0^1 \left(\int_0^t K^2(s,t)\, ds\right) dt = 2\sum_{n=1}^{\infty} \frac{1}{\lambda_n^2} \\ &= 2(d_1^2 - d_2), \end{aligned} \tag{2.99}$$

where λ_n $(n = 1, 2, \ldots)$ are eigenvalues of $K(s,t)$ repeated as many times as their multiplicities, and d_1 and d_2 are coefficients in the series expansion for $D(\lambda)$ expressed in (2.5) (see also Theorem 2.4) as

$$D(\lambda) = 1 + \sum_{n=1}^{\infty} \frac{(-1)^n d_n}{n!} \lambda^n = 1 - d_1\lambda + \frac{d_2}{2}\lambda^2 - \cdots.$$

We divide our subsequent discussions into five sections, Sections 2.4.1 through 2.4.5, according to the type of differential equation as classified in the previous section.

2.4.1 *Homogeneous Differential Equation*

In this section we deal with the case where the integral equation

$$f(t) = \lambda \int_0^1 K(s,t) f(s)\, ds \tag{2.100}$$

is equivalent to the second-order homogeneous differential equation

$$f''(t) + (\lambda - \delta) f(t) = 0 \tag{2.101}$$

with two boundary conditions. Then we find a condition for λ to be an eigenvalue and a candidate for $D(\lambda)$ by ensuring that it admits an infinite product expansion and determine the multiplicity of each eigenvalue. In this section, we consider examples [1] through [10] of related kernels.

[1] $K(s,t) = 1 - \max(s,t)$.

This kernel was fully discussed in the last section. It was shown that the integral equation (2.100) is equivalent to the differential equation $f''(t) + \lambda f(t) = 0$ with two boundary conditions $f(1) = 0$, $f'(0) = 0$. The general solution is given by $f(t) = c_1 \cos\sqrt{\lambda}t + c_2 \sin\sqrt{\lambda}t$, where c_1 and c_2 are arbitrary constants. Then we have the following homogeneous equation in $\boldsymbol{c} = (c_1, c_2)'$:

$$\begin{pmatrix} \cos\sqrt{\lambda} & \sin\sqrt{\lambda} \\ 0 & \sqrt{\lambda} \end{pmatrix} \begin{pmatrix} c_1 \\ c_2 \end{pmatrix} = M(\lambda)\,\boldsymbol{c} = \boldsymbol{0}.$$

The eigenfunction $f(t)$ must be nonvanishing, which only occurs when $\boldsymbol{c} \neq \boldsymbol{0}$. Thus we have $|M(\lambda)| = \sqrt{\lambda}\cos\sqrt{\lambda} = 0$. Then λ ($\neq 0$) is an eigenvalue if and only if $\cos\sqrt{\lambda} = 0$, and it is ensured that

$$D(\lambda) = \cos\sqrt{\lambda} \tag{2.102}$$

by checking that $\text{rank}(M(\lambda_n)) = 1$ for each eigenvalue $\lambda_n = (n - 1/2)^2\pi^2$, and $D(\lambda)$ admits an infinite product expansion, where every zero of $D(\lambda)$ is of order unity.

It follows from Theorem 2.3 and (2.97) that $\mathrm{E}\left(e^{i\theta S}\right) = (D(2i\theta))^{-1/2} = \left(\cos\sqrt{2i\theta}\right)^{-1/2}$, where S is the statistic defined in (2.96). Since $D(\lambda) = 1 - \lambda/2 + \lambda^2/24 - \cdots$, we have $\mathrm{E}(S) = 1/2$ from (2.98) and $\mathrm{Var}(S) = 2(1/4 - 1/12) = 1/3$ from (2.99).

The orthonormal eigenfunction corrsponding to λ_n is given by

$$f_n(t) = \sqrt{2}\cos\left(n - \frac{1}{2}\right)\pi t \quad (n = 1, 2, \ldots),$$

and it follows from Mercer's theorem that

$$1 - \max(s,t) = \sum_{n=1}^{\infty} \frac{2\cos(n - 1/2)\pi s\,\cos(n - 1/2)\pi t}{(n - 1/2)^2\pi^2}.$$

[2] $K(s,t) = \min(s,t)$.

This kernel yields the same FD as [1] because $K(1-s, 1-t) = \min(1-s, 1-t) = 1 - \max(s,t)$. It also holds that

$$\begin{aligned}
\int_0^1 W^2(t)\,dt &= \int_0^1\int_0^1 [1-\max(s,t)]\,dW(s)\,dW(t) \\
&\stackrel{\mathcal{D}}{=} \int_0^1\int_0^1 \mathrm{Cov}(W(s),W(t))\,dW(s)\,dW(t) \\
&= \int_0^1\int_0^1 \min(s,t)\,dW(s)\,dW(t) \\
&= \int_0^1 (W(1)-W(t))^2\,dt.
\end{aligned}$$

For illustrative purposes, however, we derive the FD of $K(s,t) = \min(s,t)$ disregarding the above facts. The integral equation

$$f(t) = \lambda \int_0^1 \min(s,t) f(s)\,ds = \lambda \left[\int_0^t s f(s)\,ds + t\int_t^1 f(s)\,ds\right]$$

yields the differential equation

$$f''(t) + \lambda f(t) = 0 \quad \Leftrightarrow \quad f(t) = c_1 \cos\sqrt{\lambda}t + c_2 \sin\sqrt{\lambda}t$$

and the boundary conditions

$$f(0) = 0, \quad f(1) = \lambda\int_0^1 s f(s)\,ds, \quad f'(0) = \lambda\int_0^1 f(s)\,ds, \quad f'(1) = 0.$$

The same differential equation emerges, but the boundary conditions are different from [1]. We choose $f(0) = f'(1) = 0$, which yields

$$\begin{pmatrix} 1 & 0 \\ -\sin\sqrt{\lambda} & \cos\sqrt{\lambda} \end{pmatrix}\begin{pmatrix} c_1 \\ c_2 \end{pmatrix} = M(\lambda)\,\mathbf{c} = \mathbf{0}, \quad |M(\lambda)| = \cos\sqrt{\lambda}.$$

Proceeding in the same way as in [1], it is found that $D(\lambda) = \cos\sqrt{\lambda}$, so that the statistic S has the same $\mathrm{E}\left(e^{i\theta S}\right)$, $\mathrm{E}(S)$, and $\mathrm{Var}(S)$ as [1], whereas the orthonormal eigenfunction corresponding to $\lambda_n = (n-1/2)^2\pi^2$ is given by

$$f_n(t) = \sqrt{2}\sin\left(n - \frac{1}{2}\right)\pi t.$$

Mercer's theorem gives $f_n(s)f_n(t) = g_n(1-s)g_n(1-t)$, where $g_n(t) = \sqrt{2}\cos(n-1/2)\pi t$ is the eigenfunction associated with the kernel $1-\max(s,t)$. It follows that

$$\min(s,t) = \sum_{n=1}^{\infty} \frac{2\sin(n-1/2)\pi s \, \sin(n-1/2)\pi t}{(n-1/2)^2\pi^2}.$$

[3] $K(s,t) = \min(s,t) - st = K(1-s, 1-t)$.

The integral equation

$$\begin{aligned} f(t) &= \lambda \int_0^1 [\min(s,t) - st] f(s)\, ds \\ &= \lambda \left[\int_0^t s f(s)\, ds + t \int_t^1 f(s)\, ds - t \int_0^1 s f(s)\, ds \right] \end{aligned}$$

gives

$$\begin{aligned} & f'(t) = \lambda \left[\int_t^1 f(s)\, ds - \int_0^1 s f(s)\, ds \right], \quad f''(t) + \lambda f(t) = 0, \\ & f(0) = 0, \quad f(1) = 0, \\ & f'(0) = \lambda \int_0^1 (1-s) f(s)\, ds, \quad f'(1) = -\lambda \int_0^1 s f(s)\, ds. \end{aligned}$$

It can be shown (Exercise 2.4.1) that the differential equation with the boundary conditions $f(0) = f(1) = 0$ is equivalent to the above integral equation. The same differential equation as [1] and [2] emerges with different boundary conditions, which gives

$$\begin{pmatrix} 1 & 0 \\ \cos\sqrt{\lambda} & \sin\sqrt{\lambda} \end{pmatrix} \begin{pmatrix} c_1 \\ c_2 \end{pmatrix} = M(\lambda)\, \boldsymbol{c} = \boldsymbol{0}, \quad |M(\lambda)| = \sin\sqrt{\lambda}.$$

The condition that $|M(\lambda)| = \sin\sqrt{\lambda} = 0$ is necessary and sufficient for λ ($\neq 0$) to be an eigenvalue, where $\mathrm{rank}(M(\lambda_n)) = 1$ for every eigenvalue $\lambda_n = n^2\pi^2$.

Let us put $h(z) = \sin z / z$, which is an even function with $h(0) = 1$ and

$$h'(z) = -\frac{1}{z^2} \sin z + \frac{1}{z} \cos z, \quad h'(0) = 0, \quad \frac{h'(z)}{h(z)} = \frac{\cos z}{\sin z} - \frac{1}{z}.$$

The zeros of $h(z)$ are all simple, and we denote them by $\pm a_n = \pm n\pi$ ($n = 1, 2, \ldots$). Since it was shown in the discussions following (2.42) that $\cos z / \sin z$ is bounded on each side of squares C_m in the complex plane with vertices $(2m + 1/2)\pi(\pm 1 \pm i)$ ($m = 1, 2, \ldots$), $h'(z)/h(z)$ is also bounded on each side of C_m as $m \to \infty$. It follows from Theorem 2.13 that $h(z)$ admits an infinite product expansion given by

$$h(z) = \prod_{n=1}^{\infty} \left(1 - \frac{z^2}{a_n^2}\right) = \prod_{n=1}^{\infty} \left(1 - \frac{z^2}{n^2\pi^2}\right).$$

Thus it is ensured that

$$D(\lambda) = \frac{\sin\sqrt{\lambda}}{\sqrt{\lambda}} \tag{2.103}$$

and $\mathrm{E}\left(e^{i\theta S}\right) = \left(\sin\sqrt{2i\theta}/\sqrt{2i\theta}\right)^{-1/2}$. Since $D(\lambda) = 1-\lambda/6+\lambda^2/120-\cdots$, we have $\mathrm{E}(S) = 1/6$ and $\mathrm{Var}(S) = 2(1/36 - 1/60) = 1/45$.

The orthonormal eigenfunction corresponding to λ_n is given by $f_n(t) = \sqrt{2}\sin n\pi t$, and it follows that

$$\min(s,t) - st = \sum_{n=1}^{\infty} \frac{2\sin n\pi s\,\sin n\pi t}{n^2\pi^2}.$$

[4] $K(s,t) = \dfrac{1}{3} - \max(s,t) + \dfrac{s^2+t^2}{2}.$

This kernel should give the same FD as [3] because

$$\begin{aligned}
&\int_0^1 (W(t) - tW(1))^2\,dt \\
&= \int_0^1\int_0^1 \left[\frac{1}{3} - \max(s,t) + \frac{s^2+t^2}{2}\right] dW(s)\,dW(t) \\
&\overset{\mathcal{D}}{=} \int_0^1\int_0^1 \mathrm{Cov}(W(s) - sW(1), W(t) - tW(1))\,dW(s)\,dW(t) \\
&= \int_0^1\int_0^1 [\min(s,t) - st]\,dW(s)\,dW(t) \\
&= \int_0^1 \left(W(t) - \int_0^1 W(s)\,ds\right)^2 dt.
\end{aligned}$$

Disregarding the above facts, let us derive the FD of the present kernel. The integral equation with this kernel yields

$$f'(t) = \lambda\left[-\int_0^t f(s)\,ds + t\int_0^1 f(s)\,ds\right], \quad f''(t)+\lambda f(t) = \lambda a, \tag{2.104}$$

where

$$a = \int_0^1 f(t)\,dt, \quad f'(0) = f'(1) = 0. \tag{2.105}$$

The general solution to the nonhomogeneous differential equation in (2.104) is

$$f(t) = c_1\cos\sqrt{\lambda}t + c_2\sin\sqrt{\lambda}t + a. \tag{2.106}$$

Given this solution, it is recognized that the three conditions in (2.105) hold for any a. However, it can be shown (Exercise 2.4.2) that the differential equation in (2.104) with the conditions in (2.105) only leads to the integral equation with the present kernel when $a = 0$, which we assume in subsequent discussions. Thus the differential equation in (2.104) becomes homogeneous. The two boundary conditions yield

$$\begin{pmatrix} 0 & \sqrt{\lambda} \\ -\sqrt{\lambda}\sin\sqrt{\lambda} & \sqrt{\lambda}\cos\sqrt{\lambda} \end{pmatrix} \begin{pmatrix} c_1 \\ c_2 \end{pmatrix} = M(\lambda)\,\boldsymbol{c} = \boldsymbol{0},$$

which gives $|M(\lambda)| = \lambda \sin\sqrt{\lambda}$, and we obtain the same FD as [3]. The reason for $a = 0$ may also be explained as follows. Noting that $c_2 = 0$, so that the eigenfunction corresponding to the eigenvalue λ_n is given by $f_n(t) = c_1 \cos n\pi t + a$, it follows from the orthogonality of $\{f_n(t)\}$ that a must be 0.

The orthonormal eigenfunction corresponding to $\lambda_n = n^2\pi^2$ is given by $f_n(t) = \sqrt{2}\cos n\pi t$, and it holds that

$$\frac{1}{3} - \max(s,t) + \frac{s^2+t^2}{2} = \sum_{n=1}^{\infty} \frac{2\cos n\pi s\,\cos n\pi t}{n^2\pi^2}.$$

[5] $K(s,t) = 1 - \max(s,t) + b.$

The case where $b = 0$ was dealt with in [1]. The integral equation with the present kernel is equivalent (Exercise 2.4.3) to

$$f''(t) + \lambda f(t) = 0, \quad f(1) = \lambda b \int_0^1 f(s)\,ds, \quad f'(0) = 0.$$

Solving for $f(t)$ we have, from the two boundary conditions, the homogeneous equation $M(\lambda)\,\boldsymbol{c} = \boldsymbol{0}$, where $\boldsymbol{c} = (c_1, c_2)'$ and

$$M(\lambda) = \begin{pmatrix} \cos\sqrt{\lambda} - b\sqrt{\lambda}\sin\sqrt{\lambda} & \sin\sqrt{\lambda} + b\sqrt{\lambda}(\cos\sqrt{\lambda} - 1) \\ 0 & \sqrt{\lambda} \end{pmatrix}.$$

The eigenvalues are given by the solution to $\cos\sqrt{\lambda} - b\sqrt{\lambda}\sin\sqrt{\lambda} = 0$. Evidently $\mathrm{rank}(M(\lambda_n)) = 1$ for each eigenvalue λ_n, so that the multiplicity of λ_n is unity for all n. Then we obtain, as a candidate for the FD,

$$\tilde{D}(\lambda) = \cos\sqrt{\lambda} - b\sqrt{\lambda}\sin\sqrt{\lambda}. \tag{2.107}$$

We dealt with $h(z) = \cos z - bz\sin z$ in (2.41), and it was ensured that $h(z)$ admits an infinite product expansion of the form

$$h(\sqrt{\lambda}) = \tilde{D}(\lambda) = \prod_{n=1}^{\infty}\left(1 - \frac{\lambda}{a_n^2}\right), \tag{2.108}$$

where $a_m^2 \neq a_n^2$ for $m \neq n$. Then we conclude that $\tilde{D}(\lambda)$ in (2.107) is the FD of the present kernel. Since

$$\tilde{D}(\lambda) = 1 - \left(\frac{1}{2} + b\right)\lambda + \left(\frac{1}{24} + \frac{b}{6}\right)\lambda^2 - \cdots,$$

it holds that $\mathrm{E}(S) = 1/2 + b$ and $\mathrm{Var}(S) = 2((1/2+b)^2 - (1/12 + b/3)) = 1/3 + 2b^2 + 4b/3$.

It can be checked (Exercise 2.4.4) that, when b is nonnegative, every eigenvalue is positive, whereas all eigenvalues except one are positive when b is negative. Thus a_n^2 $(n = 1, 2, \ldots)$ in (2.108) are all positive for $b \geq 0$, whereas only one a_n^2 is negative for $b < 0$.

The present kernel will arise in Chapter 7 when we deal with the unit root distribution

$$P\left(\frac{\int_0^1 W(t)\,dW(t)}{\int_0^1 W^2(t)\,dt} \leq x\right) = P\left(x\int_0^1 W^2(t)\,dt - \int_0^1 W(t)\,dW(t) \geq 0\right)$$

$$= P\left(\int_0^1\int_0^1 K(s,t;x)\,dW(s)\,dW(t) + \frac{1}{2} \geq 0\right),$$

where $K(s,t;x) = x\,[1 - \max(s,t)] - 1/2$. The FD of $K(s,t;x)$ can easily be derived by replacing λ and b in (2.107) by λx and $-1/(2x)$, respectively, to give

$$D(\lambda;x) = \cos\sqrt{\lambda x} + \frac{\sqrt{\lambda x}}{2x}\sin\sqrt{\lambda x} = \cos\sqrt{\lambda x} + \frac{\lambda}{2}\frac{\sin\sqrt{\lambda x}}{\sqrt{\lambda x}}. \qquad (2.109)$$

Moments of the unit root distribution will be discussed in Section 7.1 together with other types of unit root distributions presented subsequently.

[6] $K(s,t) = \min(s,t) - st + b_1 + b_2(s+t).$

The integral equation with the present kernel leads to

$$f(t) = c_1\cos\sqrt{\lambda}t + c_2\sin\sqrt{\lambda}t,$$

$$f(0) = \lambda\int_0^1 (b_1 + b_2 s)f(s)\,ds, \quad f(1) = \lambda\int_0^1 (b_1 + b_2 + b_2 s)f(s)\,ds,$$

which yields $M(\lambda)\,\boldsymbol{c} = \boldsymbol{0}$, where $M(\lambda) = \left[M_{ij}(\lambda)\right]$ with

$$M_{11}(\lambda) = 1 + A(\lambda), \quad M_{12}(\lambda) = B(\lambda),$$

$$M_{21}(\lambda) = \cos\sqrt{\lambda} - b_2\sqrt{\lambda}\sin\sqrt{\lambda} + A(\lambda),$$

$$M_{22}(\lambda) = (b_1 + b_2)\sqrt{\lambda}\cos\sqrt{\lambda} - b_2\sin\sqrt{\lambda} - b_1\sqrt{\lambda},$$

$$A(\lambda) = b_2(1 - \cos\sqrt{\lambda}) - (b_1 + b_2)\sqrt{\lambda}\sin\sqrt{\lambda},$$
$$B(\lambda) = (b_1 + b_2)\sqrt{\lambda}\cos\sqrt{\lambda} - b_2\sin\sqrt{\lambda} - b_1\sqrt{\lambda},$$
$$|M(\lambda)| = \sqrt{\lambda}\left((1 + \lambda b_2^2)\frac{\sin\sqrt{\lambda}}{\sqrt{\lambda}} + 2(b_1 + b_2 + b_2^2)(\cos\sqrt{\lambda} - 1)\right).$$

By checking Condition (ii) in Theorem 2.12, we can deduce (Exercise 2.4.5) that the FD is given by

$$D(\lambda) = (1 + \lambda b_2^2)\frac{\sin\sqrt{\lambda}}{\sqrt{\lambda}} + 2(b_1 + b_2 + b_2^2)(\cos\sqrt{\lambda} - 1) \tag{2.110}$$
$$= 1 - \left(\frac{1}{6} + b_1 + b_2\right)\lambda + \left(\frac{1}{120} + \frac{b_1}{12} + \frac{b_2}{12} - \frac{b_2^2}{12}\right)\lambda^2 - \cdots .$$

It follows that $\mathrm{E}(S) = 1/6 + b_1 + b_2$ and $\mathrm{Var}(S) = 2((1/6 + b_1 + b_2)^2 - (1/60 + b_1/6 + b_2/6 - b_2^2/6)) = 2(1/240 + (1/12 + b_1 + b_2)^2 + b_2^2/6)$.

The present kernel will arise in Chapter 7 when we deal with the mean-corrected unit root distribution

$$P\left(\frac{\int_0^1 \bar{W}(t)\,dW(t)}{\int_0^1 \bar{W}^2(t)\,dt} \le x\right) = P\left(x\int_0^1 \bar{W}^2(t)\,dt - \int_0^1 \bar{W}(t)\,dW(t) \ge 0\right)$$
$$= P\left(\int_0^1\int_0^1 K(s,t;x)\,dW(s)\,dW(t) + \frac{1}{2} \ge 0\right),$$

where

$$\bar{W}(t) = W(t) - \int_0^1 W(s)\,ds, \quad K(s,t;x) = x[\min(s,t) - st] + \frac{1}{2}(1 - s - t).$$

The FD of $K(s,t;x)$ can easily be derived by replacing λ, b_1, and b_2 in (2.110) by λx, $1/(2x)$, and $-1/(2x)$, respectively, to give

$$D(\lambda;x) = \left(1 + \frac{\lambda}{4x}\right)\frac{\sin\sqrt{\lambda x}}{\sqrt{\lambda x}} + \frac{1}{2x^2}(\cos\sqrt{\lambda x} - 1). \tag{2.111}$$

[7] $K(s,t) = \dfrac{1}{2\gamma}(e^{\gamma(s+t)} - e^{\gamma|s-t|})$.

This kernel was introduced in Section 1.4 to describe the quadratic functional of the O–U process. For $dX(t) = \gamma X(t)\,dt + dW(t)$ with $X(0) = 0$, we have (Exercise 2.4.6)

$$\int_0^1 X^2(t)\,dt = \frac{1}{2\gamma}\int_0^1\int_0^1 \left(e^{\gamma(2-s-t)} - e^{\gamma|s-t|}\right) dW(s)\,dW(t)$$

$$\overset{\mathcal{D}}{=} \int_0^1\int_0^1 \mathrm{Cov}(X(s), X(t))\,dW(s)\,dW(t) \tag{2.112}$$

$$= \frac{1}{2\gamma}\int_0^1\int_0^1 \left(e^{\gamma(s+t)} - e^{\gamma|s-t|}\right) dW(s)\,dW(t). \tag{2.113}$$

Note that $K(s,t) = \mathrm{Cov}(X(s), X(t))$ in the present case, so that it is positive definite and yields

$$f''(t) + (\lambda - \gamma^2)f(t) = 0, \quad f(0) = 0, \quad f'(1) = \gamma f(1). \tag{2.114}$$

Note also that $(e^{\gamma(2-s-t)} - e^{\gamma|s-t|})/(2\gamma) = K(1-s, 1-t)$. It can be shown (Exercise 2.4.7) that (2.114) is equivalent to the integral equation with the present kernel.

Let us assume that $\lambda \neq \gamma^2$. Then the general solution to the differential equation in (2.114) is given by

$$f(t) = c_1 \cos\mu t + c_2 \sin\mu t, \qquad \mu = \sqrt{\lambda - \gamma^2},$$

and the two boundary conditions in (2.114) yield $M(\lambda)\,\boldsymbol{c} = \boldsymbol{0}$, where

$$|M(\lambda)| = \begin{vmatrix} 1 & 0 \\ \gamma\cos\mu + \mu\sin\mu & \gamma\sin\mu - \mu\cos\mu \end{vmatrix} = \gamma\sin\mu - \mu\cos\mu$$
$$= -\mu\left(\cos\mu - \gamma\frac{\sin\mu}{\mu}\right).$$

Then we find

$$\tilde{D}(\lambda) = e^{\gamma}\left(\cos\mu - \gamma\frac{\sin\mu}{\mu}\right) \tag{2.115}$$

as a candidate for the FD, with $\tilde{D}(0) = e^{\gamma}(\cos i\gamma - \gamma\sin i\gamma/(i\gamma)) = e^{\gamma}\, e^{-\gamma} = 1$, where we can deduce (Exercise 2.4.8) that $\tilde{D}(\lambda)$ admits an infinite product expansion.

Suppose that $\lambda = \gamma^2$. Then it follows from (2.114) that $f(t) = c_1 t$ and $c_1 = \gamma c_1$, where $c_1 \neq 0$. Thus γ must be unity for λ to be an eigenvalue, and $\lambda = 1$ is the only eigenvalue. It holds that $\tilde{D}(1) = 0$ when $\lambda = \gamma^2$ and $\gamma = 1$. Thus $\tilde{D}(\lambda)$ in (2.115) is the FD of the present kernel.

To compute E(S) and Var(S), we could use the series expansion for $\tilde{D}(\lambda)$, but here we compute these as

$$\mathrm{E}(S) = \int_0^1 K(t,t)\,dt = \frac{1}{2\gamma}\int_0^1 \left(e^{2\gamma t} - 1\right)dt = \frac{1}{2\gamma}\left(\frac{e^{2\gamma}-1}{2\gamma} - 1\right),$$

$$\mathrm{Var}(S) = 4\int_0^1 \left(\int_0^t K^2(s,t)\,ds\right)dt = \frac{1}{\gamma^2}\int_0^1\left(\int_0^t \left(e^{\gamma(s+t)} - e^{\gamma(t-s)}\right)^2 ds\right)dt$$

$$= \frac{1}{2\gamma^2}\left(\frac{e^{4\gamma}}{4\gamma^2} + \frac{e^{2\gamma}}{\gamma^2} - \frac{2e^{2\gamma}}{\gamma} - \frac{5}{4\gamma^2} - \frac{1}{\gamma}\right).$$

[8] $K(s,t) = \dfrac{1}{2\gamma}\left(e^{\gamma(s+t)} - e^{\gamma|s-t|}\right) + be^{\gamma(s+t)}.$

This kernel is slightly different from [7], and it is of some interest to see the effect of the difference on the FD. We are led to

$$f''(t) + (\lambda - \gamma^2)f(t) = 0, \quad f(0) = b\lambda\int_0^1 e^{\gamma s}f(s)\,ds, \quad f'(1) = \gamma f(1).$$

When $\lambda \neq \gamma^2$, we have $f(t) = c_1\cos\mu t + c_2\sin\mu t$, where $\mu = \sqrt{\lambda - \gamma^2}$. The boundary conditions yield $M(\lambda)\,\boldsymbol{c} = \boldsymbol{0}$, where

$$|M(\lambda)| = \begin{vmatrix} \gamma(1+b\gamma) & -b\gamma\mu \\ -\gamma\cos\mu - \mu\sin\mu & \mu\cos\mu - \gamma\sin\mu \end{vmatrix}$$

$$= \gamma\mu\left(\cos\mu - (\gamma + b\lambda)\frac{\sin\mu}{\mu}\right).$$

Then we find (Exercise 2.4.9)

$$\tilde{D}(\lambda) = e^{\gamma}\left(\cos\mu - (\gamma + b\lambda)\frac{\sin\mu}{\mu}\right) = e^{\gamma}\left(\cos\mu - (\gamma + b(\gamma^2+\mu^2))\frac{\sin\mu}{\mu}\right) \tag{2.116}$$

as a candidate for the FD. We can show that γ^2 is an eigenvalue, that is,

$$\tilde{D}(\gamma^2) = e^{\gamma}(1 - \gamma - b\gamma^2) = 0.$$

In fact, for $\lambda = \gamma^2$, the differential equation is $f''(t) = 0$, so that $f(t) = c_1 + c_2 t$ and

$$f(0) = b\lambda\int_0^1 e^{\gamma s}f(s)\,ds \Leftrightarrow (b\gamma(e^{\gamma}-1) - 1)c_1 + b((\gamma-1)e^{\gamma}+1)c_2 = 0,$$

$$f'(1) = \gamma f(1) \Leftrightarrow \gamma c_1 + (\gamma - 1)c_2 = 0.$$

Then it follows that $\lambda = \gamma^2$ is an eigenvalue if and only if

$$\begin{vmatrix} b\gamma(e^\gamma - 1) - 1 & b\left((\gamma - 1)e^\gamma + 1\right) \\ \gamma & \gamma - 1 \end{vmatrix} = 1 - \gamma - b\gamma^2 = 0.$$

Since this is equivalent to $\tilde{D}(\gamma^2) = e^\gamma(1 - \gamma - b\gamma^2) = 0$, $\lambda = \gamma^2$ is an eigenvalue.

The present kernel will arise in Chapter 7 when we deal with the near-unit root distribution

$$P\left(\frac{\int_0^1 X(t)\,dX(t)}{\int_0^1 X^2(t)\,dt} \le x\right) = P\left(x\int_0^1 X^2(t)\,dt - \int_0^1 X(t)\,dX(t) \ge 0\right)$$

$$= P\left(\int_0^1\int_0^1 K(s,t;x)\,dW(s)\,dW(t) + \frac{1}{2} \ge 0\right),$$

where $dX(t) = \gamma X(t)\,dt + dW(t)$ with $X(0) = 0$ and

$$K(1-s, 1-t; x) = \frac{x}{2\gamma}\left(e^{\gamma(s+t)} - e^{\gamma|s-t|}\right) - \frac{1}{2}\,e^{\gamma(s+t)}.$$

The FD of this kernel can easily be derived by replacing λ and b in (2.116) by λx and $-1/(2x)$, respectively, to give

$$D(\lambda; x) = e^\gamma\left(\cos\nu - \left(\gamma - \frac{\lambda}{2}\right)\frac{\sin\nu}{\nu}\right), \qquad \nu = \sqrt{\lambda x - \gamma^2}. \tag{2.117}$$

[9] $K(s,t) = \frac{1}{4}[1 - 2|s-t|] = \min(s,t) - \frac{s+t}{2} + \frac{1}{4}.$

The present kernel was introduced in (1.72) together with its FD without proof when we discussed Lévy's stochastic area. It can be shown (Exercise 2.4.10) that the integral equation with this kernel is equivalent to

$$f''(t) + \lambda f(t) = 0, \quad f(0) + f(1) = 0, \quad f'(0) + f'(1) = 0, \tag{2.118}$$

which leads to the homogeneous equation $M(\lambda)\,\boldsymbol{c} = \boldsymbol{0}$, where $\boldsymbol{c} = (c_1, c_2)'$ and

$$M(\lambda) = \begin{pmatrix} 1 + \cos\sqrt{\lambda} & \sin\sqrt{\lambda} \\ -\sqrt{\lambda}\sin\sqrt{\lambda} & \sqrt{\lambda}(1 + \cos\sqrt{\lambda}) \end{pmatrix}, \qquad |M(\lambda)| = 4\sqrt{\lambda}\left(\cos\frac{\sqrt{\lambda}}{2}\right)^2.$$

Then λ $(\neq 0)$ is an eigenvalue if and only if $|M(\lambda)| = 0$, that is, $\cos(\sqrt{\lambda}/2) = 0$. Thus each eigenvalue is given by $\lambda_n = (2n-1)^2\pi^2$

($n = 1, 2, \ldots$). Since $M(\lambda_n) = 0$ for every λ_n, the multiplicity of λ_n is two because of Theorem 2.14 and (2.55), so that there are two linearly independent eigenfunctions corresponding to each eigenvalue. Then we obtain

$$D(\lambda) = \left(\cos\frac{\sqrt{\lambda}}{2}\right)^2 = \prod_{n=1}^{\infty}\left(1 - \frac{\lambda}{((2n-1)\pi)^2}\right)^2$$

as the FD of the present kernel. It now holds that

$$\begin{aligned} \mathrm{E}\left(e^{i\theta S}\right) &= \mathrm{E}\left[\exp\left\{i\theta\int_0^1\int_0^1 \frac{1}{4}\left[1 - 2\,|s-t|\,\right] dW(s)\,dW(t)\right\}\right] \\ &= \mathrm{E}\left[\exp\left\{i\theta\int_0^1 \left(W(t) - \frac{1}{2}W(1)\right)^2 dt\right\}\right] \\ &= (D(2i\theta))^{-1/2} = \left(\cos\frac{\sqrt{2i\theta}}{2}\right)^{-1}. \end{aligned} \tag{2.119}$$

Since $D(\lambda) = 1 - \lambda/4 + \lambda^2/48 - \cdots$, we have $\mathrm{E}(S) = 1/4$ and $\mathrm{Var}(S) = 2(1/16 - 1/24) = 1/24$. Note here that, if we use the formulas

$$\mathrm{E}(S) = \sum_{n=1}^{\infty}\frac{1}{\lambda_n}, \qquad \mathrm{Var}(S) = 2\sum_{n=1}^{\infty}\frac{1}{\lambda_n^2},$$

$\lambda_n = (2n-1)^2\pi^2$ must be repeated twice in the above sums; otherwise $\mathrm{E}(S)$ and $\mathrm{Var}(S)$ reduce to $1/8$ and $1/48$, respectively.

The two linearly independent orthonormal eigenfunctions corresponding to λ_n are given (Exercise 2.4.11) by

$$f_{n1}(t) = \sqrt{2}\cos(2n-1)\pi t, \qquad f_{n2}(t) = \sqrt{2}\sin(2n-1)\pi t.$$

Thus it follows from Mercer's theorem that

$$\begin{aligned} \frac{1}{4}\left[1 - 2|s-t|\right] &= \sum_{n=1}^{\infty}\frac{2}{(2n-1)^2\pi^2}[\cos(2n-1)\pi s\,\cos(2n-1)\pi t \\ &\qquad + \sin(2n-1)\pi s\,\sin(2n-1)\pi t] \\ &= \sum_{n=1}^{\infty}\frac{2\cos(2n-1)\pi(s-t)}{(2n-1)^2\pi^2}, \end{aligned}$$

which may also be derived from the Fourier expansion for $|x|$ ($|x| \le 1$) given by

$$|x| = \frac{1}{2} - 4\sum_{n=1}^{\infty}\frac{\cos(2n-1)\pi x}{(2n-1)^2\pi^2} \quad \Leftrightarrow \quad \frac{1}{4}\left[1 - 2|x|\right] = 2\sum_{n=1}^{\infty}\frac{\cos(2n-1)\pi x}{(2n-1)^2\pi^2}.$$

As was mentioned above, we discussed Lévy's stochastic area in Section 1.4, which is expressed as

$$Q = \frac{1}{2}\int_0^1 \Big(W_1(t)\,dW_2(t) - W_2(t)\,dW_1(t)\Big),$$

where the conditional distribution of Q given $\{W_1(t)\}$ was shown to be

$$Q\Big|\Big\{W_1(t)\Big\} \quad \sim \quad \mathrm{N}\left(0,\ \frac{1}{4}\int_0^1\int_0^1 [1-2|s-t|]\ dW_1(s)\,dW_1(t)\right).$$

The c.f. of Q was also presented without proof, which can now be easily derived from (2.119). We obtain

$$\begin{aligned}
\mathrm{E}\Big(e^{i\theta Q}\Big) &= \mathrm{E}\Big[\mathrm{E}\Big[e^{i\theta Q}\Big|\Big\{W_1(t)\Big\}\Big]\Big] = \mathrm{E}\left[-\frac{\theta^2}{2}\mathrm{Var}\Big(Q\Big|\Big\{W_1(t)\Big\}\Big)\right] \\
&= \mathrm{E}\left[\exp\left\{-\frac{\theta^2}{2}\int_0^1\int_0^1 \frac{1}{4}\Big[1-2\,|s-t|\,\Big]dW_1(s)\,dW_1(t)\right\}\right] \\
&= \Big(D(-\theta^2)\Big)^{-1/2} = \left(\cos\frac{\sqrt{-\theta^2}}{2}\right)^{-1} = \left(\cosh\frac{\theta}{2}\right)^{-1}. \qquad (2.120)
\end{aligned}$$

The distribution of Lévy's stochastic area is symmetric since its c.f. is real, and has $\mathrm{Var}(Q) = 1/4$ since $(\cosh(\theta/2))^{-1} = 1 - \theta^2/8 + 5\theta^4/384 - \cdots$. The probability density $f(x)$ of Q can be explicitly derived by the *inversion formula* as

$$f(x) = \frac{1}{2\pi}\int_{-\infty}^{\infty} e^{-i\theta x}\mathrm{E}\left(e^{i\theta Q}\right)d\theta = \frac{1}{\pi}\int_0^{\infty}\frac{\cos\theta x}{\cosh(\theta/2)}\,d\theta = \frac{1}{\cosh \pi x},$$

where this last result comes from Gradshteyn and Ryzhik (1965, p.540).

[10] $K(s,t) = \min(s,t) - st + \frac{1}{2}(s^2+t^2-s-t) + \frac{1}{12}.$

The present kernel was introduced in (2.21) together with its FD without proof. This kernel was used in Watson (1961) in connection with goodness-of-fit tests on a circle. It holds (Exercise 2.4.12) that

$$\begin{aligned}
\int_0^1\int_0^1 K(s,t)\,dW(s)\,dW(t) &= \int_0^1\int_0^1 \mathrm{Cov}(\tilde{X}(s),\tilde{X}(t))\,dW(s)\,dW(t) \\
&= \int_0^1 \tilde{X}^2(t)\,dt,
\end{aligned}$$

where

$$\tilde{X}(t) = \tilde{W}(t) - \int_0^1 \tilde{W}(u)\,du, \quad \tilde{W}(t) = W(t) - tW(1).$$

Note that $\{\tilde{W}(t)\}$ is the Brownian bridge (Bb), whereas $\{\tilde{X}(t)\}$ is the demeaned Bb. It is recognized that the kernel $K(s,t)$ is positive definite.

The integral equation with the kernel $K(s,t)$ leads to

$$f''(t) + \lambda f(t) = \lambda a, \quad a = \int_0^1 f(s)\,ds, \quad f(0) = f(1), \quad f'(0) = f'(1). \tag{2.121}$$

It must hold that $a = 0$ for the equations in (2.121) to be equivalent to the integral equation. Thus the general solution to the differential equation is given by $f(t) = c_1 \cos\sqrt{\lambda}t + c_2 \sin\sqrt{\lambda}t$, and it follows from the two boundary conditions in (2.121) that $M(\lambda)\,\boldsymbol{c} = \boldsymbol{0}$, where $\boldsymbol{c} = (c_1, c_2)'$ and

$$M(\lambda) = \begin{pmatrix} 1 - \cos\sqrt{\lambda} & -\sin\sqrt{\lambda} \\ \sin\sqrt{\lambda} & 1 - \cos\sqrt{\lambda} \end{pmatrix}, \qquad |M(\lambda)| = 4\sin^2\frac{\sqrt{\lambda}}{2}.$$

Eigenvalues are given by $\lambda_n = 4n^2\pi^2$ $(n = 1, 2, \ldots)$, which yields $M(\lambda_n) = 0$, so that the multiplicity of each eigenvalue is two. Thus we obtain, as the FD of K,

$$D(\lambda) = \left(\frac{\sin\sqrt{\lambda}/2}{\sqrt{\lambda}/2}\right)^2 \tag{2.122}$$
$$= 1 - \frac{1}{12}\lambda + \frac{1}{360}\lambda^2 - \cdots.$$

Then we have $\mathrm{E}\left(e^{i\theta S}\right) = (D(2i\theta))^{-1/2} = \left(\sin\sqrt{i\theta/2}/\sqrt{i\theta/2}\right)^{-1}$. It also holds that $\mathrm{E}(S) = 1/12$ and $\mathrm{Var}(S) = 1/360$, which was earlier obtained in Section 2.2.

Since eigenfunctions corresponding to $\lambda_n = 4n^2\pi^2$ are given by

$$f_n(t) = c_1 \cos\sqrt{\lambda_n}t + c_2 \sin\sqrt{\lambda_n}t = c_1 \cos 2n\pi t + c_2 \sin 2n\pi t,$$

where c_1 and c_2 are arbitrary constants, two linearly independent orthonormal eigenfunctions corresponding to λ_n are given by

$$f_{n1}(t) = \sqrt{2}\cos 2n\pi t, \qquad f_{n2}(t) = \sqrt{2}\sin 2n\pi t.$$

Thus the present kernel $K(s,t)$ can be expanded as

$$\begin{aligned} K(s,t) &= -\frac{1}{2}|s-t| + \frac{1}{2}(s-t)^2 + \frac{1}{12} \\ &= \sum_{n=1}^{\infty} \frac{2(\cos 2n\pi s\ \cos 2n\pi t + \sin 2n\pi s\ \sin 2n\pi t)}{4n^2\pi^2} \\ &= \sum_{n=1}^{\infty} \frac{\cos 2n\pi(s-t)}{2n^2\pi^2}. \end{aligned}$$

This last expression can also be derived from the Fourier expansions for $|x|$ and x^2 ($|x| \le 1$) given by

$$|x| = \frac{1}{2} - 4\sum_{n=1}^{\infty} \frac{\cos(2n-1)\pi x}{(2n-1)^2\pi^2}, \quad x^2 = \frac{1}{3} + 4\sum_{n=1}^{\infty}(-1)^n \frac{\cos n\pi x}{n^2\pi^2}.$$

2.4.2 Bessel's Differential Equation

In this section we deal with Bessel's differential equation introduced in Case 2.2 of the last section, which is given by

$$f''(t) - \frac{2\alpha - 1}{t} f'(t) + \left(\beta^2\gamma^2 t^{2\gamma-2} + \frac{\alpha^2 - \nu^2\gamma^2}{t^2}\right) f(t) = 0. \tag{2.123}$$

When ν is a noninteger, the general solution is given by

$$f(t) = t^{\alpha}\left(c_1 J_\nu(\beta t^\gamma) + c_2 J_{-\nu}(\beta t^\gamma)\right), \tag{2.124}$$

where c_1 and c_2 are arbitrary constants, and $J_\nu(z)$ is the Bessel function of the first kind defined in (2.58) as

$$J_\nu(z) = \left(\frac{z}{2}\right)^\nu \sum_{k=0}^{\infty} \frac{(-z^2/4)^k}{k!\,\Gamma(\nu+k+1)} = \frac{(z/2)^\nu}{\Gamma(\nu+1)} \prod_{n=1}^{\infty}\left(1 - \frac{z^2}{a_n^2}\right), \tag{2.125}$$

where $a_1 < a_2 < \cdots$ are positive zeros of $J_\nu(z)$, and the infinite product expansion is given in Watson (1958, p.498). The following relations are also useful (Bowman, 2010):

$$J_\nu(z) = \frac{z}{2\nu}\left(J_{\nu-1}(z) + J_{\nu+1}(z)\right), \quad J_\nu'(z) = \frac{1}{2}\left(J_{\nu-1}(z) - J_{\nu+1}(z)\right), \tag{2.126}$$

$$\left(z^\nu J_\nu(z)\right)' = z^\nu J_{\nu-1}(z), \quad \left(z^{-\nu} J_\nu(z)\right)' = -z^{-\nu} J_{\nu+1}(z), \tag{2.127}$$

$$J_{-1/2}(z) = \sqrt{\frac{2}{\pi z}}\cos z, \qquad J_{1/2}(z) = \sqrt{\frac{2}{\pi z}}\sin z.$$

Below we consider examples [11] through [17] of related kernels.

[11] $K(s,t) = \frac{1}{2m+1}\left[1 - (\max(s,t))^{2m+1}\right] \quad (m > -1/2).$

This kernel was introduced earlier in Section 1.3, where it was indicated that

$$\int_0^1 t^{2m}\, W^2(t)\, dt = \int_0^1\int_0^1 \frac{1}{2m+1}\left[1 - (\max(s,t))^{2m+1}\right] dW(s)\, dW(t)$$
$$\stackrel{\mathcal{D}}{=} \int_0^1\int_0^1 s^m t^m \min(s,t)\, dW(s)\, dW(t). \tag{2.128}$$

The related statistic was dealt with by MacNeill (1974) and Nabeya and Tanaka (1988) in connection with testing for parameter constancy. MacNeill (1974) assumed $m > -1$ rather than $m > -1/2$, although the kernel is not continuous when $-1 < m \leq -1/2$. Here we do not go into such complexity but continue to assume that the kernel is continuous, so that $m > -1/2$.

The integral equation is given by

$$f(t) = \frac{\lambda}{2m+1}\left[\int_0^1 f(s)\,ds - t^{2m+1}\int_0^t f(s)\,ds - \int_t^1 s^{2m+1} f(s)\,ds\right],$$

which yields

$$f'(t) = -\lambda t^{2m}\int_0^t f(s)\,ds, \quad f''(t) = -2\lambda m t^{2m-1}\int_0^t f(s)\,ds - \lambda t^{2m} f(t).$$

It can be shown (Exercise 2.4.13) that the above integral equation is equivalent to

$$f''(t) - \frac{2m}{t} f'(t) + \lambda t^{2m} f(t) = 0, \quad f(1) = 0, \quad \lim_{t\to 0}\frac{f'(t)}{t^{2m}} = 0. \quad (2.129)$$

The differential equation in (2.129) is a special case of Bessel's differential equation defined in (2.123) with

$$\alpha = \frac{2m+1}{2}, \quad \beta = \frac{\sqrt{\lambda}}{m+1}, \quad \gamma = m+1, \quad \nu = \frac{2m+1}{2(m+1)},$$

and it has the general solution

$$f(t) = t^{(2m+1)/2}\left\{c_1 J_\nu\left(\frac{\sqrt{\lambda}}{m+1}t^{m+1}\right) + c_2 J_{-\nu}\left(\frac{\sqrt{\lambda}}{m+1}t^{m+1}\right)\right\}. \quad (2.130)$$

Since

$$\begin{aligned} f(t) &= c_1\left(\frac{\sqrt{\lambda}}{2(m+1)}\right)^{\nu} t^{2m+1}\left(\frac{1}{\Gamma(\nu+1)} - \frac{\lambda t^{2(m+1)}/4(m+1)^2}{\Gamma(\nu+2)} + O\left(t^{4(m+1)}\right)\right) \\ &\quad + c_2\left(\frac{\sqrt{\lambda}}{2(m+1)}\right)^{-\nu}\left(\frac{1}{\Gamma(-\nu+1)} - \frac{\lambda t^{2(m+1)}/4(m+1)^2}{\Gamma(-\nu+2)} + O\left(t^{4(m+1)}\right)\right), \\ f'(t) &= c_1\left(\frac{\sqrt{\lambda}}{2(m+1)}\right)^{\nu}\left(\frac{(2m+1)t^{2m}}{\Gamma(\nu+1)} + O\left(t^{4m+2}\right)\right) \\ &\quad + c_2\left(\frac{\sqrt{\lambda}}{2(m+1)}\right)^{-\nu}\left(-\frac{2(m+1)\lambda t^{2m+1}/4(m+1)^2}{\Gamma(-\nu+2)} + O\left(t^{4m+3}\right)\right), \end{aligned}$$

the boundary conditions in (2.129) yield $M(\lambda)\,\boldsymbol{c} = \boldsymbol{0}$, where $\boldsymbol{c} = (c_1, c_2)'$ and

$$M(\lambda) = \begin{pmatrix} J_\nu\left(\dfrac{\sqrt{\lambda}}{m+1}\right) & J_{-\nu}\left(\dfrac{\sqrt{\lambda}}{m+1}\right) \\ \left(\dfrac{\sqrt{\lambda}}{2(m+1)}\right)^\nu \dfrac{2m+1}{\Gamma(\nu+1)} & 0 \end{pmatrix}. \tag{2.131}$$

A necessary and sufficient condition for λ ($\neq 0$) to be an eigenvalue is $|M(\lambda)| = 0$, that is, $J_{-\nu}(\sqrt{\lambda}/(m+1)) = 0$. It is clear that $\mathrm{rank}(M(\lambda_n)) = 1$ for each eigenvalue λ_n, so that the multiplicity is unity for every eigenvalue. Since it is known (Watson, 1958, p.498) that the infinite product expansion is possible for $J_{-\nu}(z)$, which is

$$J_{-\nu}(z) = \frac{(z/2)^{-\nu}}{\Gamma(-\nu+1)} \prod_{n=1}^{\infty} \left(1 - \frac{z^2}{a_n^2}\right), \tag{2.132}$$

where $a_1 < a_2 < \cdots$ are positive zeros of $J_{-\nu}(z)$, we obtain the following as the FD:

$$\begin{aligned} D(\lambda) &= \prod_{n=1}^{\infty} \left(1 - \frac{\lambda}{(m+1)^2 a_n^2}\right) \\ &= \Gamma(-\nu+1)\, J_{-\nu}\left(\frac{\sqrt{\lambda}}{m+1}\right) \Big/ \left(\frac{\sqrt{\lambda}}{2(m+1)}\right)^{-\nu} \\ &= 1 - \frac{1}{2(m+1)}\lambda + \frac{1}{8(m+1)^2(2m+3)}\lambda^2 - \cdots, \end{aligned} \tag{2.133}$$

where $\nu = (2m+1)/2(m+1)$. It holds that $\mathrm{E}(S) = 1/2(m+1)$ and $\mathrm{Var}(S) = 1/(m+1)(2m+3)$.

Note that the kernel $s^m t^m \min(s,t)$ has the same FD, as was described in (2.128). It may also be noted that, when $m = 0$ so that $\nu = 1/2$, $D(\lambda)$ reduces to $\cos\sqrt{\lambda}$ since

$$\Gamma\left(\frac{1}{2}\right) = \sqrt{\pi}, \qquad J_{-1/2}(z) = \sqrt{\frac{2}{\pi z}} \cos z.$$

Conversely, $D(2m\lambda) \to 1 - \lambda$ as $m \to \infty$. Thus the statistic in (2.128) multiplied by $2m$ approaches $\chi^2(1)$ as m becomes large.

[12] $K(s,t) = s^m t^m [\min(s,t) - st] \quad (m > -1/2)$.

This kernel arises in a quadratic functional of the Bb defined by

$$S = \int_0^1 \left[t^m \{W(t) - tW(1)\} \right]^2 dt$$
$$\stackrel{\mathcal{D}}{=} \int_0^1 \int_0^1 s^m t^m [\min(s,t) - st] \, dW(s) \, dW(t). \qquad (2.134)$$

The integral equation with the present kernel is given by

$$f(t) = \lambda \left[t^m \int_0^t s^{m+1} f(s) \, ds + t^{m+1} \int_t^1 s^m f(s) \, ds - t^{m+1} \int_0^1 s^{m+1} f(s) \, ds \right],$$

which we rewrite, by putting $f(t) = t^m h(t)$, as

$$h(t) = \lambda \left[\int_0^t s^{2m+1} h(s) \, ds + t \int_t^1 s^{2m} h(s) \, ds - t \int_0^1 s^{2m+1} h(s) \, ds \right].$$

Then it can be shown that the integral equation is equivalent to

$$h''(t) + \lambda t^{2m} h(t) = 0, \qquad h(0) = h(1) = 0. \qquad (2.135)$$

The differential equation in (2.135) is a special case of Bessel's equation (2.123) with

$$\alpha = \frac{1}{2}, \quad \beta = \frac{\sqrt{\lambda}}{m+1}, \quad \gamma = m+1, \quad \nu = \frac{1}{2(m+1)},$$

which yields

$$h(t) = \sqrt{t} \left\{ c_1 J_\nu \left(\frac{\sqrt{\lambda}}{m+1} t^{m+1} \right) + c_2 J_{-\nu} \left(\frac{\sqrt{\lambda}}{m+1} t^{m+1} \right) \right\}.$$

Proceeding in the same way as [11], we obtain (Exercise 2.4.14)

$$D(\lambda) = \Gamma \left(\frac{2m+3}{2(m+1)} \right) J_{1/2(m+1)} \left(\frac{\sqrt{\lambda}}{m+1} \right) \Big/ \left(\frac{\sqrt{\lambda}}{2(m+1)} \right)^{1/2(m+1)} \qquad (2.136)$$
$$= 1 - \frac{1}{2(m+1)(2m+3)} \lambda + \frac{1}{8(m+1)^2(2m+3)(4m+5)} \lambda^2 - \cdots,$$

and

$$\mathrm{E}(S) = \frac{1}{2(m+1)(2m+3)}, \quad \mathrm{Var}(S) = \frac{1}{(m+1)(2m+3)^2(4m+5)}.$$

It can be checked that when $m = 0$, $D(\lambda)$ reduces to $\sin\sqrt{\lambda}/\sqrt{\lambda}$ since

$$\Gamma\left(\frac{3}{2}\right) = \frac{\sqrt{\pi}}{2}, \qquad J_{1/2}(z) = \sqrt{\frac{2}{\pi z}}\,\sin z.$$

It can also be checked that $D((m+1)^2\lambda) \to J_0(\sqrt{\lambda})$ as $m \to \infty$.

It is noted that the statistic S in (2.134) is related to the Anderson–Darling statistic for goodness-of-fit tests (Anderson and Darling, 1952), which is

$$\begin{aligned} W^2 &= \int_0^1 \frac{(W(t) - tW(1))^2}{t(1-t)}\,dt \\ &\overset{\mathcal{D}}{=} \int_0^1\int_0^1 \frac{1}{\sqrt{st(1-s)(1-t)}}\,[\min(s,t) - st\,]\,dW(s)\,dW(t) \\ &= \int_0^1\int_0^1 K(s,t)\,dW(s)\,dW(t), \end{aligned} \tag{2.137}$$

where

$$K(s,t) = \sqrt{\frac{s(1-t)}{t(1-s)}} \qquad (s \le t).$$

It can be seen that the present kernel $K(s,t)$ is not continuous at $(s,t) = (0,0)$ and $(1,1)$. If we disregard this fact, we have the differential equation

$$t(1-t)h''(t) + \lambda h(t) = 0, \qquad h(0) = h(1) = 0. \tag{2.138}$$

The resulting equation is the hypergeometric differential equation, and the solution is well defined for $0 < t < 1$ but is not at $t = 0$ and $t = 1$. It seems difficult to apply our procedure to obtain the FD. Instead, Anderson and Darling (1952) (see also Pettitt (1976)) derived the following, using the Ferrer associated Legendre polynomial:

$$W^2 = \int_0^1 \frac{(W(t) - tW(1))^2}{t(1-t)}\,dt \overset{\mathcal{D}}{=} \sum_{n=1}^{\infty} \frac{1}{n(n+1)}\,Z_n^2, \quad \{Z_n\} \sim \mathrm{NID}(0,1).$$

Then we obtain (Exercise 2.4.15), as the FD of $K(s,t)$ in (2.137),

$$\begin{aligned} D(\lambda) &= \prod_{n=1}^{\infty}\left(1 - \frac{\lambda}{n(n+1)}\right) = -\frac{\cos\dfrac{\pi\sqrt{1+4\lambda}}{2}}{\pi\lambda} \\ &= 1 - \lambda + \frac{12-\pi^2}{6}\lambda^2 - \cdots. \end{aligned} \tag{2.139}$$

It holds that $\mathrm{E}(W^2) = 1$ and

$$\mathrm{Var}(W^2) = 2\left(1 - \frac{12-\pi^2}{3}\right) = 2\left(\frac{\pi^2}{3} - 3\right) = 2\sum_{n=1}^{\infty} \frac{1}{n^2(n+1)^2}.$$

The probability density of W^2 will be presented in Figure 10 of Appendix B.

Returning to (2.134), it may also be of some interest to note (Exercise 2.4.16) that the kernel defined by

$$K(s,t) = \frac{1}{2m+1}\left[(\min(s,t))^{1/(2m+1)} - (st)^{1/(2m+1)}\right] \quad (m > -1/2) \tag{2.140}$$

has the same FD as the kernel $s^m t^m\,[\min(s,t) - st]$.

[13] $K(s,t) = \dfrac{1}{2m+1}\left[(\min(s,t))^{2m+1} - (st)^{2m+1}\right] \quad (m > -1/2).$

This kernel was used in Nabeya and Tanaka (1988) and arises (Exercise 2.4.17) in the statistic

$$S = \int_0^1 \left[t^m \left\{W(t) - (2m+1)\int_0^1 s^{2m} W(s)\,ds\right\}\right]^2 dt.$$

The integral equation with the present kernel is given by

$$f(t) = \frac{\lambda}{2m+1}\left[\int_0^t s^{2m+1} f(s)\,ds + t^{2m+1}\int_t^1 f(s)\,ds - t^{2m+1}\int_0^1 s^{2m+1} f(s)\,ds\right],$$

which is shown to be equivalent to

$$f''(t) - \frac{2m}{t} f'(t) + \lambda t^{2m} f(t) = 0, \quad f(0) = f(1) = 0.$$

The differential equation is the same as (2.129), but the boundary conditions are slightly different, and it can be shown (Exercise 2.4.18) that the FD is given by

$$\begin{aligned} D(\lambda) &= \Gamma(\nu+1) J_\nu\left(\frac{\sqrt{\lambda}}{m+1}\right) \Big/ \left(\frac{\sqrt{\lambda}}{2(m+1)}\right)^{\nu} && (2.141) \\ &= 1 - \frac{1}{2(m+1)(4m+3)}\lambda + \frac{1}{8(m+1)^2(4m+3)(6m+5)}\lambda^2 - \cdots, \end{aligned}$$

where $\nu = (2m+1)/2(m+1)$. We have

$$\mathrm{E}(S) = \frac{1}{2(m+1)(4m+3)}, \qquad \mathrm{Var}(S) = \frac{1}{(m+1)(4m+3)^2(6m+5)}.$$

It can be checked that when $m = 0$, $D(\lambda)$ reduces to $\sin\sqrt{\lambda}/\sqrt{\lambda}$.

[14] $K(s,t) = \dfrac{1}{2}(\max(s,t))^m \quad (m > 0).$

The present kernel is nearly definite since it is the sum of a nondegenerate negative definite kernel $-(1 - (\max(s,t))^m)/2$ and a degenerate positive definite kernel $1/2$. This kernel was introduced in connection with the Itô integral in (1.59) in Section 1.4. In fact, the stochastic differential

$$d\left(t^m W^2(t)\right) = 2t^m W(t)\,dW(t) + \left(mt^{m-1}W^2(t) + t^m\right) dt$$

yields

$$\begin{aligned}\int_0^1 t^m W(t)\,dW(t) &= \frac{1}{2}\int_0^1 \left[d\left(t^m W^2(t)\right) - \left(mt^{m-1}W^2(t) + t^m\right) dt\right] \\ &= \int_0^1\int_0^1 \frac{1}{2}(\max(s,t))^m\, dW(s)\,dW(t) - \frac{1}{2(m+1)}.\end{aligned}$$

The integral equation with the present kernel is given by

$$f(t) = \frac{\lambda}{2}\left[t^m \int_0^t f(s)\,ds + \int_t^1 s^m f(s)\,ds\right],$$

which leads to

$$\begin{aligned}f'(t) &= \frac{\lambda m t^{m-1}}{2}\int_0^t f(s)\,ds, \\ f''(t) &= \frac{\lambda m t^{m-1}}{2}\left[\frac{m-1}{t}\int_0^t f(s)\,ds + f(t)\right].\end{aligned}$$

Then it can be checked that the integral equation is equivalent to

$$f''(t) - \frac{m-1}{t}f'(t) - \frac{\lambda m}{2}t^{m-1}f(t) = 0, \quad \lim_{t\to 0}\frac{f'(t)}{t^{m-1}} = 0, \quad f'(1) = mf(1). \tag{2.142}$$

The differential equation in (2.142) is a special case of Bessel's equation (2.123) with

$$\alpha = \frac{m}{2}, \quad \beta = \frac{\sqrt{-2\lambda m}}{m+1}, \quad \gamma = \frac{m+1}{2}, \quad \nu = \frac{m}{m+1},$$

which yields

$$\begin{aligned}f(t) &= t^{m/2}\left\{c_1 J_\nu\left(\beta t^\gamma\right) + c_2 J_{-\nu}\left(\beta t^\gamma\right)\right\} \\ &= c_1 t^m\left(\frac{\beta}{2}\right)^\nu \left\{\frac{1}{\Gamma(\nu+1)} - \frac{\beta^2}{4\,\Gamma(\nu+2)}t^{m+1} + O\left(t^{2(m+1)}\right)\right\} \\ &\quad + c_2\left(\frac{\beta}{2}\right)^{-\nu}\left\{\frac{1}{\Gamma(-\nu+1)} - \frac{\beta^2}{4\,\Gamma(-\nu+2)}t^{m+1} + O\left(t^{2(m+1)}\right)\right\}.\end{aligned}$$

Thus it follows that

$$\lim_{t\to 0}\frac{f'(t)}{t^{m-1}}=c_1\left(\frac{\beta}{2}\right)^{\nu}\frac{m}{\Gamma(\nu+1)},$$

which gives $c_1=0$ from the first boundary condition in (2.142). Putting $c_1=0$ and using the second relation in (2.126), we have

$$\begin{aligned} f'(t)&=c_2\left(\frac{m}{2}t^{m/2-1}J_{-\nu}(\beta t^{\gamma})+t^{m/2}\frac{dJ_{-\nu}(\beta t^{\gamma})}{dt}\right)\\ &=c_2\left(\frac{m}{2}t^{m/2-1}J_{-\nu}(\beta t^{\gamma})+t^{m-1/2}\frac{\sqrt{-2\lambda m}}{4}\left(J_{-\nu-1}(\beta t^{\gamma})-J_{-\nu+1}(\beta t^{\gamma})\right)\right). \end{aligned}$$

Using the first relation in (2.126), the second boundary condition in (2.142) yields

$$\begin{aligned} f'(1)-mf(1)&=c_2\left(\frac{m}{2}J_{-\nu}(\beta)+\frac{\sqrt{-2\lambda m}}{4}\left(J_{-\nu-1}(\beta)-J_{-\nu+1}(\beta)\right)-mJ_{-\nu}(\beta)\right)\\ &=c_2\frac{\sqrt{-2\lambda m}}{2}J_{-\nu-1}(\beta). \end{aligned}$$

Then we obtain, as the FD of the present kernel,

$$\begin{aligned} D(\lambda)&=\Gamma(-\nu)J_{-\nu-1}\left(\frac{\sqrt{-2\lambda m}}{m+1}\right)\bigg/\left(\frac{\sqrt{-2\lambda m}}{2(m+1)}\right)^{-\nu-1} \qquad (2.143)\\ &=1-\frac{1}{2(m+1)}\lambda-\frac{m}{8(m+1)^2}\lambda^2-\cdots, \end{aligned}$$

where $\nu=m/(m+1)$. It holds that $\mathrm{E}(S)=\mathrm{Var}(S)=1/2(m+1)$.

Note that the case of $m=0$ has been excluded in the above discussion. When $m=0$, we have $K(s,t)=1/2$, whose FD is evidently $D(\lambda)=1-\lambda/2$. It can be checked (Exercise 2.4.19) that $D(\lambda)$ in (2.143) reduces to $1-\lambda/2$ when $m=0$.

[15] $K(s,t)=\dfrac{2(1-H)}{4H-1}\left[1-(\max(s,t))^{4H-1}\right](st)^{1/2-H}\quad(1/2<H<1).$

This kernel was used in Tanaka (2014) and appears in

$$S=\int_0^1 C_H^2(t)\,dt=\int_0^1\int_0^1 K(s,t)\,dW(s)\,dW(t),$$

where

$$C_H(t)=\sqrt{2(1-H)}\,t^{2H-1}\int_0^t u^{1/2-H}\,dW(u). \qquad (2.144)$$

The statistic S will be discussed in Chapter 8 when we deal with *fractional Brownian motion*. The integral equation with the present kernel is given by

$$f(t) = \frac{2(1-H)\lambda}{4H-1}\left(t^{1/2-H}\int_0^1 s^{1/2-H} f(s)\,ds - t^{3H-1/2}\int_0^t s^{1/2-H} f(s)\,ds \right.$$
$$\left. -t^{1/2-H}\int_t^1 s^{3H-1/2} f(s)\,ds\right).$$

Putting $h(t) = t^{1/2-H} f(t)$, we have

$$t^{H-1/2}h(t) = \frac{2(1-H)\lambda}{4H-1}\left(t^{1/2-H}\int_0^1 h(s)\,ds - t^{3H-1/2}\int_0^t h(s)\,ds \right.$$
$$\left. -t^{1/2-H}\int_t^1 s^{4H-1}h(s)\,ds\right),$$

so that

$$t^{2H-1}h(t) = \frac{2(1-H)\lambda}{4H-1}\left(\int_0^1 h(s)\,ds - t^{4H-1}\int_0^t h(s)\,ds - \int_t^1 s^{4H-1}h(s)\,ds\right).$$

Differentiating on both sides yields

$$(2H-1)t^{2H-2}h(t) + t^{2H-1}h'(t) = -2(1-H)\lambda t^{4H-2}\int_0^t h(s)\,ds,$$

and differentiating once more yields

$$\begin{aligned}
&(2H-1)(2H-2)t^{2H-3}h(t) + 2(2H-1)t^{2H-2}h'(t) + t^{2H-1}h''(t) \\
&= -2(1-H)(4H-2)\lambda t^{4H-3}\int_0^t h(s)\,ds - 2(1-H)\lambda t^{4H-2}h(t) \\
&= \frac{4H-2}{t}\left[(2H-1)t^{2H-2}h(t) + t^{2H-1}h'(t)\right] - 2(1-H)\lambda t^{4H-2}h(t).
\end{aligned}$$

Thus we have the differential equation

$$h''(t) + \left(2(1-H)\lambda t^{2H-1} - \frac{2H(2H-1)}{t^2}\right)h(t) = 0 \tag{2.145}$$

and the boundary conditions

$$h(1) = 0, \quad \lim_{t\to 0}\left((2H-1)t^{-2H}h(t) + t^{1-2H}h'(t)\right) = 0. \tag{2.146}$$

The differential equation (2.145) is a special case of Bessel's equation (2.123) with

$$\alpha = \frac{1}{2}, \quad \beta = \frac{\sqrt{2(1-H)\lambda}}{H+\frac{1}{2}}, \quad \gamma = H + \frac{1}{2}, \quad \nu = \frac{2H-\frac{1}{2}}{H+\frac{1}{2}},$$

which yields the general solution

$$h(t) = t^{1/2}\left(c_1 J_\nu\left(\beta t^{H+1/2}\right) + c_2 J_{-\nu}\left(\beta t^{H+1/2}\right)\right).$$

It follows from the two boundary conditions in (2.146) that $M(\lambda)\,\boldsymbol{c} = \mathbf{0}$, where $\boldsymbol{c} = (c_1,\, c_2)'$ and

$$M(\lambda) = \begin{pmatrix} J_\nu(\beta) & J_{-\nu}(\beta) \\ 1 & 0 \end{pmatrix}.$$

Then we obtain, as the FD of the present kernel,

$$\begin{aligned} D(\lambda) &= \Gamma(1-\nu)\, J_{-\nu}(\beta) \Big/ \left(\frac{\beta}{2}\right)^{-\nu} \qquad (2.147) \\ &= 1 - \frac{1}{2H+1}\lambda + \frac{1-H}{3(2H+1)^2}\lambda^2 - \cdots, \end{aligned}$$

where $\beta = \sqrt{2(1-H)\lambda}/(H+1/2)$ and $\nu = (2H-1/2)/(H+1/2)$. We have $\mathrm{E}(S) = 1/(2H+1)$ and $\mathrm{Var}(S) = 1/3(H+1/2)$.

[16] $K(s,t) = (st)^{2H-1}\,(\min(s,t))^{2-2H} \quad (1/2 < H < 1).$

This kernel should give the same FD as obtained in [15] because we have, for $C_H(t)$ in (2.144),

$$\begin{aligned} \mathrm{Cov}\,(C_H(s),\, C_H(t)) &= 2(1-H)(st)^{2H-1}\int_0^{\min(s,t)} u^{1-2H}\, du \\ &= (st)^{2H-1}\,(\min(s,t))^{2-2H}. \end{aligned}$$

Putting aside this fact, let us derive the FD of the present kernel. The corresponding integral equation is given by

$$f(t) = \lambda\left(t^{2H-1}\int_0^t sf(s)\, ds + t\int_t^1 s^{2H-1} f(s)\, ds\right),$$

which gives

$$\begin{aligned} f'(t) &= \lambda\left((2H-1)t^{2H-2}\int_0^t sf(s)\, ds + \int_t^1 s^{2H-1} f(s)\, ds\right), \\ f''(t) &= \lambda\left((2H-1)(2H-2)t^{2H-3}\int_0^t sf(s)\, ds + (2H-2)t^{2H-1} f(t)\right) \\ &= (2H-1)\left(\frac{1}{t} f'(t) - \frac{1}{t^2} f(t)\right) + (2H-2)\lambda t^{2H-1} f(t). \end{aligned}$$

Thus we have the differential equation

$$f''(t) - \frac{2H-1}{t} f'(t) + \left(2(1-H)\lambda t^{2H-1} + \frac{2H-1}{t^2}\right) f(t) = 0 \quad (2.148)$$

and the boundary conditions

$$\lim_{t\to 0} t^{2-2H} f'(t) = 0, \quad f'(1) = (2H-1)f(1). \quad (2.149)$$

The differential equation (2.148) is a special case of Bessel's equation in (2.123) with

$$\alpha = H, \quad \beta = \frac{\sqrt{2(1-H)\lambda}}{H+\frac{1}{2}}, \quad \gamma = H + \frac{1}{2}, \quad \nu = \frac{1-H}{H+\frac{1}{2}},$$

which gives the general solution

$$\begin{aligned} f(t) &= t^H \left(c_1 J_\nu(\beta t^{H+1/2}) + c_2 J_{-\nu}(\beta t^{H+1/2})\right) \\ &= c_1 t \left(\frac{\beta}{2}\right)^\nu \left(\frac{1}{\Gamma(\nu+1)} - \frac{\beta^2 t^{2H+1}/4}{\Gamma(\nu+2)} + O(t^{4H+2})\right) \\ &\quad + c_2 t^{2H-1} \left(\frac{\beta}{2}\right)^{-\nu} \left(\frac{1}{\Gamma(-\nu+1)} - \frac{\beta^2 t^{2H+1}/4}{\Gamma(-\nu+2)} + O(t^{4H+2})\right). \end{aligned}$$

We also have

$$\begin{aligned} f'(t) &= c_1 \left(\frac{\beta}{2}\right)^\nu \left(\frac{1}{\Gamma(\nu+1)} - \frac{(2H+2)\beta^2 t^{2H+1}/4}{\Gamma(\nu+2)} + O(t^{4H+2})\right) \\ &\quad + c_2 t^{2H-2} \left(\frac{\beta}{2}\right)^{-\nu} \left(\frac{2H-1}{\Gamma(-\nu+1)} - \frac{H\beta^2 t^{2H+1}}{\Gamma(-\nu+2)} + O(t^{4H+2})\right). \end{aligned}$$

It follows from the first boundary condition in (2.149) that $c_2 = 0$. Using the relation $J'_\nu(z) = (J_{\nu-1}(z) - J_{\nu+1}(z))/2$ in (2.126), we have, putting $c_2 = 0$,

$$f'(t) = c_1 \left[H t^{H-1} J_\nu(a) + t^H \beta(H+1/2) t^{H-1/2} \frac{1}{2}(J_{\nu-1}(a) - J_{\nu+1}(a))\right],$$

where $a = \beta t^{H+1/2}$. Thus the second boundary condition in (2.149) yields

$$\left[H J_\nu(\beta) + \frac{1}{2}\beta(H+1/2)(J_{\nu-1}(\beta) - J_{\nu+1}(\beta))\right] c_1 = (2H-1) J_\nu(\beta) c_1.$$

A necessary and sufficient condition for λ ($\neq 0$) to be an eigenvalue is

$$(1-H) J_\nu(\beta) + \frac{1}{2}\beta(H+1/2)(J_{\nu-1}(\beta) - J_{\nu+1}(\beta)) = 0.$$

Using the relation $J_\nu(\beta) = \beta(J_{\nu-1}(\beta) + J_{\nu+1}(\beta))/(2\nu)$ in (2.126) yields $\beta(H+1/2)J_{\nu-1}(\beta) = 0$ with $\nu - 1 = -(2H-1/2)/(H+1/2)$, from which we obtain the same FD as (2.147).

[17] $K(s,t) = \left[1 - (\max(s,t))^{4H-1}\right] (st)^{1/2-H} + b\,(st)^{1/2-H}.$

This kernel is an extended version of [15] with $1/2 < H < 1$. The corresponding integral equation is given by

$$f(t) = \lambda \left((b+1)t^{1/2-H} \int_0^1 s^{1/2-H} f(s)\,ds - t^{3H-1/2} \int_0^t s^{1/2-H} f(s)\,ds \right.$$
$$\left. - t^{1/2-H} \int_t^1 s^{3H-1/2} f(s)\,ds \right),$$

which yields, by putting $h(t) = t^{1/2-H} f(t)$,

$$t^{2H-1} h(t) = \lambda \left((b+1) \int_0^1 h(s)\,ds - t^{4H-1} \int_0^t h(s)\,ds - \int_t^1 s^{4H-1} h(s)\,ds \right).$$

Differentiating on both sides we have

$$(2H-1)t^{2H-2}h(t) + t^{2H-1}h'(t) = -(4H-1)\lambda t^{4H-2} \int_0^t h(s)\,ds.$$

Differentiating once more yields

$$\begin{aligned}
&(2H-1)(2H-2)t^{2H-3}h(t) + 2(2H-1)t^{2H-2}h'(t) + t^{2H-1}h''(t) \\
&= -(4H-1)(4H-2)\lambda t^{4H-3} \int_0^t h(s)\,ds - (4H-1)\lambda t^{4H-2}h(t) \\
&= \frac{4H-2}{t} \left((2H-1)t^{2H-2}h(t) + t^{2H-1}h'(t) \right) - (4H-1)\lambda t^{4H-2}h(t).
\end{aligned}$$

Thus we have the differential equation

$$h''(t) + \left((4H-1)\lambda t^{2H-1} - \frac{2H(2H-1)}{t^2} \right) h(t) = 0 \tag{2.150}$$

and the boundary conditions

$$\lim_{t\to 0} \left\{ (2H-1)t^{-2H}h(t) + t^{1-2H}h'(t) \right\} = 0, \tag{2.151}$$

$$\left(2H - 1 + \frac{4H-1}{b} \right) h(1) + h'(1) = 0. \tag{2.152}$$

The differential equation (2.150) is a special case of Bessel's equation in (2.123) with

$$\alpha = \frac{1}{2}, \quad \beta = \frac{\sqrt{(4H-1)\lambda}}{H+\frac{1}{2}}, \quad \gamma = H + \frac{1}{2}, \quad \nu = \frac{4H-1}{2H+1},$$

which gives the general solution

$$\begin{aligned} h(t) &= t^{1/2}\left(c_1 J_\nu(\beta t^{H+1/2}) + c_2 J_{-\nu}(\beta t^{H+1/2})\right) \\ &= c_1 t^{2H}\left(\frac{\beta}{2}\right)^{\nu}\left(\frac{1}{\Gamma(\nu+1)} + O\left(t^{2H+1}\right)\right) \\ &\quad + c_2 t^{1-2H}\left(\frac{\beta}{2}\right)^{-\nu}\left(\frac{1}{\Gamma(1-\nu)} + O\left(t^{2H+1}\right)\right). \end{aligned}$$

We also have

$$\begin{aligned} h'(t) &= 2c_1 H t^{2H-1}\left(\frac{\beta}{2}\right)^{\nu}\left(\frac{1}{\Gamma(\nu+1)} + O\left(t^{2H+1}\right)\right) \\ &\quad + c_2(1-2H)t^{-2H}\left(\frac{\beta}{2}\right)^{-\nu}\left(\frac{1}{\Gamma(1-\nu)} + O\left(t^{2H+2}\right)\right). \end{aligned}$$

It follows from the boundary condition (2.151) that $c_1 = 0$. Using the relation $J'_{-\nu}(z) = -J_{-\nu+1}(z) - \nu J_{-\nu}(z)/z$ in (2.126), we have

$$h'(t) = c_2\left[(1-2H)t^{-1/2}J_{-\nu}(\beta t^{H+1/2}) - \beta\left(H+\frac{1}{2}\right)t^H J_{1-\nu}(\beta t^{H+1/2})\right].$$

Thus the boundary condition (2.152) yields

$$\left[\frac{4H-1}{b}J_{-\nu}(\beta) - \sqrt{(4H-1)\lambda}\,J_{1-\nu}(\beta)\right]c_2 = 0.$$

A necessary and sufficient condition for λ ($\neq 0$) to be an eigenvalue is

$$G(\lambda) = J_{-\nu}(\beta) - \frac{b\sqrt{\lambda}}{\sqrt{4H-1}}J_{1-\nu}(\beta) = 0.$$

Using the relation

$$J_{-\nu}(\beta) = \frac{(\beta/2)^{-\nu}}{\Gamma(1-\nu)}\prod_{n=1}^{\infty}\left(1 - \frac{\beta^2}{a_n^2}\right),$$

where $0 < a_1 < a_2 < \cdots$ are positive zeros of $J_{-\nu}(\beta)$, we obtain, as the FD of the present kernel,

$$\begin{aligned} D(\lambda) &= \Gamma(1-\nu)\left(\frac{\beta}{2}\right)^{\nu}G(\lambda) \\ &= \Gamma(1-\nu)\left(\frac{\beta}{2}\right)^{\nu}\left(J_{-\nu}(\beta) - \frac{b\sqrt{\lambda}}{\sqrt{4H-1}}J_{1-\nu}(\beta)\right), \end{aligned} \tag{2.153}$$

where $\beta = \sqrt{(4H-1)\lambda}/(H+1/2)$ and $\nu = (4H-1)/(2H+1)$.

The present kernel is closely related to the *fractional unit root distribution* to be discussed in Chapter 8. The corresponding statistic is given by

$$S = \frac{C_H^2(1)}{2} \Bigg/ \int_0^1 C_H^2(t)\, dt\,,$$

where $C_H(t)$ is defined in (2.144). Then we have

$$P(S < x) = P\left(\int_0^1 \int_0^1 K_H(s,t;x)\, dW(s)\, dW(t) > 0\right),$$

where

$$K_H(s,t;x) = \frac{x}{4H-1}\left[1 - (\max(s,t))^{4H-1}\right](st)^{1/2-H} - \frac{1}{2}(st)^{1/2-H}. \tag{2.154}$$

The FD of this kernel can easily be derived by replacing λ and b in (2.153) by $\lambda x/(4H-1)$ and $-(4H-1)/(2x)$, respectively, as

$$D_H(\lambda; x) = \Gamma(1-\nu)\left(\frac{\kappa}{2}\right)^{\nu}\left(J_{-\nu}(\kappa) + \frac{\lambda}{2}\frac{J_{1-\nu}(\kappa)}{\sqrt{\lambda x}}\right), \tag{2.155}$$

where $\kappa = \sqrt{\lambda x}/(H+1/2)$ and $\nu = (4H-1)/(2H+1)$.

2.4.3 Nonhomogeneous Differential Equation – Part 1

In this section we deal with Case 2.3 discussed in Section 2.3.1. Here we consider two types of kernels. One is

$$K_1(s,t) = \min(s,t) - st - \psi(s)\psi(t), \quad \psi(0) = 0, \quad \psi(1) = 0, \tag{2.156}$$

where $\psi(t)$ is twice continuously differentiable on $[0,1]$. Then we are led to the second-order nonhomogeneous differential equation of the form

$$f''(t) + \lambda f(t) = -\lambda a \psi''(t), \tag{2.157}$$

where

$$a = \int_0^1 \psi(s) f(s)\, ds, \quad f(0) = 0, \quad f(1) = 0. \tag{2.158}$$

To derive the FD of $K_1(s,t)$, we can avoid dealing with the differential equation. Instead we can use Darling's formula given in Theorem 2.17 to obtain

$$D_1(\lambda) = \frac{\sin\sqrt{\lambda}}{\sqrt{\lambda}} - 2\int_0^1 \left(\int_0^t \psi'(s)\cos\sqrt{\lambda}s\, ds\right)\psi'(t)\cos\sqrt{\lambda}(1-t)\, dt. \tag{2.159}$$

The other kernel is given by

$$K_2(s,t) = 1 - \max(s,t) - \psi(s)\psi(t), \quad \psi(1) = 0, \tag{2.160}$$

which leads to the same differential equation as (2.157) with the conditions

$$a = \int_0^1 \psi(s) f(s)\, ds, \quad f(1) = 0, \quad f'(0) = -\lambda \psi'(0) a. \tag{2.161}$$

To derive the FD of $K_2(s,t)$, we can also avoid dealing with the differential equation. Instead the formula given in Theorem 2.18 yields

$$D_2(\lambda) = \cos\sqrt{\lambda} + 2\sqrt{\lambda} \int_0^1 \left(\int_0^t \psi'(s) \sin\sqrt{\lambda} s\, ds \right) \psi'(t) \cos\sqrt{\lambda}(1-t)\, dt. \tag{2.162}$$

In the following we consider examples [18] through [20] of kernels related to $K_1(s,t)$ and examples [21] through [23] of kernels related to $K_2(s,t)$. Note that the kernels $K_1(s,t)$ and $K_2(s,t)$ are special cases of the kernel discussed in the next section, where the approach based on a differential equation is used to derive the FD.

[18] $K(s,t) = \min(s,t) - st - 9/16st\left(1-s^4\right)\left(1-t^4\right).$

This kernel has $\psi(t) = 3t(1-t^4)/4$ with $\psi(0) = \psi(1) = 0$ in (2.156) and was used in Nabeya and Tanaka (1988) for testing parameter constancy. With the aid of computerized algebra applied to (2.159), it is an easy matter to obtain, as the FD of the present kernel,

$$\begin{aligned} D_1(\lambda) &= \frac{\sin\sqrt{\lambda}}{\sqrt{\lambda}} - \frac{9}{8} \int_0^1 \left(\int_0^t \left(1-5s^4\right) \cos\sqrt{\lambda} s\, ds \right) \\ &\qquad \times \left(1-5t^4\right) \cos\sqrt{\lambda}(1-t)\, dt \\ &= \frac{1350}{\lambda^4} \left[\left(-6 + 2\lambda - \frac{\lambda^2}{6} \right) \cos\sqrt{\lambda} \right. \\ &\qquad \left. + \left(6 - 4\lambda + \frac{7\lambda^2}{10} - \frac{\lambda^3}{42} \right) \frac{\sin\sqrt{\lambda}}{\sqrt{\lambda}} \right] \\ &= 1 - \frac{41}{462}\lambda + \frac{467}{168{,}168}\lambda^2 - \cdots . \end{aligned} \tag{2.163}$$

It holds that $\mathrm{E}(S) = 41/462$ and $\mathrm{Var}(S) = 3221/693{,}693$.

[19] $K(s,t) = \min(s,t) - st - 2/\pi^2 \sin\pi s \sin\pi t.$

This kernel has $\psi(t) = \sqrt{2}\sin\pi t/\pi$ with $\psi(0) = \psi(1) = 0$ in (2.156). It follows from (2.159) that

$$
\begin{aligned}
D_1(\lambda) &= \frac{\sin\sqrt{\lambda}}{\sqrt{\lambda}} - 4\int_0^1\left(\int_0^t \cos\pi s\,\cos\sqrt{\lambda}s\,ds\right)\cos\pi t\,\cos\sqrt{\lambda}(1-t)\,dt \\
&= \frac{\sin\sqrt{\lambda}}{\sqrt{\lambda}} \Big/ \left(1-\frac{\lambda}{\pi^2}\right) \qquad (2.164)\\
&= 1-\left(\frac{1}{6}-\frac{1}{\pi^2}\right)\lambda+\left(\frac{1}{120}-\frac{1}{6\pi^2}+\frac{1}{\pi^4}\right)\lambda^2-\cdots.
\end{aligned}
$$

It holds that $\mathrm{E}(S) = 1/6 - 1/\pi^2$ and $\mathrm{Var}(S) = 1/45 - 2/\pi^4$.

Because of the special structure of the present kernel, the above result can also be derived as follows. By Mercer's theorem (Theorem 2.2) it holds that

$$K_0(s,t) = \min(s,t) - st = \sum_{n=1}^{\infty}\frac{f_n(s)f_n(t)}{\lambda_n},$$

where $\lambda_n = n^2\pi^2$ and $f_n(t) = \sqrt{2}\sin n\pi t$, and the FD of $K_0(s,t)$ is given by

$$D_0(\lambda) = \frac{\sin\sqrt{\lambda}}{\sqrt{\lambda}} = \prod_{n=1}^{\infty}\left(1-\frac{\lambda}{\lambda_n}\right).$$

Since we have

$$K(s,t) = \sum_{n=1}^{\infty}\frac{f_n(s)f_n(t)}{\lambda_n} - \frac{f_1(s)f_1(t)}{\lambda_1},$$

which excludes the eigenvalue λ_1 from $D_0(\lambda)$, the corresponding FD is given by

$$D(\lambda) = \prod_{n=2}^{\infty}\left(1-\frac{\lambda}{\lambda_n}\right) = \prod_{n=1}^{\infty}\left(1-\frac{\lambda}{\lambda_n}\right)\Big/\left(1-\frac{\lambda}{\lambda_1}\right),$$

which yields the present FD.

We can use the same reasoning to establish that the FD of the kernel

$$K(s,t) = \min(s,t) - st - \sum_{j=1}^{m}\frac{2}{j^2\pi^2}\sin j\pi s\,\sin j\pi t$$

is given by

$$D(\lambda) = \frac{\sin\sqrt{\lambda}}{\sqrt{\lambda}}\Big/\prod_{j=1}^{m}\left(1-\frac{\lambda}{j^2\pi^2}\right),$$

which will also be directly established in [28].

[20] $K(s,t) = \min(s,t) - st - 2/\pi^2\,\sin^2\pi s\,\sin^2\pi t.$

This kernel has $\psi(t) = \sqrt{2}\sin^2\pi t/\pi$ with $\psi(0) = \psi(1) = 0$. Noting that $\psi'(t) = \sqrt{2}\sin 2\pi t$, it follows from (2.159) that

$$D_1(\lambda) = \left(\frac{\sin\sqrt{\lambda}}{\sqrt{\lambda}} + \frac{1-\cos\sqrt{\lambda}}{\pi^2\left(1-\frac{\lambda}{4\pi^2}\right)} \right) \Bigg/ \left(1-\frac{\lambda}{4\pi^2}\right) \tag{2.165}$$
$$= 1 - \left(\frac{1}{6} - \frac{3}{4\pi^2}\right)\lambda + \left(\frac{1}{120} - \frac{1}{12\pi^2} + \frac{5}{16\pi^4}\right)\lambda^2 - \cdots .$$

We have $\mathrm{E}(S) = 1/6 - 3/4\pi^2$ and $\mathrm{Var}(S) = 1/45 - 1/6\pi^2 - 1/8\pi^4$. It can be checked that $D_1(4\pi^2) = 0$, that is, $\lambda = 4\pi^2$ is an eigenvalue although $1-\lambda/4\pi^2$ appears in the denominator of $D_1(\lambda)$. Darling (1955) also dealt with this kernel and derived the FD, but it is incorrect.

[21] $K(s,t) = 1 - \max(s,t) - 3/4\,(1-s^2)(1-t^2)$.

This kernel has $\psi(t) = \sqrt{3}(1-t^2)/2$ with $\psi(1) = 0$ and was used in Nabeya and Tanaka (1988) for testing parameter constancy. We can use the formula (2.162) for computing the FD to obtain

$$D_2(\lambda) = \cos\sqrt{\lambda} + 6\sqrt{\lambda}\int_0^1 \left(\int_0^t s\,\sin\sqrt{\lambda}s\,ds\right) t\,\cos\sqrt{\lambda}(1-t)\,dt$$
$$= \frac{3}{\lambda}\left(\frac{\sin\sqrt{\lambda}}{\sqrt{\lambda}} - \cos\sqrt{\lambda}\right) \tag{2.166}$$
$$= 1 - \frac{1}{10}\lambda + \frac{1}{280}\lambda^2 - \cdots .$$

We have $\mathrm{E}(S) = 1/10$ and $\mathrm{Var}(S) = 1/175$.

[22] $K(s,t) = 1 - \max(s,t) - 8/\pi^2\,\cos\pi s/2\,\cos\pi t/2$.

This kernel has $\psi(t) = 2\sqrt{2}\cos(\pi t/2)/\pi$ with $\psi(1) = 0$. Noting that $\psi'(t) = -\sqrt{2}\sin(\pi t/2)$, it follows from (2.162) that

$$D_2(\lambda) = \cos\sqrt{\lambda} + 4\sqrt{\lambda}\int_0^1 \left(\int_0^t \sin\frac{\pi s}{2}\,\sin\sqrt{\lambda}s\,ds\right)\sin\frac{\pi t}{2}\,\cos\sqrt{\lambda}(1-t)\,dt$$
$$= \cos\sqrt{\lambda}\Bigg/\left(1-\frac{4\lambda}{\pi^2}\right) \tag{2.167}$$
$$= 1 - \left(\frac{1}{2} - \frac{4}{\pi^2}\right)\lambda + \left(\frac{1}{24} - \frac{2}{\pi^2} + \frac{16}{\pi^4}\right)\lambda^2 - \cdots .$$

It follows that $\mathrm{E}(S) = 1/2 - 4/\pi^2$ and $\mathrm{Var}(S) = 1/3 - 6/\pi^2 + 16/\pi^4$. The expression in (2.167) is a parallel to [19]. In fact, Mercer's theorem gives

$$K_0(s,t) = 1 - \max(s,t) = \sum_{n=1}^{\infty} \frac{f_n(s) f_n(t)}{\lambda_n},$$

where $\lambda_n = (n - 1/2)^2\pi^2$ and $f_n(t) = \sqrt{2}\cos(n - 1/2)\pi t$, and the FD of $K_0(s,t)$ is given by

$$D_0(\lambda) = \cos\sqrt{\lambda} = \prod_{n=1}^{\infty}\left(1 - \frac{\lambda}{\lambda_n}\right).$$

Since we have

$$K(s,t) = \sum_{n=1}^{\infty} \frac{f_n(s) f_n(t)}{\lambda_n} - \frac{f_1(s) f_1(t)}{\lambda_1},$$

which excludes the eigenvalue λ_1 from $D_0(\lambda)$, the corresponding FD is given by

$$D(\lambda) = \prod_{n=2}^{\infty}\left(1 - \frac{\lambda}{\lambda_n}\right) = \prod_{n=1}^{\infty}\left(1 - \frac{\lambda}{\lambda_n}\right) \Big/ \left(1 - \frac{\lambda}{\lambda_1}\right),$$

which yields the present FD.

We can use the same reasoning to establish that the FD of the kernel

$$K(s,t) = 1 - \max(s,t) - \sum_{j=1}^{m} \frac{2}{\left(j - \frac{1}{2}\right)^2 \pi^2} \cos\left(j - \frac{1}{2}\right)\pi s \cos\left(j - \frac{1}{2}\right)\pi t$$

is given by

$$D(\lambda) = \cos\sqrt{\lambda} \Big/ \prod_{j=1}^{m}\left(1 - \frac{\lambda}{\left(j - \frac{1}{2}\right)^2 \pi^2}\right),$$

which will also be directly established in [29].

[23] $K(s,t) = 1 - \max(s,t) - 8/\pi^2\ (\cos\pi s/2\ \cos\pi t/2)^2$.

This kernel has $\psi(t) = 2\sqrt{2}\cos^2(\pi t/2)/\pi$, so that $\psi(1) = 0$ and $\psi'(t) = -\sqrt{2}\sin\pi t$. It follows from (2.162) that

$$D_2(\lambda) = \cos\sqrt{\lambda} + 4\sqrt{\lambda}\int_0^1 \left(\int_0^t \sin\pi s\, \sin\sqrt{\lambda}s\, ds\right) \sin\pi t\, \cos\sqrt{\lambda}(1-t)\, dt$$

$$= \left(\cos\sqrt{\lambda} + \frac{2\sqrt{\lambda}\sin\sqrt{\lambda}}{\pi^2\left(1-\frac{\lambda}{\pi^2}\right)}\right) \Big/ \left(1-\frac{\lambda}{\pi^2}\right) \tag{2.168}$$

$$= 1 - \left(\frac{1}{2}-\frac{3}{\pi^2}\right)\lambda + \left(\frac{1}{24}-\frac{5}{6\pi^2}+\frac{5}{\pi^4}\right)\lambda^2 - \cdots .$$

We have $D_2(\pi^2) = -1/4$, so that $\lambda = \pi^2$ is not an eigenvalue, unlike the FD in [20]. It follows that $\mathrm{E}(S) = 1/2 - 3/\pi^2$ and $\mathrm{Var}(S) = 1/3 - 8/3\pi^2 - 2/\pi^4$.

2.4.4 Nonhomogeneous Differential Equation – Part 2

In this section we consider the integral equation with the kernel defined by

$$K(s,t) = K_0(s,t) + \sum_{k=1}^{q} \xi_k(s)\eta_k(t), \tag{2.169}$$

where $K_0(s,t) = \min(s,t) - st$ or $1 - \max(s,t)$, and $\xi_k(s)$ and $\eta_k(t)$ $(k = 1, \ldots, q)$ are deterministic functions that satisfy the following:

(i) $\xi_k(s)$ and $\eta_k(t)$ $(k = 1, \ldots, q)$ are continuous and each set is linearly independent in the space $C[0,1]$.
(ii) $\eta_k''(t)$ $(k = 1, \ldots, q)$ are linearly independent.

Then we are led to the second-order nonhomogeneous differential equation of the form

$$f''(t) + \lambda f(t) = \lambda \sum_{k=1}^{q} a_k \eta_k''(t), \tag{2.170}$$

where

$$a_k = \int_0^1 \xi_k(s) f(s)\, ds \qquad (k = 1, \ldots, q), \tag{2.171}$$

and we have two appropriate boundary conditions.

The general solution to (2.170) is

$$f(t) = c_1 \cos\sqrt{\lambda}t + c_2 \sin\sqrt{\lambda}t + \sum_{k=1}^{q} a_k g_k(t), \tag{2.172}$$

where $g_k(t)$ is a special solution of

$$g_k''(t) + \lambda g_k(t) = \lambda \eta_k''(t). \tag{2.173}$$

The q equations in (2.171) and two appropriate boundary conditions yield a system of $q+2$ linear homogeneous equations $M(\lambda)\boldsymbol{c} = \boldsymbol{0}$ in $\boldsymbol{c} = (a_1, \ldots, a_q, c_1, c_2)'$. Then we compute $|M(\lambda)|$ to obtain a candidate for the FD, determining the multiplicity ℓ_n of each eigenvalue λ_n by $\ell_n = q + 2 - \mathrm{rank}(M(\lambda_n))$ described in (2.68). Next, we consider four examples [24] through [27] of related kernels.

We also consider the kernel expressed as

$$K(s,t) = K_0(s,t) - \sum_{j=1}^{m} \psi_j(s)\psi_j(t),$$

where $\psi_j(t)$ $(j = 1, \ldots, m)$ are continuously differentiable. For $K_0(s,t) = \min(s,t) - st$ and $\psi_j(0) = \psi_j(1) = 0$, Theorem 2.20 is applied to deal with examples [25], [27], and [28], whereas, for $K_0(s,t) = 1 - \max(s,t)$ and $\psi_j(1) = 0$, Theorem 2.21 is applied to examples [26] and [29]. The computation of the FD in these examples is seen to be more efficient than the general method explained above.

[24] $K(s,t) = \min(s,t) - st - 3st(1-s)(1-t)$.

The FD of this kernel can be obtained easily by Darling's formula given in (2.159) and was earlier presented in (1.79) as

$$\begin{aligned} D(\lambda) &= \frac{12}{\lambda}\left(\frac{2(1-\cos\sqrt{\lambda})}{\lambda} - \frac{\sin\sqrt{\lambda}}{\sqrt{\lambda}}\right) \\ &= 1 - \frac{1}{15}\lambda + \frac{1}{560}\lambda^2 - \cdots . \end{aligned} \tag{2.174}$$

It holds that $\mathrm{E}(S) = 1/15$ and $\mathrm{Var}(S) = 11/6300$.

Nonetheless we derive the FD by the present method for demonstration purposes. We rewrite the kernel as

$$K(s,t) = \min(s,t) - st + \xi(s)\eta(t),$$

where $\xi(s) = s(1-s)$ and $\eta(t) = -3t(1-t)$. The integral equation

$$f(t) = \lambda \int_0^1 K(s,t) f(s)\, ds = \lambda \left(\int_0^1 [\min(s,t) - st]\, f(s)\, ds + a\eta(t) \right)$$

leads to the differential equation

$$f''(t) + \lambda f(t) = \lambda a \eta''(t) = 6\lambda a, \tag{2.175}$$

where $\eta''(t) = 6$ and

$$a = \int_0^1 s(1-s)f(s)\,ds, \quad f(0) = 0, \quad f(1) = 0. \tag{2.176}$$

The general solution to (2.175) is

$$f(t) = c_1 \cos\sqrt{\lambda}t + c_2 \sin\sqrt{\lambda}t + 6a.$$

Substituting this into the three conditions in (2.176), we have $M(\lambda)\boldsymbol{c} = \boldsymbol{0}$, where $\boldsymbol{c} = (a, c_1, c_2)'$ and

$$M(\lambda) = \begin{pmatrix} 6 & 1 & 0 \\ 6 & \cos\sqrt{\lambda} & \sin\sqrt{\lambda} \\ 0 & -\dfrac{1+\cos\sqrt{\lambda}}{\lambda} + \dfrac{2\sin\sqrt{\lambda}}{\lambda\sqrt{\lambda}} & \dfrac{2(1-\cos\sqrt{\lambda})}{\lambda\sqrt{\lambda}} - \dfrac{\sin\sqrt{\lambda}}{\lambda} \end{pmatrix}.$$

A necessary and sufficient condition for λ ($\neq 0$) to be an eigenvalue is $|M(\lambda)| = 0$, where

$$|M(\lambda)| = \frac{2}{\sqrt{\lambda}}\left(\frac{\sin\sqrt{\lambda}}{\sqrt{\lambda}} - \frac{2(1-\cos\sqrt{\lambda})}{\lambda}\right) = -\frac{\sqrt{\lambda}}{6} + O(\lambda\sqrt{\lambda}) \qquad (\lambda \to 0).$$

Then we find, as the candidate for the FD,

$$\tilde{D}(\lambda) = -\frac{6}{\sqrt{\lambda}}|M(\lambda)| = \frac{12}{\lambda}\left(\frac{2(1-\cos\sqrt{\lambda})}{\lambda} - \frac{\sin\sqrt{\lambda}}{\sqrt{\lambda}}\right).$$

To ensure that $\tilde{D}(\lambda)$ is the FD of the present kernel, we first note that $\operatorname{rank}(M(\lambda_n)) = 2$ for each eigenvalue λ_n, so that the multiplicity is unity for each λ_n. Moreover, consider

$$h(z) = \frac{12}{z^4}\left(2(1-\cos z) - z\sin z\right),$$

where $z = \sqrt{\lambda}$. Then it can be checked that $h'(z)/h(z)$ is bounded on the squares C_m ($m = 1, 2, \ldots$) with the vertices at $(2m + 1/2)\pi(\pm 1 \pm i)$. Since the function $h(z)$ is even and analytic for all z with $h(0) = 1$ and $h'(0) = 0$, it follows from Theorem 2.13 that $h(z)$ has an infinite product expansion (2.39) with $\lambda_n = a_n^2$ and $\ell_n = 1$, where $\pm a_n$ ($n = 1, 2, \ldots$) are simple zeros of $h(z)$, which proves that $\tilde{D}(\lambda)$ really is the FD of the present kernel.

[25] $K(s,t) = \min(s,t) - st - 2st(1-s)(1-t)(4 - 5s - 5t + 10st)$.

This kernel is positive definite and is used in Nabeya and Tanaka (1988). Note that $K(s,t)$ can be rewritten as

$$K(s,t) = \min(s,t) - st + \sum_{k=1}^{2} \xi_k(s)\, \eta_k(t),$$

where

$$\xi_1(s) = s(1-s)(4-5s), \quad \eta_1(t) = -2t(1-t),$$
$$\xi_2(s) = s(1-s)(1-2s), \quad \eta_2(t) = 10t^2(1-t).$$

The integral equation with the present kernel is given by

$$f(t) = \lambda \left(\int_0^1 [\min(s,t) - st] f(s)\, ds + a_1\eta_1(t) + a_2\eta_2(t) \right),$$

which leads to the differential equation

$$f''(t) + \lambda f(t) = \lambda \left(a_1\eta_1''(t) + a_2\eta_2''(t) \right) = \lambda \left(4a_1 + 20(1-3t)a_2 \right), \quad (2.177)$$

where $\eta_1''(t) = 4$, $\eta_2''(t) = 20(1-3t)$, and

$$a_1 = \int_0^1 \xi_1(s) f(s)\, ds, \quad a_2 = \int_0^1 \xi_2(s) f(s)\, ds, \quad f(0) = 0, \quad f(1) = 0. \quad (2.178)$$

The general solution to (2.177) is given by

$$f(t) = c_1 \cos\sqrt{\lambda}t + c_2 \sin\sqrt{\lambda}t + 4a_1 + 20(1-3t)a_2.$$

Substituting this into the four conditions in (2.178), we have $M(\lambda)\,\boldsymbol{c} = \boldsymbol{0}$, where $\boldsymbol{c} = (a_1, a_2, c_1, c_2)'$ and

$$M(\lambda) = \begin{pmatrix} 4 & 20 & 1 & 0 \\ 4 & -40 & \cos\sqrt{\lambda} & \sin\sqrt{\lambda} \\ 0 & 0 & M_{33}(\lambda) & M_{34}(\lambda) \\ 0 & 0 & M_{43}(\lambda) & M_{44}(\lambda) \end{pmatrix},$$

$$M_{33}(\lambda) = \frac{1}{\lambda^2}\left(30 - 4\lambda - (30-\lambda)\cos\sqrt{\lambda} - 12\sqrt{\lambda}\sin\sqrt{\lambda}\right),$$
$$M_{34}(\lambda) = \frac{1}{\lambda^2}\left(18\sqrt{\lambda} + 12\sqrt{\lambda}\cos\sqrt{\lambda} - (30-\lambda)\sin\sqrt{\lambda}\right),$$
$$M_{43}(\lambda) = \frac{1}{\lambda^2}\left(12 - \lambda - (12-\lambda)\cos\sqrt{\lambda} - 6\sqrt{\lambda}\sin\sqrt{\lambda}\right),$$
$$M_{44}(\lambda) = \frac{1}{\lambda^2}\left(6\sqrt{\lambda} + 6\sqrt{\lambda}\cos\sqrt{\lambda} - (12-\lambda)\sin\sqrt{\lambda}\right).$$

We obtain

$$|M(\lambda)| = \frac{720}{\lambda^3\sqrt{\lambda}}\left(4(6+\lambda) - 8(3-\lambda)\cos\sqrt{\lambda} - \sqrt{\lambda}(24-\lambda)\sin\sqrt{\lambda}\right)$$
$$= \sqrt{\lambda} + O(\lambda) \qquad (\lambda \to 0).$$

Then we ensure (Exercise 2.4.20) that the FD is given by

$$
\begin{aligned}
D(\lambda) &= \frac{|M(\lambda)|}{\sqrt{\lambda}} \\
&= \frac{720}{\lambda^4}\left(4(6+\lambda) - 8(3-\lambda)\cos\sqrt{\lambda} - \sqrt{\lambda}(24-\lambda)\sin\sqrt{\lambda}\right) \qquad (2.179) \\
&= 1 - \frac{3}{70}\lambda + \frac{1}{1260}\lambda^2 - \cdots .
\end{aligned}
$$

It follows that $\mathrm{E}(S) = 3/70$ and $\mathrm{Var}(S) = 11/22{,}050$.

The FD in (2.179) can also be obtained from Theorem 2.20. For this purpose the present kernel may be rewritten as

$$
\begin{aligned}
K(s,t) &= \min(s,t) - st - 2st(1-s)(1-t)(-1+5st+5(1-s)(1-t)) \\
&= \min(s,t) - st - \sum_{j=1}^{3} \psi_j(s)\psi_j(t), \qquad (2.180)
\end{aligned}
$$

$$
\psi_1(t) = \sqrt{2}it(1-t), \quad \psi_2(t) = \sqrt{10}t^2(1-t), \quad \psi_3(t) = \sqrt{10}t(1-t)^2.
$$

Note that $\psi_j(0) = \psi_j(1) = 0$ for $j = 1, 2, 3$. Then Theorem 2.20 gives us

$$
\begin{aligned}
P_{jj}(\lambda) &= 1 - \frac{2\sqrt{\lambda}}{\sin\sqrt{\lambda}} \int_0^1 \left(\int_0^t \psi_j'(s)\cos\sqrt{\lambda}s\, ds \right) \psi_j'(t)\cos\sqrt{\lambda}(1-t)\, dt, \\
P_{jk}(\lambda) &= -\frac{\sqrt{\lambda}}{\sin\sqrt{\lambda}} \int_0^1 \left(\int_0^t \psi_j'(s)\cos\sqrt{\lambda}s\cos\sqrt{\lambda}(1-t)\, ds \right. \\
&\qquad \left. + \int_t^1 \psi_j'(s)\cos\sqrt{\lambda}t\cos\sqrt{\lambda}(1-s)\, ds \right) \psi_k'(t)\, dt,
\end{aligned}
$$

where $j \neq k$, from which we obtain, using computerized algebra,

$$
\begin{aligned}
P_{11}(\lambda) &= \frac{5}{3} + \frac{8}{\lambda} + \frac{16(\cos\sqrt{\lambda}-1)}{\lambda\sqrt{\lambda}\sin\sqrt{\lambda}}, \\
P_{22}(\lambda) &= -\frac{1}{3} - \frac{40}{\lambda} + \frac{360}{\lambda^2} - \frac{40(5\cos\sqrt{\lambda}+4)}{\lambda\sqrt{\lambda}\sin\sqrt{\lambda}},
\end{aligned}
$$

$$
\begin{aligned}
P_{33}(\lambda) &= P_{22}(\lambda), \\
P_{12}(\lambda) &= i\left[-\frac{\sqrt{5}}{3} - \frac{4\sqrt{5}}{\lambda} - \frac{8\sqrt{5}(\cos\sqrt{\lambda}-1)}{\lambda\sqrt{\lambda}\sin\sqrt{\lambda}} \right], \\
P_{13}(\lambda) &= P_{12}(\lambda), \\
P_{23}(\lambda) &= -\frac{1}{3} + \frac{20}{\lambda} - \frac{360}{\lambda^2} + \frac{40(4\cos\sqrt{\lambda}+5)}{\lambda\sqrt{\lambda}\sin\sqrt{\lambda}}.
\end{aligned}
$$

Then we have

$$|P(\lambda)| = \frac{720}{\lambda^3\sqrt{\lambda}\sin\sqrt{\lambda}}\left(4(6+\lambda) - 8(3-\lambda)\cos\sqrt{\lambda} - \sqrt{\lambda}(24-\lambda)\sin\sqrt{\lambda}\right)$$

and arrive at the FD given in (2.179).

We note in passing that $\psi_j(t)$ in (2.180) can be defined in a different way. For example, rewriting "$-1+5st+5(1-s)(1-t)$" as "$9-5(1+s)(1+t)+15st$," we can put

$$\psi_1(t) = 3\sqrt{2}(t-t^2), \quad \psi_2(t) = -\sqrt{10}i(t-t^3), \quad \psi_3(t) = \sqrt{30}(t^2-t^3). \tag{2.181}$$

Then it follows (Exercise 2.4.21) from Theorem 2.20 that the kernel $K(s,t)$ with $\psi_j(t)$ defined in (2.181) gives the FD in (2.179).

[26] $K(s,t) = 1 - \max(s,t) + \frac{2}{9}(1-s)(1-t)\Big(5(s^2+t^2-8s^2t^2 + st(s+t)) - 4(1+s)(1+t)\Big).$

This kernel is also used in Nabeya and Tanaka (1988) and can be rewritten as

$$K(s,t) = 1 - \max(s,t) + \sum_{k=1}^{2} \xi_k(s)\,\eta_k(t),$$

$$\xi_1(s) = (1-s)(1+s-8s^2), \quad \eta_1(t) = \frac{10}{9}(1-t)t^2,$$

$$\xi_2(s) = (1-s)(-4-4s+5s^2), \quad \eta_2(t) = \frac{2}{9}(1-t^2).$$

The integral equation with the present kernel leads to the differential equation

$$f''(t) + \lambda f(t) = \lambda\left(a_1\eta_1''(t) + a_2\eta_2''(t)\right) = \frac{4\lambda}{9}\left(5(1-3t)a_1 - a_2\right), \tag{2.182}$$

where $\eta_1''(t) = 20(1-3t)/9$, $\eta_2''(t) = -4/9$, and

$$a_1 = \int_0^1 \xi_1(s)f(s)\,ds, \quad a_2 = \int_0^1 \xi_2(s)f(s)\,ds, \quad f(1) = 0, \quad f'(0) = 0. \tag{2.183}$$

The general solution to (2.182) is given by

$$f(t) = c_1\cos\sqrt{\lambda}t + c_2\sin\sqrt{\lambda}t + \frac{20}{9}(1-3t)a_1 - \frac{4}{9}a_2.$$

Substituting this into the four conditions in (2.183), we have $M(\lambda)\boldsymbol{c} = \boldsymbol{0}$, where $\boldsymbol{c} = (a_1, a_2, c_1, c_2)'$ and

$$M(\lambda) = \begin{pmatrix} -\dfrac{40}{9} & -\dfrac{4}{9} & \cos\sqrt{\lambda} & \sin\sqrt{\lambda} \\ -\dfrac{20}{3} & 0 & 0 & \sqrt{\lambda} \\ 0 & 0 & M_{33}(\lambda) & M_{34}(\lambda) \\ 0 & 0 & M_{43}(\lambda) & M_{44}(\lambda) \end{pmatrix},$$

$$M_{33}(\lambda) = \frac{1}{\lambda^2}\left(48 - 6(8-\lambda)\cos\sqrt{\lambda} - 30\sqrt{\lambda}\sin\sqrt{\lambda}\right),$$
$$M_{34}(\lambda) = \frac{1}{\lambda^2}\left(\sqrt{\lambda}(18+\lambda) + 30\sqrt{\lambda}\cos\sqrt{\lambda} - 6(8-\lambda)\sin\sqrt{\lambda}\right),$$
$$M_{43}(\lambda) = \frac{1}{\lambda^2}\left(-30 + 3(10+\lambda)\cos\sqrt{\lambda} + 12\sqrt{\lambda}\sin\sqrt{\lambda}\right),$$
$$M_{44}(\lambda) = \frac{1}{\lambda^2}\left(-18\sqrt{\lambda} - 4\lambda\sqrt{\lambda} - 12\sqrt{\lambda}\cos\sqrt{\lambda} + 3(10+\lambda)\sin\sqrt{\lambda}\right).$$

We obtain

$$\begin{aligned} |M(\lambda)| &= \frac{80}{\lambda^3\sqrt{\lambda}}\left(12(2+\lambda) - (24-\lambda^2)\cos\sqrt{\lambda} - 4\sqrt{\lambda}(6+\lambda)\sin\sqrt{\lambda}\right) \\ &= \sqrt{\lambda} + O(\lambda) \qquad (\lambda \to 0), \end{aligned}$$

and the FD of the present kernel is shown to be

$$\begin{aligned} D(\lambda) &= \frac{80}{\lambda^4}\left(12(2+\lambda) - (24-\lambda^2)\cos\sqrt{\lambda} - 4\sqrt{\lambda}(6+\lambda)\sin\sqrt{\lambda}\right) \qquad (2.184) \\ &= 1 - \frac{11}{210}\lambda + \frac{13}{11{,}340}\lambda^2 - \cdots. \end{aligned}$$

It holds that $\mathrm{E}(S) = 11/210$ and $\mathrm{Var}(S) = 179/198{,}450$.

The present kernel can also be rewritten as

$$\begin{aligned} K(s,t) &= 1 - \max(s,t) - \frac{2}{9}(1-s)(1-t) \\ &\quad \times \left(9(1+s)(1+t) - 5(1+s+s^2)(1+t+t^2) + 45s^2t^2\right) \\ &= 1 - \max(s,t) - \sum_{j=1}^{3}\psi_j(s)\psi_j(t), \qquad (2.185) \end{aligned}$$

$$\psi_1(t) = \sqrt{2}(1-t^2), \quad \psi_2(t) = -\frac{\sqrt{10}}{3}i(1-t^3), \quad \psi_3(t) = \sqrt{10}(t^2-t^3)$$

with $\psi_j(1)=0$ for $j=1,2,3$. Then Theorem 2.21 gives us, using computerized algebra,

$$
\begin{aligned}
P_{11}(\lambda) &= -\frac{5}{3} - \frac{8}{\lambda} + \frac{8\sin\sqrt{\lambda}}{\lambda\sqrt{\lambda}\cos\sqrt{\lambda}},\\
P_{22}(\lambda) &= 3 + \frac{40}{3\lambda} - \frac{40}{\lambda^2} - \frac{40}{\lambda^2\sqrt{\lambda}\cos\sqrt{\lambda}}\left((1+\lambda)\sin\sqrt{\lambda} - 2\sqrt{\lambda}\right),\\
P_{33}(\lambda) &= -\frac{1}{3} - \frac{40}{\lambda} + \frac{120}{\lambda^2} + \frac{40}{\lambda^2\sqrt{\lambda}\cos\sqrt{\lambda}}\left((9+4\lambda)\sin\sqrt{\lambda} - 12\sqrt{\lambda}\right),\\
P_{12}(\lambda) &= \sqrt{5}i\left[1 + \frac{4}{\lambda} - \frac{8}{\lambda^2} + \frac{8}{\lambda^2\sqrt{\lambda}}\left(\sqrt{\lambda}\sin\sqrt{\lambda} - 1\right)\right],\\
P_{13}(\lambda) &= -\frac{\sqrt{5}}{3}\left(1 + \frac{12}{\lambda} - \frac{72}{\lambda^2}\right) + \frac{8\sqrt{5}}{\lambda^2\cos\sqrt{\lambda}}\left(2\sqrt{\lambda}\sin\sqrt{\lambda} - 3\right),\\
P_{23}(\lambda) &= i\left[1 + \frac{20}{\lambda} - \frac{80}{\lambda^2} - \frac{40}{\lambda^2\sqrt{\lambda}\cos\sqrt{\lambda}}\left((2\lambda+3)\sin\sqrt{\lambda} - 5\sqrt{\lambda}\right)\right],
\end{aligned}
$$

from which we obtain

$$
|P(\lambda)| = \frac{80}{\lambda^4\cos\sqrt{\lambda}}\left(12(2+\lambda) - (24-\lambda^2)\cos\sqrt{\lambda} - 4\sqrt{\lambda}(6+\lambda)\sin\sqrt{\lambda}\right)
$$

and arrive at the FD given in (2.184).

[27] $K(s,t) = \min(s,t) - st - 2/\pi^2 \sin^2\pi s \sin^2\pi t - 1/2\pi^2 \sin 2\pi s \sin 2\pi t.$

This kernel is an extended version of [20] considered in Section 2.4.3 and is used for goodness-of-fit tests (see Section 4.2).

We present two methods for obtaining the FD of the present kernel. One is based on a differential equation and the other on Theorem 2.20. The former method rewrites the kernel as

$$
\begin{aligned}
K(s,t) &= \min(s,t) - st + \sum_{k=1}^{2} \xi_k(s)\,\eta_k(t),\\
\xi_1(s) &= \sin^2\pi s, \quad \eta_1(t) = -\frac{2}{\pi^2}\sin^2\pi t,\\
\xi_2(s) &= \sin 2\pi s, \quad \eta_2(t) = -\frac{1}{2\pi^2}\sin 2\pi t.
\end{aligned}
$$

The integral equation with the present kernel leads to

$$
f''(t) + \lambda f(t) = a_1\eta_1''(t) + a_2\eta_2''(t), \tag{2.186}
$$

where $\eta_1''(t) = -4\cos 2\pi t$, $\eta_2''(t) = 2\sin 2\pi t$, and

$$a_1 = \int_0^1 \xi_1(s) f(s)\, ds,\ a_2 = \int_0^1 \xi_2(s) f(s)\, ds, \quad f(0) = 0, \quad f(1) = 0. \tag{2.187}$$

The general solution to (2.186) is given by

$$f(t) = c_1 \cos\sqrt{\lambda}t + c_2 \sin\sqrt{\lambda}t + \sum_{k=1}^{2} a_k g_k(t),$$

where $g_k(t)$ is a special solution to

$$g_k''(t) + \lambda g_k(t) = \lambda \eta_k''(t).$$

When $\lambda \neq 4\pi^2$, operational calculus with the differential operator D gives

$$g_1(t) = \frac{-4\lambda\cos 2\pi t}{D^2+\lambda} = \frac{-4\lambda\cos 2\pi t}{-4\pi^2+\lambda}, \quad g_2(t) = \frac{2\lambda\sin 2\pi t}{D^2+\lambda} = \frac{2\lambda\sin 2\pi t}{-4\pi^2+\lambda},$$

so that

$$f(t) = c_1 \cos\sqrt{\lambda}t + c_2 \sin\sqrt{\lambda}t + \frac{2\lambda}{4\pi^2-\lambda}(2a_1\cos 2\pi t - a_2 \sin 2\pi t).$$

Substituting $f(t)$ into the four conditions in (2.187) yields $M(\lambda)\,\boldsymbol{c} = \boldsymbol{0}$, where $\boldsymbol{c} = (a_1, a_2, c_1, c_2)'$ and $M(\lambda)$ is equal to

$$\begin{pmatrix} \dfrac{4\lambda}{4\pi^2-\lambda} & 0 & 1 & 0 \\[2ex] \dfrac{4\lambda}{4\pi^2-\lambda} & 0 & \cos\sqrt{\lambda} & \sin\sqrt{\lambda} \\[2ex] \dfrac{-4\pi^2}{4\pi^2-\lambda} & 0 & M_{33}(\lambda) & M_{34}(\lambda) \\[2ex] 0 & \dfrac{-4\pi^2}{4\pi^2-\lambda} & \dfrac{\sqrt{\lambda}}{\pi}M_{34}(\lambda) & -\dfrac{\sqrt{\lambda}}{\pi}M_{33}(\lambda) \end{pmatrix},$$

$$M_{33}(\lambda) = \frac{2\pi^2\sin\sqrt{\lambda}}{\sqrt{\lambda}(4\pi^2-\lambda)}, \quad M_{34}(\lambda) = \frac{2\pi^2(1-\cos\sqrt{\lambda})}{\sqrt{\lambda}(4\pi^2-\lambda)}.$$

We obtain

$$\begin{aligned} |M(\lambda)| &= \left(\sin\sqrt{\lambda} + \frac{\sqrt{\lambda}(1-\cos\sqrt{\lambda})}{\pi^2\left(1-\dfrac{\lambda}{4\pi^2}\right)} \right) \Bigg/ \left(1-\frac{\lambda}{4\pi^2}\right)^2 \\ &= \sqrt{\lambda} + O(\lambda) \qquad (\lambda \to 0). \end{aligned}$$

When $\lambda = 4\pi^2$, operational calculus gives

$$\frac{1}{D^2 + 4\pi^2} \cos 2\pi t = \frac{t \sin 2\pi t}{4\pi}, \qquad \frac{1}{D^2 + 4\pi^2} \sin 2\pi t = -\frac{t \cos 2\pi t}{4\pi},$$

so that

$$f(t) = c_1 \cos 2\pi t + c_2 \sin 2\pi t - 4\pi a_1 t \sin 2\pi t - 2\pi a_2 t \cos 2\pi t.$$

The four conditions in (2.187) yield $M\boldsymbol{c} = \boldsymbol{0}$, where $\boldsymbol{c} = (a_1, a_2, c_1, c_2)'$ and

$$M = \begin{pmatrix} -\frac{1}{4} & \frac{\pi}{4} & -\frac{1}{4} & 0 \\ -\pi & -\frac{3}{4} & 0 & \frac{1}{2} \\ 0 & 0 & 1 & 0 \\ 0 & -2\pi & 1 & 0 \end{pmatrix},$$

which yields $|M| = -\pi/4 \neq 0$. Thus $\lambda = 4\pi^2$ cannot be an eigenvalue. Then the FD of the present kernel is shown to be

$$D(\lambda) = \left(\frac{\sin\sqrt{\lambda}}{\sqrt{\lambda}} + \frac{1 - \cos\sqrt{\lambda}}{\pi^2 \left(1 - \frac{\lambda}{4\pi^2}\right)} \right) \Bigg/ \left(1 - \frac{\lambda}{4\pi^2}\right)^2 \tag{2.188}$$

$$= 1 - \left(\frac{1}{6} - \frac{1}{\pi^2}\right)\lambda + \left(\frac{1}{120} - \frac{1}{8\pi^2} + \frac{9}{16\pi^4}\right)\lambda^2 - \cdots.$$

It holds that $\mathrm{E}(S) = 1/6 - 1/\pi^2$, $\mathrm{Var}(S) = 1/45 - 1/6\pi^2 - 1/4\pi^4$ and $D(4\pi^2) = -1/8$.

It is recognized that the present FD just excludes the eigenvalue $\lambda = 4\pi^2$ from the FD in [20].

The other method for computing $D(\lambda)$ in (2.188) is based on Theorem 2.20 and follows the same lines as in [25]. The present kernel is expressed as

$$K(s,t) = \min(s,t) - st - \psi_1(s)\psi_1(t) - \psi_2(s)\psi_2(t),$$

$$\psi_1(t) = \frac{\sqrt{2}}{\pi} \sin^2 \pi t, \quad \psi_1'(t) = \sqrt{2} \sin 2\pi t,$$

$$\psi_2(t) = \frac{1}{\sqrt{2}\pi} \sin 2\pi t, \quad \psi_2'(t) = \sqrt{2} \cos 2\pi t.$$

Note that $\psi_j(0) = \psi_j(1) = 0$ for $j = 1, 2$. Then Theorem 2.20 yields

$$\begin{aligned}
P_{11}(\lambda) &= 1 - \frac{2\sqrt{\lambda}}{\sin\sqrt{\lambda}} \int_0^1 \left(\int_0^t \psi_1'(s) \cos\sqrt{\lambda}s\, ds \right) \psi_1'(t) \cos\sqrt{\lambda}(1-t)\, dt \\
&= \frac{1}{1 - \dfrac{\lambda}{4\pi^2}} \left[1 + \frac{\sqrt{\lambda}}{\sin\sqrt{\lambda}} \frac{1 - \cos\sqrt{\lambda}}{\pi^2 \left(1 - \frac{\lambda}{4\pi^2}\right)} \right], \\
P_{22}(\lambda) &= 1 - \frac{2\sqrt{\lambda}}{\sin\sqrt{\lambda}} \int_0^1 \left(\int_0^t \psi_2'(s) \cos\sqrt{\lambda}s\, ds \right) \psi_2'(t) \cos\sqrt{\lambda}(1-t)\, dt \\
&= \frac{1}{1 - \dfrac{\lambda}{4\pi^2}},
\end{aligned}$$

and $P_{12}(\lambda) = 0$. Thus we obtain $D(\lambda) = \sin\sqrt{\lambda}\, P_{11}(\lambda) P_{22}(\lambda)/\sqrt{\lambda}$, which leads to (2.188).

[28] $$\begin{aligned}
K(s,t) &= \min(s,t) - st - \sum_{j=1}^{m} \frac{2 \sin j\pi s \, \sin j\pi t}{j^2\pi^2} \\
&= \sum_{j=m+1}^{\infty} \frac{2 \sin j\pi s \, \sin j\pi t}{j^2\pi^2}.
\end{aligned}$$

This kernel is an extended version of [19]. We derive the FD of the present kernel using Theorem 2.20. Let us put

$$K(s,t) = \min(s,t) - st - \sum_{j=1}^{m} \psi_j(s)\psi_j(t),$$

$$\psi_j(t) = \frac{\sqrt{2}\sin j\pi t}{j\pi}, \quad \psi_j'(t) = \sqrt{2}\cos j\pi t.$$

Note that $\psi_j(0) = \psi_j(1) = 0$ for $j = 1, \ldots, m$. Using Theorem 2.20 we have

$$\begin{aligned}
P_{jj}(\lambda) &= 1 - \frac{2\sqrt{\lambda}}{\sin\sqrt{\lambda}} \int_0^1 \left(\int_0^t \psi_j'(s) \cos\sqrt{\lambda}s\, ds \right) \psi_j'(t) \cos\sqrt{\lambda}(1-t)\, dt \\
&= 1 - \frac{2\sqrt{\lambda}}{\sin\sqrt{\lambda}} \times \frac{2\sqrt{\lambda}\sin\sqrt{\lambda}}{4(\lambda - j^2\pi^2)} \\
&= \frac{1}{1 - \dfrac{\lambda}{j^2\pi^2}} \qquad (j = 1, \ldots, m),
\end{aligned}$$

$$
\begin{aligned}
P_{jk}(\lambda) &= -\frac{\sqrt{\lambda}}{\sin\sqrt{\lambda}} \int_0^1 \left(\int_0^t \psi_j'(s) \cos\sqrt{\lambda}s \, \cos\sqrt{\lambda}(1-t)\, ds \right. \\
&\qquad \left. + \int_t^1 \psi_j'(s) \cos\sqrt{\lambda}t \, \cos\sqrt{\lambda}(1-s)\, ds \right) \psi_k'(t)\, dt \\
&= 0 \qquad (j \neq k).
\end{aligned}
$$

Thus it follows from Theorem 2.20 that the FD of the present kernel is given by

$$
\begin{aligned}
D(\lambda) &= \frac{\sin\sqrt{\lambda}}{\sqrt{\lambda}} \prod_{j=1}^{m} P_{jj}(\lambda) = \frac{\sin\sqrt{\lambda}}{\sqrt{\lambda}} \Big/ \prod_{j=1}^{m} \left(1 - \frac{\lambda}{j^2\pi^2}\right) \\
&= \prod_{j=m+1}^{\infty} \left(1 - \frac{\lambda}{j^2\pi^2}\right) = 1 - d_1\lambda + d_2\lambda^2/2 - \cdots, \\
d_1 &= \frac{1}{6} - \frac{1}{\pi^2} \sum_{j=1}^{m} \frac{1}{j^2}, \\
d_2 &= \frac{1}{60} - \frac{1}{3\pi^2} \sum_{j=1}^{m} \frac{1}{j^2} + \frac{1}{\pi^4} \left\{ \left(\sum_{j=1}^{m} \frac{1}{j^2} \right)^2 + \sum_{j=1}^{m} \frac{1}{j^4} \right\}.
\end{aligned}
$$

It can be checked that $d_1 \to 0$ and $d_2 \to 0$ as $m \to \infty$.

[29] $$
\begin{aligned}
K(s,t) &= 1 - \max(s,t) - \sum_{j=1}^{m} \frac{2\cos(j-1/2)\pi s \, \cos(j-1/2)\pi t}{(j-1/2)^2\pi^2} \\
&= \sum_{j=m+1}^{\infty} \frac{2\cos(j-1/2)\pi s \, \cos(j-1/2)\pi t}{(j-1/2)^2\pi^2}.
\end{aligned}
$$

This kernel is an extended version of [22]. We derive the FD of the present kernel using Theorem 2.21. Let us put

$$
\begin{aligned}
K(s,t) &= 1 - \max(s,t) - \sum_{j=1}^{m} \psi_j(s)\psi_j(t), \\
\psi_j(t) &= \frac{\sqrt{2}\cos(j-1/2)\pi t}{(j-1/2)\pi}, \qquad \psi_j'(t) = -\sqrt{2}\sin(j-1/2)\pi t.
\end{aligned}
$$

Note that $\psi_j(1) = 0$ for $j = 1, \ldots, m$. Using Theorem 2.21 we have

$$
\begin{aligned}
P_{jj}(\lambda) &= 1 + \frac{2\sqrt{\lambda}}{\cos\sqrt{\lambda}} \int_0^1 \left(\int_0^t \psi_j'(s) \sin\sqrt{\lambda} s\, ds \right) \psi_j'(t) \cos\sqrt{\lambda}(1-t)\, dt \\
&= 1 + \frac{2\sqrt{\lambda}}{\cos\sqrt{\lambda}} \times \frac{-2\sqrt{\lambda}\cos\sqrt{\lambda}}{4\lambda - (2j-1)^2\pi^2} \\
&= \frac{1}{1 - \dfrac{\lambda}{(j-1/2)^2\pi^2}} \qquad (j = 1, \ldots, m),
\end{aligned}
$$

$$
\begin{aligned}
P_{jk}(\lambda) &= \frac{\sqrt{\lambda}}{\cos\sqrt{\lambda}} \int_0^1 \left(\int_0^t \psi_j'(s) \sin\sqrt{\lambda} s\, \cos\sqrt{\lambda}(1-t)\, ds \right. \\
&\qquad \left. + \int_t^1 \psi_j'(s) \sin\sqrt{\lambda} t\, \cos\sqrt{\lambda}(1-s)\, ds \right) \psi_k'(t)\, dt \\
&= 0 \qquad (j \neq k).
\end{aligned}
$$

Thus it follows from Theorem 2.21 that the FD of the present kernel is given by

$$
\begin{aligned}
D(\lambda) &= \cos\sqrt{\lambda} \prod_{j=1}^m P_{jj}(\lambda) = \cos\sqrt{\lambda} \Big/ \prod_{j=1}^m \left(1 - \frac{\lambda}{(j-1/2)^2\pi^2} \right) \\
&= \prod_{j=m+1}^{\infty} \left(1 - \frac{\lambda}{(j-1/2)^2\pi^2} \right) = 1 - d_1\lambda + d_2\lambda^2/2 - \cdots,
\end{aligned}
$$

where

$$
d_1 = \frac{1}{2} - \frac{1}{\pi^2} \sum_{j=1}^m \frac{1}{(j-1/2)^2},
$$

$$
d_2 = \frac{1}{12} - \frac{1}{\pi^2} \sum_{j=1}^m \frac{1}{(j-1/2)^2} + \frac{1}{\pi^4} \left\{ \left(\sum_{j=1}^m \frac{1}{(j-1/2)^2} \right)^2 + \sum_{j=1}^m \frac{1}{(j-1/2)^4} \right\}.
$$

It can be checked that $d_1 \to 0$ and $d_2 \to 0$ as $m \to \infty$.

2.4.5 Higher-Order Differential Equation

In this section we deal with Case 2.5 discussed in Section 2.3.1. We are concerned with g-fold integrated Bm defined by

$$
\begin{aligned}
F_g(t) &= \int_0^t F_{g-1}(u)\, du = \int_0^t \int_0^{t_1} \cdots \int_0^{t_{g-1}} W(t_g)\, dt_g\, dt_{g-1} \cdots dt_1 \\
&= \int_0^t \frac{(t-s)^g}{g!}\, dW(s) \qquad (g = 1, 2, \ldots),
\end{aligned}
$$

and consider its quadratic functional

$$\begin{aligned}S_g &= \int_0^1 F_g^2(t)\,dt \\ &= \int_0^1\int_0^1 \frac{1}{(g!)^2}\left(\int_{\max(s,t)}^1 ((u-s)(u-t))^g\,du\right) dW(s)\,dW(t) \\ &\overset{\mathcal{D}}{=} \int_0^1\int_0^1 \frac{1}{(g!)^2}\left(\int_0^{\min(s,t)} ((s-u)(t-u))^g\,du\right) dW(s)\,dW(t). \quad (2.189)\end{aligned}$$

Here we deal with the last expression, which is slightly simpler than the second for our purpose. Note that S_g is well defined for any nonnegative integer g.

As was explained in Case 2.5 of Section 2.3.1, the integral equation with the kernel in (2.189) is equivalent to the homogeneous differential equation of order $2g+2$

$$f^{(2g+2)}(t) + (-1)^g \lambda f(t) = 0 \quad (2.190)$$

with $2g+2$ boundary conditions

$$f^{(j)}(0) = 0, \quad f^{(g+1+j)}(1) = 0 \quad (j = 0, 1, \ldots, g). \quad (2.191)$$

Let us denote by x_j $(j = 1, \ldots, 2g+2)$ the roots of the characteristic equation of (2.190). Then we have

$$x_j = \begin{cases} \lambda^{1/(2g+2)}\left(\cos\frac{(j-1)\pi}{g+1} + i\sin\frac{(j-1)\pi}{g+1}\right) & (g:\text{ odd}), \\ \lambda^{1/(2g+2)}\left(\cos\frac{(2j-1)\pi}{2g+2} + i\sin\frac{(2j-1)\pi}{2g+2}\right) & (g:\text{ even}), \end{cases} \quad (2.192)$$

where $x_{g+1+k} = -x_k$ for $k = 1, \ldots, g+1$. Then the general solution to (2.190) is given by

$$\begin{aligned}f(t) &= \sum_{j=1}^{2g+2} b_j e^{x_j t} = \sum_{k=1}^{g+1}\left(b_k e^{x_k t} + b_{g+1+k} e^{-x_k t}\right) \\ &= \sum_{k=1}^{g+1}\left(b_k e^{i(-i x_k t)} + b_{g+1+k} e^{i(i x_k t)}\right) \\ &= \sum_{k=1}^{g+1}\left(c_{2k-1}\cos(i y_k t) + c_{2k}\sin(i y_k t)\right), \quad (2.193)\end{aligned}$$

where b_j and c_j $(j = 1, \ldots, 2g+2)$ are arbitrary constants, and $y_1, \ldots, y_{g+1}$ are distinct elements of $x_1, \ldots, x_{2g+2}$ for which $y_k \neq -y_\ell$.

The computational burden for computing the FD associated with S_g becomes more serious as g gets large. The computation of the FD for $g > 5$ remains to be done. However, it is found that the smallest eigenvalue λ_1 becomes more dominant as g gets large, and the upper bound B_2 for $1/\lambda_1$ described in Theorem 2.8 is useful, where

$$B_2 = \sqrt{\frac{1}{\ell}\int_0^1\int_0^1 K^2(s,t)\,ds\,dt} > \frac{1}{\lambda_1} \tag{2.194}$$

with ℓ equal to the multiplicity of λ_1.

Following, we consider examples [30] through [34] of related kernels, and at the end of this section we tabulate the upper bound B_2 up to $g = 10$, which is compared with available $1/\lambda_1$.

[30] $K(s,t) = \displaystyle\int_0^{\min(s,t)} (s-u)(t-u)\,du.$

This is the kernel for $g = 1$ in (2.189), where $K(s,t) = s^2(-s+3t)/6$ for $s \le t$. The differential equation is given by $f^{(4)}(t) - \lambda f(t) = 0$. Putting $A = \lambda^{1/4}$, it follows from (2.192) that the solutions to $x^4 - \lambda = 0$ are

$$x_1 = A, \quad x_2 = iA, \quad x_3 = -A, \quad x_4 = -iA.$$

Choosing $iy_1 = ix_4 = A$ and $iy_2 = ix_1 = iA$, the general solution to the differential equation is given from (2.193) by

$$f(t) = c_1\cos At + c_2\sin At + c_3\cos(iAt) + c_4\sin(iAt).$$

The four boundary conditions in (2.191) yield

$$\begin{aligned}
f(0) &= c_1 + c_3 = 0,\\
f'(0)/A &= c_2 + c_4 i = 0,\\
f''(1)/A^2 &= -c_1\cos A - c_2\sin A + c_3\cos iA + c_4\sin iA = 0,\\
f'''(1)/A^3 &= c_1\sin A - c_2\cos A - c_3 i\sin iA + c_4 i\cos iA = 0.
\end{aligned}$$

Then we have $M(\lambda)\,\boldsymbol{c} = \boldsymbol{0}$, where $\boldsymbol{c} = (c_1, c_2, c_3, c_4)'$ and

$$M(\lambda) = \begin{pmatrix} 1 & 0 & 1 & 0\\ 0 & 1 & 0 & i\\ -\cos A & -\sin A & \cos iA & \sin iA\\ \sin A & -\cos A & -i\sin iA & i\cos iA \end{pmatrix}.$$

A necessary and sufficient condition for $\lambda\ (\neq 0)$ to be an eigenvalue is that $|M(\lambda)| = 2i(1 + \cos A\cos iA) = 0$. Since $\mathrm{rank}(M(\lambda_n)) = 3$ for each eigenvalue λ_n, the multiplicity is unity for each eigenvalue. Then we find, as the candidate for the FD of the present kernel,

$$
\begin{aligned}
\tilde{D}(\lambda) &= \frac{|M(\lambda)|}{4i} = \frac{1}{2}\left(1+\cos A \cos iA\right) \\
&= \frac{1}{2}\left(1+\cos \lambda^{1/4} \cosh \lambda^{1/4}\right) \\
&= 1 - \frac{1}{12}\lambda + \frac{1}{5040}\lambda^2 - \cdots .
\end{aligned}
\tag{2.195}
$$

To ensure that $\tilde{D}(\lambda)$ is the FD, consider

$$h(z) = \frac{1}{2}\left(1+\cos z \, \cosh z\right),$$

where $z = \lambda^{1/4}$. It can be checked that every zero of $h(z)$ is simple. We have

$$\frac{h'(z)}{h(z)} = \frac{-\sin z \, \cosh z + \cos z \, \sinh z}{1+\cos z \, \cosh z} = \frac{-\sin z + \cos z \, \tanh z}{\cos z + 1/\cosh z},$$

which is bounded on the squares C_m $(m = 1, 2, \ldots)$ with the vertices at $2m\pi(\pm 1 \pm i)$. Since $h(z)$ is even and analytic for all z with $h(0) = 1$ and $h'(0) = 0$, it follows from Theorem 2.13 that, denoting by a_n $(n = 1, 2, \ldots)$ positive zeros of $h(z)$, $h(z)$ has an infinite product expansion (2.39) of the form

$$h(z) = \prod_{n=1}^{\infty}\left(1 - \frac{z^2}{a_n^2}\right) = \prod_{n=1}^{\infty}\left(1 - \frac{\sqrt{\lambda}}{a_n^2}\right).$$

Note here that $\lambda_n = a_n^4$. We now ensure that the FD of the present kernel is given by (2.195). The series expansion for $\tilde{D}(\lambda)$ gives $\mathrm{E}(S_1) = 1/12$ and $\mathrm{Var}(S_1) = 11/840$.

The smallest eigenvalue is the smallest zero of $\tilde{D}(\lambda)$ and is given by $\lambda_1 = 12.36236$. Since

$$\sum_{n=1}^{\infty} \frac{1}{\lambda_n} = \int_0^1 K(t,t)\, dt = \mathrm{E}(S_1) = \frac{1}{12},$$

we have $(1/\lambda_1)/(\sum_{n=1}^{\infty} 1/\lambda_n) = 12/12.36236 = 0.9707$, which shows a high contribution of the smallest eigenvalue to the total sum. This was earlier reported in Table 2.1 of Section 2.2. Theorem 2.8 suggests an upper bound for $1/\lambda_1 = 0.080891$, given by

$$B_2 = \sqrt{\frac{2}{\ell}\int_0^1 \left(\int_0^t K^2(s,t)\, ds\right) dt} = \sqrt{\frac{1}{2}\mathrm{Var}(S_1)} = 0.080917,$$

which is seen to be quite close to $1/\lambda_1$. Note here that ℓ must be unity; otherwise the inequaltiy (2.194) does not hold.

[31] $K(s,t) = 1/4 \int_0^{\min(s,t)} ((s-u)(t-u))^2\, du.$

This is the kernel for $g = 2$ in (2.189), where $K(s,t) = s^3(s^2-5st+10t^2)/120$ for $s \le t$. The differential equation is given by $f^{(6)}(t) + \lambda f(t) = 0$. Putting $A = \lambda^{1/6}$, it follows from (2.192) that the six solutions to $x^6 + \lambda = 0$ are $x_1 = A(\sqrt{3}+i)/2$, $x_2 = iA$, $x_3 = A(-\sqrt{3}+i)/2$, $x_4 = -x_1$, $x_5 = -x_2$, $x_6 = -x_3$. Putting $\omega = (1+\sqrt{3}i)/2$ and choosing $iy_1 = ix_5 = A$, $iy_2 = ix_6 = A\omega$, $iy_3 = ix_1 = A\omega^2$, the general solution to the differential equation is given from (2.193) by

$$f(t) = c_1 \cos At + c_2 \sin At + c_3 \cos A\omega t + c_4 \sin A\omega t + c_5 \cos A\omega^2 t + c_6 \sin A\omega^2 t.$$

The six boundary conditions in (2.191) yield $M(\lambda)\boldsymbol{c} = \boldsymbol{0}$, where $\boldsymbol{c} = (c_1, c_2, \ldots, c_6)'$ and $M(\lambda)$ is equal to

$$\begin{pmatrix} 1 & 0 & 1 & 0 & 1 & 0 \\ 0 & 1 & 0 & \omega & 0 & \omega^2 \\ -1 & 0 & -\omega^2 & 0 & \omega & 0 \\ \sin A & -\cos A & -\sin A\omega & \cos A\omega & \sin A\omega^2 & -\cos A\omega^2 \\ \cos A & \sin A & -\omega\cos A\omega & -\omega \sin A\omega & \omega^2 \cos A\omega^2 & \omega^2 \sin A\omega^2 \\ -\sin A & \cos A & \omega^2 \sin A\omega & -\omega^2 \cos A\omega & \omega \sin A\omega^2 & -\omega \cos A\omega^2 \end{pmatrix}.$$

We have $|M(\lambda)| = 27D(\lambda)$, where $D(\lambda)$ is the FD of the present kernel given by

$$\begin{aligned} D(\lambda) &= \frac{1}{9}\Big[3\cos A \cos A\omega \cos A\omega^2 + 2\Big(\cos A \sin A\omega \sin A\omega^2 \\ &\quad - \cos A\omega \sin A \sin A\omega^2 + \cos A\omega^2 \sin A \sin A\omega\Big) \\ &\quad + 2\Big(\cos A + \cos A\omega + \cos A\omega^2\Big)\Big] \\ &= \frac{1}{9}\Big[2\Big(1 + \cos\lambda^{1/6} + \cos\lambda^{1/6}\omega + \cos\lambda^{1/6}\omega^2\Big) \\ &\quad + \cos\lambda^{1/6} \cos\lambda^{1/6}\omega \cos\lambda^{1/6}\omega^2\Big] \\ &= 1 - \frac{1}{120}\lambda + \frac{19}{26{,}611{,}200}\lambda^2 - \cdots. \end{aligned} \tag{2.196}$$

It holds that $\mathrm{E}(S_2) = 1/120$ and $\mathrm{Var}(S_2) = 181/1{,}330{,}560$.

The smallest eigenvalue is obtained as $\lambda_1 = 121.25906$. Since

$$\sum_{n=1}^{\infty} \frac{1}{\lambda_n} = \int_0^1 K(t,t)\, dt = \mathrm{E}(S_2) = \frac{1}{120},$$

we have $(1/\lambda_1)/(\sum_{n=1}^{\infty} 1/\lambda_n) = 120/121.25906 = 0.98962$, which shows a higher contribution of the smallest eigenvalue to the total sum than the case of $g = 1$. This was also reported in Table 2.1. An upper bound for $1/\lambda_1 = 0.00824681$ is given by $B_2 = \sqrt{\mathrm{Var}(S_2)/2} = 0.00824721$, which tells us that the distribution of S_2 in (2.189) may be approximated by that of $B_2\, Z^2$, where $Z \sim \mathrm{N}(0, 1)$.

[32] $K(s,t) = 1/36 \displaystyle\int_0^{\min(s,t)} ((s-u)(t-u))^3\, du.$

This is the kernel for $g = 3$ in (2.189), where $K(s,t) = s^4(-s^3+7s^2t-21st^2+35t^3)/5040$ for $s \le t$. The differential equation is given by $f^{(8)}(t)-\lambda f(t) = 0$. Putting $A = \lambda^{1/8}$, it follows from (2.192) that the eight solutions to $x^8 - \lambda = 0$ are $x_1 = A$, $x_2 = A(1+i)/\sqrt{2}$, $x_3 = iA$, $x_4 = A(-1+i)/\sqrt{2}$, $x_5 = -x_1$, $x_6 = -x_2$, $x_7 = -x_3$, $x_8 = -x_4$. Putting $\omega = (1+i)/\sqrt{2}$ and choosing $iy_1 = ix_7 = A$, $iy_2 = ix_8 = A\omega$, $iy_3 = ix_1 = A\omega^2$, $iy_4 = ix_2 = A\omega^3$, the general solution to the differential equation is given from (2.193) by

$$\begin{aligned} f(t) &= c_1 \cos At + c_2 \sin At + c_3 \cos A\omega t + c_4 \sin A\omega t \\ &\quad + c_5 \cos A\omega^2 t + c_6 \sin A\omega^2 t + c_7 \cos A\omega^3 t + c_8 \sin A\omega^3 t, \end{aligned}$$

and the eight boundary conditions in (2.191) yield $M(\lambda)\boldsymbol{c} = \boldsymbol{0}$, where $\boldsymbol{c} = (c_1, c_2, \ldots, c_8)'$. The jth row M_j of $M(\lambda)$ is given by

$$\begin{aligned} M_1 &= (1,0,1,0,1,0,1,0), \\ M_2 &= (0,1,0,\omega,0,\omega^2,0,\omega^3), \\ M_3 &= (-1,0,-\omega^2,0,-\omega^4,0,-\omega^6,0) = (-1,0,-\omega^2,0,1,0,i,0), \\ M_4 &= (0,-1,0,-\omega^3,0,-\omega^6,0,-\omega^9) = (0,-1,0,-\omega^3,0,i,0,-\omega), \\ M_5 &= (\cos A, \sin A, -\cos A\omega, -\sin A\omega, \cos A\omega^2, \sin A\omega^2, \\ &\qquad -\cos A\omega^3, -\sin A\omega^3), \\ M_6 &= (-\sin A, \cos A, \omega \sin A\omega, -\omega \cos A\omega, -\omega^2 \sin A\omega^2, \omega^2 \cos A\omega^2, \\ &\qquad \omega^3 \sin A\omega^3, -\omega^3 \cos A\omega^3), \\ M_7 &= (-\cos A, -\sin A, i \cos A\omega, i \sin A\omega, \cos A\omega^2, \sin A\omega^2, \\ &\qquad -\omega^2 \cos A\omega^3, -\omega^2 \sin A\omega^3), \\ M_8 &= (\sin A, -\cos A, -\omega^3 \sin A\omega, \omega^3 \cos A\omega, -\omega^2 \sin A\omega^2, \omega^2 \cos A\omega^2, \\ &\qquad -\omega \sin A\omega^3, \omega \cos A\omega^3). \end{aligned}$$

We obtain $|M(0)| = 256i$ and the FD

$$\begin{aligned} D(\lambda) &= |M(\lambda)|/|M(0)| \\ &= \frac{1}{16}\Big[3\cos a\,\cos b\,\cos c\,\cos d + 2\Big(\cos a\,\cos b + \cos b\,\cos c \\ &\qquad + \cos c\,\cos d + \cos d\,\cos a\Big) + \cos a\,\cos c + \cos b\,\cos d + 3\Big] \\ &\quad + \frac{\sqrt{2}}{16}\Big[\sin a\,\sin b\,(1 + \cos c\,\cos d) + \sin b\,\sin c\,(1 + \cos d\,\cos a) \\ &\qquad + \sin c\,\sin d\,(1 + \cos a\,\cos b) - \sin d\,\sin a\,(1 + \cos b\,\cos c)\Big] \\ &= 1 - \frac{1}{2016}\lambda + \frac{257}{201{,}180{,}672{,}000}\lambda^2 - \cdots, \end{aligned}$$

where $a = \lambda^{1/8}$, $b = a\omega$, $c = a\omega^2$, $d = a\omega^3$. We have $\mathrm{E}(S_3) = 1/2016$ and $\mathrm{Var}(S_3) = 3499/7{,}185{,}024{,}000$.

The smallest eigenvalue is obtained as $\lambda_1 = 2026.5755$. Since

$$\sum_{n=1}^{\infty} \frac{1}{\lambda_n} = \int_0^1 K(t,t)\,dt = \mathrm{E}(S_3) = \frac{1}{2016},$$

we have $(1/\lambda_1)/(\sum_{n=1}^{\infty} 1/\lambda_n) = 2016/2026.5755 = 0.99478$, which shows a more dominant contribution of the smallest eigenvalue to the total sum than the case of $g = 2$. An upper bound for $1/\lambda_1 = 0.0004934432$ is

$$B_2 = \sqrt{\frac{1}{2}\,\mathrm{Var}(S_3)} = 0.0004934497,$$

which coincides with $1/\lambda_1$ up to five digits. The distribution of S_3 in (2.189) can be well approximated by $B_2\,Z^2$, where $Z \sim \mathrm{N}(0, 1)$.

[33] $K(s,t) = 1/576 \displaystyle\int_0^{\min(s,t)} ((s-u)(t-u))^4\,du.$

This is the kernel for $g = 4$ in (2.189), where $K(s,t) = s^5(s^4 - 9s^3t + 36s^3t^2 - 84st^3 + 126t^4)/362{,}880$ for $s \le t$. The differential equation is given by $f^{(10)}(t) + \lambda f(t) = 0$. Putting $A = \lambda^{1/10}$, it follows from (2.192) that the 10 solutions x_j $(j = 1, \ldots, 10)$ to $x^{10} + \lambda = 0$ are given by

$$x_j = A\left(\cos\frac{(2j-1)\pi}{10} + i\sin\frac{(2j-1)\pi}{10}\right) = A\left(\cos\frac{\pi}{10} + i\sin\frac{\pi}{10}\right)^{2j-1},$$

where

$$\cos\frac{\pi}{10} = \frac{\sqrt{5+\sqrt{5}}}{2\sqrt{2}}, \quad \sin\frac{\pi}{10} = \frac{\sqrt{5}-1}{4}.$$

Putting $\omega = (\sqrt{5}+1)/4+i\sqrt{5-\sqrt{5}}/2\sqrt{2}$ and choosing $iy_1 = ix_8 = A$, $iy_2 = ix_9 = A\omega$, $iy_3 = ix_{10} = A\omega^2$, $iy_4 = ix_1 = A\omega^3$, $iy_5 = ix_2 = A\omega^4$, the general solution to the differential equation is given from (2.193) by

$$f(t) = \sum_{k=1}^{5} \left(c_{2k-1} \cos A\omega^{k-1}t + c_{2k} \sin A\omega^{k-1}t \right).$$

The 10 boundary conditions in (2.191) yield a 10-dimensional homogeneous equation $M(\lambda)\,\boldsymbol{c} = \boldsymbol{0}$ in $\boldsymbol{c} = (c_1, \ldots, c_{10})'$. Evaluating $|M(\lambda)|$ we can obtain the FD. The final expression, however, contains as many as 100 terms; so we omit it.

The smallest eigenvalue is obtained as $\lambda_1 = 52{,}002.83876$. Since

$$\sum_{n=1}^{\infty} \frac{1}{\lambda_n} = \int_0^1 K(t,t)\,dt = \int_0^1 \frac{1}{5184}t^9\,dt = \frac{1}{51{,}840},$$

we have $(1/\lambda_1)/(\sum_{n=1}^{\infty} 1/\lambda_n) = 51{,}840/52{,}002.83876 = 0.99687$. The smallest eigenvalue is quite dominant. An upper bound for $1/\lambda_1 = 1.922972 \times 10^{-5}$ is

$$B_2 = \sqrt{2\int_0^1 \left(\int_0^t K^2(s,t)\,ds \right) dt} = 1.922981 \times 10^{-5}.$$

[34] $K(s,t) = 1/14{,}400 \displaystyle\int_0^{\min(s,t)} ((s-u)(t-u))^5\,du.$

This is the kernel for $g = 5$ in (2.189), where $K(s,t) = s^6(-s^5 + 11s^4t - 55s^3t^2 + 165s^2t^3 - 330st^4 + 462t^5)/39{,}916{,}800$ for $s \le t$. The differential equation is given by $f^{(12)} - \lambda f(t) = 0$. Putting $A = \lambda^{1/12}$, it follows from (2.192) that the 12 solutions x_j $(j = 1, \ldots, 12)$ to $x^{12} - \lambda = 0$ are given by

$$x_j = A\left(\cos\frac{(j-1)\pi}{6} + i\sin\frac{(j-1)\pi}{6}\right) = A\left(\cos\frac{\pi}{6} + i\sin\frac{\pi}{6}\right)^{j-1}.$$

Putting $\omega = (\sqrt{3}+i)/2$ and choosing $iy_1 = ix_{10} = A$, $iy_2 = ix_{11} = A\omega$, $iy_3 = ix_{12} = A\omega^2$, $iy_4 = ix_1 = A\omega^3$, $iy_5 = ix_2 = A\omega^4$, $iy_6 = ix_3 = A\omega^5$, the general solution to the differential equation is given from (2.193) by

$$f(t) = \sum_{k=1}^{6} \left(c_{2k-1} \cos A\omega^{k-1}t + c_{2k} \sin A\omega^{k-1}t \right).$$

The 12 boundary conditions in (2.191) yield a 12-dimensional homogeneous equation $M(\lambda)\,\boldsymbol{c} = \boldsymbol{0}$ in $\boldsymbol{c} = (c_1, c_2, \ldots, c_{12})'$.

We obtained the expression for $D(\lambda) = |M(\lambda)|/|M(0)|$, which we omit here because of the lengthy expression of as many as 150 terms. The smallest eigenvalue is obtained as $\lambda_1 = 1{,}904{,}774.734$. Since

$$\sum_{n=1}^{\infty} \frac{1}{\lambda_n} = \int_0^1 K(t,t)\,dt = \int_0^1 \frac{1}{158{,}400} t^{11}\,dt = \frac{1}{1{,}900{,}800},$$

we have $(1/\lambda_1)/(\sum_{n=1}^{\infty} 1/\lambda_n) = 1{,}900{,}800/1{,}904{,}774.734 = 0.99791$. An upper bound for $1/\lambda_1 = 5.249965 \times 10^{-7}$ is

$$B_2 = \sqrt{2\int_0^1 \left(\int_0^t K^2(s,t)\,ds\right) dt} = 5.249976 \times 10^{-7}.$$

In this section we obtained the FD of the kernel associated with S_g in (2.189) up to $g = 5$, although the FDs for $g = 4, 5$ were omitted because of lengthy expressions. It was found that, as the integrated order g gets large, the smallest eigenvalue becomes more dominant, and the distribution may be approximated by

$$\begin{aligned} S_g &= \int_0^1 F_g^2(t)\,dt \overset{\mathcal{D}}{=} \int_0^1 \int_0^1 K_g(s,t)\,dW(s)\,dW(t) \\ &\approx \frac{1}{\lambda_1(g)} Z^2 < B_2(g)\, Z^2, \end{aligned}$$

where $\lambda_1(g)$ is the smallest eigenvalue of $K_g(s,t) = \mathrm{Cov}(F_g(s), F_g(t))$ and $Z \sim \mathrm{N}(0,1)$, and $B_2(g)$ is an upper bound for $1/\lambda_1(g)$ defined by

$$B_2(g) = \sqrt{2\int_0^1 \left(\int_0^t K_g^2(s,t)\,ds\right) dt} = \sqrt{\frac{1}{2}\mathrm{Var}(S_g)}.$$

Table 2.2 presents $\mathrm{E}(S_g) = \sum_{n=1}^{\infty} 1/\lambda_n(g)$, $1/\lambda_1(g)$, $B_2(g)$, $r_1(g) = (1/\lambda_1(g))/\mathrm{E}(S_g)$, and $r_2(g) = B_2(g)/\mathrm{E}(S_g)$ for $g = 0, 1, 2, 3, 4, 5$. It is seen from the table that

(a) The total sum $\mathrm{E}(S_g)$, the inverse of the smallest eigenvalue $1/\lambda_1(g)$, and its uppper bound $B_2(g)$ become smaller quickly as g gets large.
(b) $1/\lambda_1(g)$ and $B_2(g)$ become closer to $\mathrm{E}(S_g)$ with g.
(c) It is expected that the distribution of S_g can be approximated just by using the smallest eigenvalue and its upper bound (see Figure 2.1).

Table 2.2 *Smallest eigenvalues and related statistics for* $g = 0, 1, 2, 3, 4, 5$.

g	$\mathrm{E}(S_g)$	$1/\lambda_1(g)$	$B_2(g)$	$r_1(g)$	$r_2(g)$
0	1/2	$4.053/10$	$4.082/10$	0.8106	0.8165
1	1/12	$8.0891/10^2$	$8.0917/10^2$	0.9707	0.9710
2	1/120	$8.24681/10^3$	$8.24721/10^3$	0.9896	0.9897
3	1/2,016	$4.934432/10^4$	$4.934497/10^4$	0.9948	0.9948
4	1/51,840	$1.922972/10^5$	$1.922981/10^5$	0.9969	0.9969
5	1/1,900,800	$5.249965/10^7$	$5.249976/10^7$	0.9979	0.9979

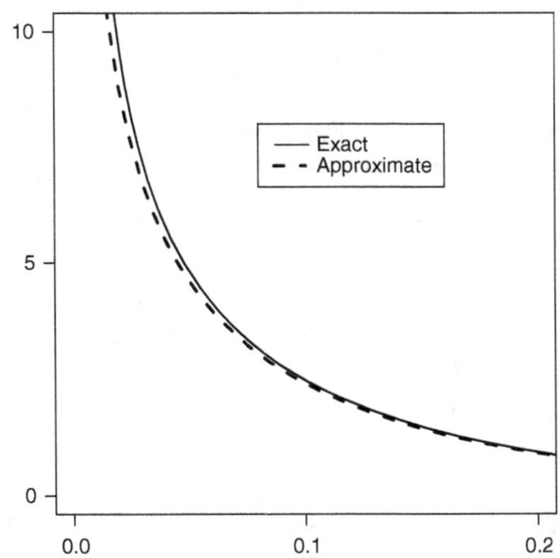

Figure 2.1 Densities of $S_1 = \int_0^1 F_1^2(t)\,dt = \int_0^1 \left\{ \int_0^t W(s)\,ds \right\}^2 dt.$

Figure 2.1 draws the exact and approximate probability densities of S_g for $g = 1$, where the former is given by the inversion formula as

$$f(x) = \frac{1}{2\pi} \int_{-\infty}^{\infty} e^{-i\theta x} \Big(D(2i\theta) \Big)^{-1/2} d\theta,$$

where $D(\lambda)$ is the FD in (2.195), and the latter is the density of $Z^2/\lambda_1(g)$ with $g = 1$ and $Z \sim \mathrm{N}(0, 1)$. It is seen that both densities are monotone, and the approximation based on the smallest eigenvalue approximates the exact density from below and is fairly good, as was anticipated. The density based on the upper bound $B_2(g)$ with $g = 1$ is very close to the approximate density, so it is not shown here. The approximation becomes better as g gets large, although this is not shown here either. We note in passing that the distribution of S_g with $g = 0$ is not monotone; its density is presented in Figure 1 of Appendix B.

Exercises for Section 2.4

Exercise 2.4.1 Show that

$$f''(t) + \lambda f(t) = 0, \quad f(0) = 0, \quad f(1) = 0$$

lead to the integral equation

$$f(t) = \lambda \int_0^1 [\min(s,t) - st] f(s)\, ds.$$

Exercise 2.4.2 Show that the conditions given by

$$f''(t) + \lambda f(t) = \lambda a, \quad a = \int_0^1 f(s)\, ds, \quad f'(0) = 0, \quad f'(1) = 0$$

lead to the integral equation

$$f(t) = \lambda \int_0^1 \left[\frac{1}{3} - \max(s,t) + \frac{s^2 + t^2}{2} \right] f(s)\, ds$$

only when $a = 0$.

Exercise 2.4.3 Show that the integral equation

$$f(t) = \lambda \int_0^1 [1 - \max(s,t) + b] f(s)\, ds$$

is equivalent to

$$f''(t) + \lambda f(t) = 0, \quad f(1) = \lambda b \int_0^1 f(s)\, ds, \quad f'(0) = 0.$$

Exercise 2.4.4 Prove that the eigenvalues of $K(s,t) = 1 - \max(s,t) + b$ are all positive when b is nonnegative, whereas all eigenvalues except one are positive when b is negative.

Exercise 2.4.5 Prove that the FD of $K(s,t) = \min(s,t) - st + b_1 + b_2(s+t)$ is given by

$$D(\lambda) = \left(1 + \lambda b_2^2\right) \frac{\sin\sqrt{\lambda}}{\sqrt{\lambda}} + 2\left(b_1 + b_2 + b_2^2\right)(\cos\sqrt{\lambda} - 1).$$

Exercise 2.4.6 Show that, for $dX(t) = \gamma\, X(t)\, dt + dW(t)$ with $X(0) = 0$,

$$\int_0^1 X^2(t)\, dt = \frac{1}{2\gamma} \int_0^1 \int_0^1 \left(e^{\gamma(2-s-t)} - e^{\gamma|s-t|}\right) dW(s)\, dW(t)$$
$$\stackrel{\mathcal{D}}{=} \frac{1}{2\gamma} \int_0^1 \int_0^1 \left(e^{\gamma(s+t)} - e^{\gamma|s-t|}\right) dW(s)\, dW(t).$$

Exercise 2.4.7 Show that the integral equation

$$f(t) = \frac{\lambda}{2\gamma} \int_0^1 \left(e^{\gamma(s+t)} - e^{\gamma|s-t|}\right) f(s)\, ds$$

is equivalent to

$$f''(t) + (\lambda - \gamma^2) f(t) = 0, \quad f(0) = 0, \quad f'(1) = \gamma f(1).$$

Exercise 2.4.8 Prove that

$$D(\lambda) = e^{\gamma} \left(\cos \mu - \gamma \frac{\sin \mu}{\mu}\right), \quad \mu = \sqrt{\lambda - \gamma^2},$$

admits an infinite product expansion.

Exercise 2.4.9 Prove that the FD of $K(s,t) = (e^{\gamma(s+t)} - e^{\gamma|s-t|})/(2\gamma) + be^{\gamma(s+t)}$ is given by

$$D(\lambda) = e^{\gamma} \left(\cos \mu - (\gamma + b\lambda) \frac{\sin \mu}{\mu}\right), \quad \mu = \sqrt{\lambda - \gamma^2}.$$

Exercise 2.4.10 For the integral equation

$$f(t) = \lambda \int_0^1 K(s,t) f(s)\, ds, \quad K(s,t) = \frac{1}{4}\left[1 - 2|s-t|\right],$$

show that $K(s,t)$ is positive definite and that this integral equation is equivalent to

$$f''(t) + \lambda f(t) = 0, \quad f(0) + f(1) = 0, \quad f'(0) + f'(1) = 0.$$

Exercise 2.4.11 Show that every eigenvalue λ_n of $K(s,t) = [1 - 2|s-t|]/4$ has two linearly independent orthonormal eigenfunctions given by

$$f_{n1}(t) = \sqrt{2} \cos(2n-1)\pi t, \qquad f_{n2}(t) = \sqrt{2} \sin(2n-1)\pi t.$$

Exercise 2.4.12 For the kernel $K(s,t) = \min(s,t) - st + (s^2 + t^2 - s - t)/2 + 1/12$, show that

$$\int_0^1 \int_0^1 K(s,t)\, dW(s)\, dW(t)$$
$$= \int_0^1 \int_0^1 \mathrm{Cov}(\tilde{X}(s), \tilde{X}(t))\, dW(s)\, dW(t) = \int_0^1 \tilde{X}^2(t)\, dt,$$

where

$$\tilde{X}(t) = \tilde{W}(t) - \int_0^1 \tilde{W}(u)\,du, \quad \tilde{W}(t) = W(t) - tW(1).$$

Exercise 2.4.13 Show that the integral equation

$$f(t) = \int_0^1 \frac{1}{2m+1}\left[1 - (\max(s,t))^{2m+1}\right] f(s)\,ds, \quad m > -1/2$$

is equivalent to

$$f''(t) - \frac{2m}{t} f'(t) + \lambda t^{2m} f(t) = 0, \quad f(1) = 0, \quad \lim_{t\to 0} \frac{f'(t)}{t^{2m}} = 0.$$

Exercise 2.4.14 Show that the FD of $K(s,t) = s^m t^m\,[\min(s,t) - st]$ for $m > -1/2$ is given by

$$D(\lambda) = \Gamma\left(\frac{2m+3}{2(m+1)}\right) J_{1/2(m+1)}\left(\frac{\sqrt{\lambda}}{m+1}\right) \Bigg/ \left(\frac{\sqrt{\lambda}}{2(m+1)}\right)^{1/2(m+1)}.$$

Exercise 2.4.15 Prove that

$$\prod_{n=1}^{\infty}\left(1 - \frac{\lambda}{n(n+1)}\right) = -\frac{\cos\dfrac{\pi\sqrt{1+4\lambda}}{2}}{\pi\lambda}.$$

Exercise 2.4.16 Show that the FD of

$$K(s,t) = \frac{1}{2m+1}\left[(\min(s,t))^{1/(2m+1)} - (st)^{1/(2m+1)}\right], \quad m > -1/2$$

is the same as that of the kernel given in Exercise 2.4.14.

Exercise 2.4.17 Prove the following relation:

$$\begin{aligned} S &= \int_0^1 \left[t^m \left\{W(t) - (2m+1)\int_0^1 s^{2m} W(s)\,ds\right\}\right]^2 dt \\ &= \int_0^1\int_0^1 \frac{1}{2m+1}\left[(\min(s,t))^{2m+1} - (st)^{2m+1}\right] dW(s)\,dW(t), \end{aligned}$$

where $m > -1/2$.

Exercise 2.4.18 Show that the FD of

$$K(s,t) = \frac{1}{2m+1}\left[(\min(s,t))^{2m+1} - (st)^{2m+1}\right], \quad m > -1/2$$

is given by

$$D(\lambda) = \Gamma\left(\frac{4m+3}{2(m+1)}\right) J_\nu\left(\frac{\sqrt{\lambda}}{m+1}\right) \Bigg/ \left(\frac{\sqrt{\lambda}}{2(m+1)}\right)^{\nu}, \quad \nu = \frac{2m+1}{2(m+1)}.$$

Exercise 2.4.19 Show that the FD defined by

$$D(\lambda) = \Gamma(-\nu) J_{-\nu-1}\left(\frac{\sqrt{-2\lambda m}}{m+1}\right) \Big/ \left(\frac{\sqrt{-2\lambda m}}{2(m+1)}\right)^{-\nu-1}, \quad \nu = \frac{m}{m+1}$$

reduces to $1 - \lambda/2$ when $m = 0$.

Exercise 2.4.20 Ensure that the FD of

$$K(s,t) = \min(s,t) - st - 2st(1-s)(1-t)(4 - 5s - 5t + 10st)$$

is given by

$$D(\lambda) = \frac{720}{\lambda^4}\left(4(6+\lambda) - 8(3-\lambda)\cos\sqrt{\lambda} - \sqrt{\lambda}(24-\lambda)\sin\sqrt{\lambda}\right).$$

Exercise 2.4.21 Using Theorem 2.20, derive the FD of the following kernel:

$$K(s,t) = \min(s,t) - st - \sum_{j=1}^{3} \psi_j(s)\psi_j(t),$$

where

$$\psi_1(t) = \sqrt{2}it(1-t), \quad \psi_2(t) = \sqrt{10}t^2(1-t), \quad \psi_3(t) = \sqrt{10}t(1-t)^2.$$

3

Integral Equations and the Resolvent

In the last chapter we were concerned with purely quadratic functionals of Bm and discussed how to derive the FD associated with such functionals by presenting various examples. In this chapter we extend the class of functionals to quadratic plus linear or bilinear functionals of Bm. In this case, it turns out that the FD is not sufficient to obtain the c.f. of such functionals, and we need to deal with the resolvent appearing in the theory of nonhomogeneous integral equations of Fredholm type. We explain the resolvent from a statistical point of view. It is recognized that the computation of the resolvent itself is unnecessary to compute the c.f. but that we need only compute a function derived from the resolvent. We illustrate by some statistical examples how to compute the c.f. of those statistics.

3.1 Resolvent

We have so far discussed how to obtain the FD associated with purely quadratic functionals of Bm. In this chapter we extend the class of our statistics to consider quadratic plus linear or bilinear functionals of Bm. The statistic we first deal with is

$$S_1 = \int_0^1 (Y(t) + m(t))^2 \, dt, \tag{3.1}$$

where $Y(t)$ is a zero-mean Gaussian process with $\mathrm{Cov}(Y(s), Y(t)) = K(s,t)$, and $m(t)$ is a continuous and deterministic function that does not vanish identically. Note that $K(s,t)$ is positive definite. The statistic S_1 was dealt with in Nabeya (1992).

We also deal with statistics of the form

$$S_2 = \int_0^1 \int_0^1 K(s,t)\,dW(s)\,dW(t) + 2a \int_0^1 K(0,t)\,dW(t) + b, \tag{3.2}$$

$$S_3 = \int_0^1 (Y(t) + Zm(t))^2\,dt, \tag{3.3}$$

$$S_4 = \int_0^1 \int_0^1 K(s,t)\,dW(s)\,dW(t) + 2aZ \int_0^1 K(0,t)\,dW(t) + bZ^2, \tag{3.4}$$

where a and b are constants, and Z follows N(0, 1) and is independent of $\{Y(t)\}$ and $\{W(t)\}$. The statistics S_1 and S_2 are quadratic plus linear functionals of Bm, whereas S_3 and S_4 are quadratic plus bilinear functionals of Bm. It should be noted that the process $X(t) = Y(t) + Zm(t)$ in S_3 is a zero-mean Gaussian process with $\mathrm{Cov}(X(s), X(t)) = K(s,t) + m(s)m(t)$. Thus the c.f. of S_3 may be derived from the method discussed in Chapter 2, but we present here an alternative method as an extension of S_1. Note also that a restrictive assumption is imposed on the linear part of S_2 and the bilinear part of S_4 with $K(0,t)$. This will enable us to derive the c.f. of S_2 and S_4 explicitly. The statistics S_2, S_3, and S_4 were dealt with in Tanaka (1990a).

It turns out that, to derive the c.f. of the above statistics, we need some knowledge of the Fredholm theory on nonhomogeneous integral equations, which we describe briefly (see Courant and Hilbert (1953) and Hochstadt (1973) for more details). Let us consider

$$f(t) = \lambda \int_0^1 K(s,t) f(s)\,ds + g(t), \tag{3.5}$$

where $g(t)$ is continuous, whereas $K(s,t)$ is symmetric, continuous, and nearly definite.

Given $K(s,t)$ and $g(t)$, we try to solve the integral equation (3.5) for $f(t)$. Consider the iterated kernels $K^{(j)}(s,t)$ $(j = 2, 3, \ldots)$ defined in (2.28) as

$$K^{(j)}(s,t) = \int_0^1 K^{(j-1)}(s,u) K(u,t)\,du = \sum_{n=1}^{\infty} \frac{1}{\lambda_n^j} f_n(s) f_n(t), \tag{3.6}$$

where $K^{(1)}(s,t) = K(s,t)$, λ_n $(n = 1, 2, \ldots)$ are eigenvalues of $K(s,t)$, and $f_n(t)$ is the orthonormal eigenfunction corresponding to λ_n. Then we obtain

$$
\begin{aligned}
f(t) &= g(t) + \lambda \int_0^1 K(s,t) \left(g(s) + \lambda \int_0^1 K(u,s) f(u)\, du \right) ds \\
&= g(t) + \lambda \int_0^1 K(s,t) g(s)\, ds + \lambda^2 \int_0^1 K^{(2)}(s,t) f(s)\, ds \\
&= g(t) + \lambda \int_0^1 K(s,t) g(s)\, ds + \lambda^2 \int_0^1 K^{(2)}(s,t) g(s)\, ds \\
&\quad + \lambda^3 \int_0^1 K^{(3)}(s,t) f(s)\, ds \\
&= g(t) + \lambda \int_0^1 \Gamma(s,t;\lambda) g(s)\, ds,
\end{aligned}
$$

where

$$
\Gamma(s,t;\lambda) = \sum_{j=1}^{\infty} \lambda^{j-1} K^{(j)}(s,t). \tag{3.7}
$$

The symmetric function $\Gamma(s,t;\lambda)$ is called the *resolvent* of $K(s,t)$. It is known (Courant and Hilbert, 1953) that the infinite series in (3.7) converges uniformly for $|\lambda| < \min_{0 \le s,t \le 1} (1/|K(s,t)|)$.

Using again the expression for $K^{(j)}(s,t)$ given in (3.6), we obtain, for $|\lambda| < \min_{1 \le n}(|\lambda_n|)$,

$$
\begin{aligned}
\Gamma(s,t;\lambda) &= \sum_{j=1}^{\infty} \lambda^{j-1} K^{(j)}(s,t) = \sum_{j=1}^{\infty} \lambda^{j-1} \sum_{n=1}^{\infty} \frac{1}{\lambda_n^j} f_n(s) f_n(t) \\
&= \sum_{n=1}^{\infty} \sum_{j=1}^{\infty} \frac{1}{\lambda_n} \left(\frac{\lambda}{\lambda_n} \right)^{j-1} f_n(s) f_n(t) \\
&= \sum_{n=1}^{\infty} \frac{1}{\lambda_n - \lambda} f_n(s) f_n(t).
\end{aligned} \tag{3.8}
$$

It is mentioned in Courant and Hilbert (1953) that (3.8) provides the analytic continuation of the resolvent into the entire complex λ-plane and that the resolvent is a meromorphic function of λ that possesses simple poles at the eigenvalues.

It also holds that

$$
\begin{aligned}
\Gamma(s,t;\lambda) &= K(s,t) + \sum_{j=1}^{\infty} \lambda^j K^{(j+1)}(s,t) \\
&= K(s,t) + \lambda \sum_{j=1}^{\infty} \lambda^{j-1} \int_0^1 K^{(j)}(s,u) K(u,t)\, du \\
&= K(s,t) + \lambda \int_0^1 \Gamma(s,u;\lambda) K(u,t)\, du.
\end{aligned} \tag{3.9}
$$

Thus we have obtained three expressions (3.7), (3.8), and (3.9) for the resolvent $\Gamma(s,t;\lambda)$, which is summarized in the following theorem.

Theorem 3.1 *Let $\{\lambda_n\}$ be a sequence of eigenvalues of a symmetric, continuous, and nearly definite kernel $K(s,t)$ and $\{f_n(t)\}$ be an orthonormal sequence of eigenfunctions corresponding to $\{\lambda_n\}$. Then the resolvent $\Gamma(s,t;\lambda)$ of $K(s,t)$ is given by*

$$\begin{aligned}\Gamma(s,t;\lambda) &= \sum_{j=1}^{\infty} \lambda^{j-1} K^{(j)}(s,t) \\ &= \sum_{n=1}^{\infty} \frac{1}{\lambda_n - \lambda} f_n(s) f_n(t) \\ &= K(s,t) + \lambda \int_0^1 \Gamma(s,u;\lambda) K(u,t)\, du, \end{aligned} \tag{3.10}$$

where $K^{(j)}(s,t)$ is the iterated kernel defined in (3.6).

It is quite hard to compute $\Gamma(s,t;\lambda)$ for any s and t following the above definitions. Our purpose, however, is to compute the c.f.s of the statistics S_k $(k = 1,2,3,4)$ given above, and we can avoid the computation of $\Gamma(s,t;\lambda)$ for any s and t to derive the c.f.s, which will be discussed subsequently.

In Section 3.2 we deal with the statistic S_1 in (3.1). The statistics S_2 in (3.2), S_3 in (3.3), and S_4 in (3.4) are discussed in Sections 3.3, 3.4, and 3.5, respectively.

3.2 Computation with the Resolvent – Case 1

In this section we consider the statistic S_1 in (3.1) given by

$$S_1 = \int_0^1 (Y(t) + m(t))^2\, dt, \tag{3.11}$$

where $\{Y(t)\}$ is a zero-mean Gaussian process with $\mathrm{Cov}(Y(s),\, Y(t)) = K(s,t)$, and $m(t)$ is a continuous and deterministic function.

To derive the distribution of S_1 we use the Karhunen–Loève expansion (Theorem 2.11) given by

$$Y(t) = \sum_{n=1}^{\infty} \frac{f_n(t)}{\sqrt{\lambda_n}} Z_n,$$

where $\{Z_n\} \sim \mathrm{NID}(0,1)$ and $\{\lambda_n\}$ is a sequence of eigenvalues of the positive definite kernel $K(s,t) = \mathrm{Cov}(Y(s),\, Y(t))$, repeated as many times as their multiplicities, and $\{f_n(t)\}$ is an orthonormal sequence of eigenfunctions corresponding to $\{\lambda_n\}$.

Then we obtain (Exercise 3.2.1)

$$S_1 \stackrel{\mathcal{D}}{=} \int_0^1 \left\{ \sum_{n=1}^{\infty} \frac{f_n(t)}{\sqrt{\lambda_n}} Z_n + m(t) \right\}^2 dt = \sum_{n=1}^{\infty} T_n + \int_0^1 m^2(t)\, dt, \tag{3.12}$$

where

$$T_n = \frac{1}{\lambda_n} Z_n^2 + \frac{2\, c_n}{\sqrt{\lambda_n}} Z_n, \quad c_n = \int_0^1 f_n(t) m(t)\, dt. \tag{3.13}$$

Since

$$\begin{aligned}
\mathrm{E}\left(e^{i\theta T_n}\right) &= \mathrm{E}\left[\exp\left\{i\theta\left(\frac{1}{\lambda_n} Z_n^2 + \frac{2c_n}{\sqrt{\lambda_n}} Z_n\right)\right\}\right] \\
&= \frac{1}{\sqrt{2\pi}} \int_{-\infty}^{\infty} \exp\left[-\frac{1}{2}\left\{x^2 - 2i\theta\left(\frac{1}{\lambda_n} x^2 + \frac{2c_n}{\sqrt{\lambda_n}} x\right)\right\}\right] dx \\
&= \left(1 - \frac{2i\theta}{\lambda_n}\right)^{-1/2} \exp\left\{\frac{-2\theta^2 c_n^2}{\lambda_n - 2i\theta}\right\},
\end{aligned}$$

we can deduce that

$$\begin{aligned}
\mathrm{E}\left(e^{i\theta S_1}\right) &= \mathrm{E}\left[\exp\left\{i\theta\left(\sum_{n=1}^{\infty} T_n + \int_0^1 m^2(t)\, dt\right)\right\}\right] \\
&= \left(\prod_{n=1}^{\infty} \mathrm{E}\left(e^{i\theta T_n}\right)\right) \exp\left\{i\theta \int_0^1 m^2(t)\, dt\right\} \\
&= \prod_{n=1}^{\infty}\left(1 - \frac{2i\theta}{\lambda_n}\right)^{-1/2} \exp\left\{i\theta \int_0^1 m^2(t)\, dt - 2\theta^2 \sum_{n=1}^{\infty} \frac{c_n^2}{\lambda_n - 2i\theta}\right\} \\
&= (D(2i\theta))^{-1/2} \exp\left\{i\theta \int_0^1 m^2(t)\, dt - 2\theta^2 \sum_{n=1}^{\infty} \frac{c_n^2}{\lambda_n - 2i\theta}\right\},
\end{aligned}$$

where $D(\lambda)$ is the FD of $K(s,t)$. Using (3.13), we have

$$c_n^2 = \int_0^1 \int_0^1 f_n(s) f_n(t) m(s) m(t)\, ds\, dt,$$

which gives, because of the second relation of the resolvent in (3.10),

$$\begin{aligned}
\sum_{n=1}^{\infty} \frac{c_n^2}{\lambda_n - 2i\theta} &= \sum_{n=1}^{\infty} \frac{1}{\lambda_n - 2i\theta} \int_0^1 \int_0^1 f_n(s) f_n(t) m(s) m(t)\, ds\, dt \\
&= \int_0^1 \int_0^1 \Gamma(s,t;2i\theta) m(s) m(t)\, ds\, dt.
\end{aligned} \tag{3.14}$$

Here we can avoid the computation of the integral in (3.14), which involves the resolvent. In fact, multiplying by $m(s)$ on the both sides of the last relation for $\Gamma(s,t;\lambda)$ in (3.10) and integrating with respect to s, we are led to

$$h(t;\lambda) = \int_0^1 \Gamma(s,t;\lambda)m(s)\,ds \tag{3.15}$$

$$= \int_0^1 K(s,t)m(s)\,ds + \lambda \int_0^1 K(s,t)h(s;\lambda)\,ds. \tag{3.16}$$

It now follows from (3.14) and (3.15) that

$$\sum_{n=1}^{\infty} \frac{c_n^2}{\lambda_n - \lambda} = \int_0^1 \int_0^1 \Gamma(s,t;\lambda)m(s)m(t)\,ds\,dt = \int_0^1 h(t;\lambda)m(t)\,dt. \tag{3.17}$$

Thus we can avoid the computation of the integral in (3.14) if we can find the solution $h(t;\lambda)$ to the nonhomogeneous integral equation (3.16).

We can now establish the following theorem (Nabeya, 1992).

Theorem 3.2 *Let us consider*

$$S_1 = \int_0^1 (Y(t) + m(t))^2\,dt,$$

where $\{Y(t)\}$ is a zero-mean Gaussian process with the covariance function given by $K(s,t) = Cov(Y(s),\ Y(t))$, and $m(t)$ is a continuous and deterministic function. Then the c.f. of S_1 is given by

$$E\left(e^{i\theta S_1}\right) = (D(\lambda))^{-1/2} \exp\left[\frac{\lambda}{2}\int_0^1 m^2(t)\,dt + \frac{\lambda^2}{2}\int_0^1 h(t;\lambda)\,m(t)\,dt\right], \tag{3.18}$$

where $\lambda = 2i\theta$ and $D(\lambda)$ is the FD of $K(s,t)$, and $h(t;\lambda)$ is the solution of the equation in (3.16).

The mean and variance of S_1 can be easily obtained by differentiating the c.f. of S_1. Alternatively we have

Corollary 3.2 *For the statistic S_1 defined in Theorem 3.2, it holds that*

$$\mathrm{E}(S_1) = \int_0^1 K(t,t)\,dt + \int_0^1 m^2(t)\,dt = d_1 + \int_0^1 m^2(t)\,dt,$$

$$\mathrm{Var}(S_1) = 4\int_0^1 \left(\int_0^t K^2(s,t)\,ds\right)dt + 8\int_0^1 \left(\int_0^t m(s)K(s,t)\,ds\right)m(t)\,dt$$

$$= 2(d_1^2 - d_2) + 4\int_0^1 h(t;0)m(t)\,dt,$$

where d_1 and d_2 are coefficients in the series expansion for $D(\lambda)$ expressed as $1 - d_1\lambda + d_2\lambda^2/2 - \cdots$.

We present below various examples for the computation of the c.f. of S_1 together with $\mathrm{E}(S_1)$ and $\mathrm{Var}(S_1)$.

Example 3.3 As the first example, let us consider

$$S_1 = \int_0^1 (W(t) + a + bt)^2 \, dt. \tag{3.19}$$

The integral equation (3.16) with $K(s,t) = \mathrm{Cov}(W(s),\ W(t)) = \min(s,t)$ and $m(s) = a + bs$ leads to

$$\begin{aligned} h(t;\lambda) &= \int_0^t s\, m(s)\, ds + t \int_t^1 m(s)\, ds \\ &\quad + \lambda \left(\int_0^t s\, h(s;\lambda)\, ds + t \int_t^1 h(s;\lambda)\, ds \right), \\ h'(t;\lambda) &= \int_t^1 m(s)\, ds + \lambda \int_t^1 h(s;\lambda)\, ds \end{aligned}$$

and is equivalent to

$$h''(t;\lambda) + \lambda h(t;\lambda) = -(a+bt), \qquad h(0;\lambda) = h'(1;\lambda) = 0.$$

The general solution is given by

$$h(t;\lambda) = c_1 \cos \sqrt{\lambda}\, t + c_2 \sin \sqrt{\lambda}\, t - \frac{1}{\lambda}(a+bt).$$

It follows from the boundary conditions $h(0;\lambda) = h'(1;\lambda) = 0$ that we can determine c_1 and c_2 uniquely as

$$c_1 = \frac{a}{\lambda}, \qquad c_2 = \frac{a\sqrt{\lambda} \sin \sqrt{\lambda} + b}{\lambda \sqrt{\lambda} \cos \sqrt{\lambda}}$$

and obtain

$$\begin{aligned} &\frac{\lambda}{2} \int_0^1 m^2(t)\, dt + \frac{\lambda^2}{2} \int_0^1 h(t;\lambda) m(t)\, dt \\ &= \frac{\lambda^2}{2} \int_0^1 \left(c_1 \cos \sqrt{\lambda}\, t + c_2 \sin \sqrt{\lambda}\, t \right) (a+bt)\, dt \\ &= \frac{\lambda^2}{2} \left[-\frac{b(2a+b)}{\lambda^2} + 2ab \frac{\sec \sqrt{\lambda}}{\lambda^2} + \left(a^2\lambda + b^2\right) \frac{\tan \sqrt{\lambda}}{\lambda^2 \sqrt{\lambda}} \right]. \end{aligned}$$

Then Theorem 3.2 gives

$$\mathrm{E}\left(e^{i\theta S_1}\right) = \left(\cos \sqrt{\lambda}\right)^{-1/2} \exp\left[-\frac{b(2a+b)}{2} + ab \sec \sqrt{\lambda} + \frac{a^2\lambda + b^2}{2} \frac{\tan \sqrt{\lambda}}{\sqrt{\lambda}} \right], \tag{3.20}$$

where $\lambda = 2i\theta$. It follows from Corollary 3.2 that

$$\begin{aligned}
\mathrm{E}(S_1) &= \frac{1}{2} + \int_0^1 (a+bt)^2\,dt = \frac{1}{2} + a^2 + ab + \frac{1}{3}b^2,\\
\mathrm{Var}(S_1) &= 4\int_0^1 \left(\int_0^t s^2\,ds\right) dt + 8\int_0^1 \left(\int_0^t s(a+bs)\,ds\right)(a+bt)dt\\
&= 2(d_1^2 - d_2) + 4\int_0^1 \frac{1}{6}t(-3at + 6a - bt^2 + 3b)(a+bt)\,dt\\
&= \frac{1}{3} + \frac{4}{3}a^2 + \frac{5}{3}ab + \frac{8}{15}b^2.
\end{aligned}$$

When $a = b = 0$, the distribution of S_1 is far from normal, but it can be shown (Exercise 3.2.2) that, as $|ab| \to \infty$,

$$S_1 \Rightarrow \mathrm{N}\left(a^2 + ab + \frac{1}{3}b^2,\ \frac{4}{3}a^2 + \frac{5}{3}ab + \frac{8}{15}b^2\right).$$

Figure 3.1 draws probability densities of $(S_1 - \mathrm{E}(S_1))/\sqrt{\mathrm{Var}(S_1)}$ for various values of a and b. It is seen that the distribution tends to normality even if $|a|$ and $|b|$ are relatively small.

Note that, when $b = 0$, we have

$$\begin{aligned}
S_1 &= \int_0^1 (W(t) + a)^2\,dt\\
&= \int_0^1\int_0^1 K(s,t)\,dW(s)\,dW(t) + 2a\int_0^1 K(0,t)\,dW(t) + a^2,
\end{aligned}$$

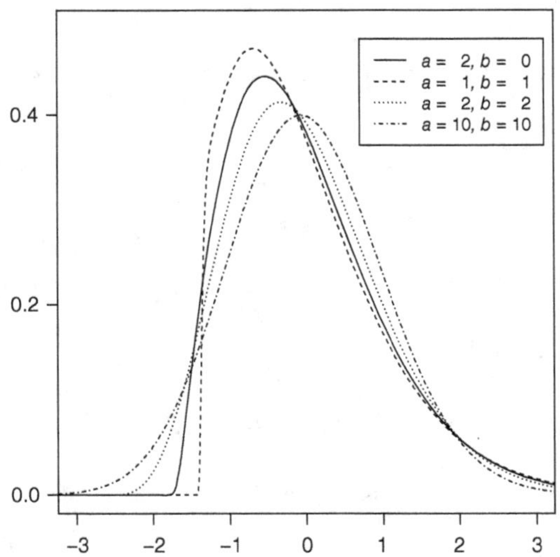

Figure 3.1 Probability densities of the normalized S_1 in (3.19).

where $K(s,t) = 1 - \max(s,t)$, which is of the same form as S_2 in (3.2), and the c.f. of S_1 can be obtained by the approach discussed in the next section.

Example 3.4 Let us consider

$$S_1 = \int_0^1 (Y(t) + a + bt)^2 \, dt, \tag{3.21}$$

where $\{Y(t)\}$ is the Bb with $K(s,t) = \mathrm{Cov}(Y(s), Y(t)) = \min(s,t) - st$. The integral equation (3.16) leads to

$$h''(t;\lambda) + \lambda h(t;\lambda) = -(a + bt), \quad h(0;\lambda) = h(1;\lambda) = 0.$$

Proceeding in the same way as in Example 3.3, we obtain (Exercise 3.2.3)

$$h(t;\lambda) = \frac{a}{\lambda} \cos\sqrt{\lambda}t + \left(\frac{a+b}{\lambda \sin\sqrt{\lambda}} - \frac{a}{\lambda}\cot\sqrt{\lambda}\right)\sin\sqrt{\lambda}t - \frac{a+bt}{\lambda},$$

which yields

$$\mathrm{E}\left(e^{i\theta S_1}\right) = \left(\frac{\sin\sqrt{\lambda}}{\sqrt{\lambda}}\right)^{-1/2} \exp\left[\frac{b^2}{2} + a(a+b)\sqrt{\lambda}\,\mathrm{cosec}\sqrt{\lambda} \right.$$
$$\left. -\sqrt{\lambda}\left(a(a+b) + \frac{b^2}{2}\right)\cot\sqrt{\lambda}\right], \tag{3.22}$$

where $\lambda = 2i\theta$. Since $h(t;\lambda) \to t(-3at + 3a - bt^2 + b)/6$ as $\lambda \to 0$, we have, from Corollary 3.2,

$$\mathrm{E}(S_1) = \frac{1}{6} + \int_0^1 (a+bt)^2 \, dt = \frac{1}{6} + a^2 + ab + \frac{1}{3}b^2,$$

$$\mathrm{Var}(S_1) = 2(d_1^2 - d_2) + 4\int_0^1 \frac{1}{6} t(-3at + 3a - bt^2 + b)(a+bt)\, dt$$
$$= \frac{1}{45} + \frac{1}{3}a^2 + \frac{1}{3}ab + \frac{4}{45}b^2,$$

and it holds that, as $|ab| \to \infty$,

$$S_1 \Rightarrow \mathrm{N}\left(a^2 + ab + \frac{1}{3}b^2, \frac{1}{3}a^2 + \frac{1}{3}ab + \frac{4}{45}b^2\right).$$

Example 3.5 The third example is concerned with

$$S_1 = \int_0^1 \left(X(t) + ae^{\gamma t}\right)^2 dt, \tag{3.23}$$

where $\{X(t)\}$ is the O–U process defined by

$$dX(t) = \gamma X(t)\, dt + dW(t), \quad X(0) = 0 \quad \Leftrightarrow \quad X(t) = e^{\gamma t}\int_0^t e^{-\gamma s}\, dW(s).$$

Thus

$$Y(t) = X(t) + ae^{\gamma t} = ae^{\gamma t} + e^{\gamma t} \int_0^t e^{-\gamma s}\, dW(s)$$

is the O–U process starting from $Y(0) = a$.

It follows from (2.113) that

$$K(s,t) = \mathrm{Cov}(X(s), X(t)) = \mathrm{E}(X(s)X(t)) = \frac{e^{\gamma(s+t)} - e^{\gamma|s-t|}}{2\gamma},$$

and the FD of $K(s,t)$ is given in (2.115) as

$$D(\lambda) = e^{\gamma} \left(\cos\mu - \gamma \frac{\sin\mu}{\mu} \right),$$

where $\mu = \sqrt{\lambda - \gamma^2}$. The integral equation (3.16) leads to

$$h''(t;\lambda) + \mu^2 h(t;\lambda) = -ae^{\gamma t}, \quad h(0;\lambda) = 0, \quad h'(1;\lambda) = \gamma h(1;\lambda).$$

Operational calculus with the differential operator D yields the general solution

$$h(t;\lambda) = c_1 \cos\mu t + c_2 \sin\mu t - \frac{ae^{\gamma t}}{D^2 + \mu^2} = c_1 \cos\mu t + c_2 \sin\mu t - \frac{ae^{\gamma t}}{\gamma^2 + \mu^2}.$$

The two boundary conditions yield

$$c_1 = \frac{a}{\lambda}, \quad c_2 = \frac{a(\mu \sin\mu + \gamma\cos\mu)}{\lambda(\mu\cos\mu - \gamma\sin\mu)},$$

and, using computerized algebra, we obtain, for $m(t) = ae^{\gamma t}$,

$$\frac{\lambda}{2}\int_0^1 m^2(t)\, dt + \frac{\lambda^2}{2}\int_0^1 h(t;\lambda)m(t)\, dt = \frac{a^2\lambda \sin\mu/\mu}{2(\cos\mu - \gamma\sin\mu/\mu)}.$$

Then we arrive at

$$\mathrm{E}\left(e^{i\theta S_1}\right) = (D(2i\theta))^{-1/2} \exp\left[\frac{ia^2\theta \sin\nu/\nu}{\cos\nu - \gamma\sin\nu/\nu}\right], \tag{3.24}$$

where $\nu = \sqrt{2i\theta - \gamma^2}$. It follows from Corollary 3.2 that

$$\mathrm{E}(S_1) = \int_0^1 \left(K(t,t) + m^2(t)\right) dt = \frac{1}{2\gamma}\left(\frac{e^{2\gamma} - 1}{2\gamma} - 1\right) + \frac{a^2(e^{2\gamma} - 1)}{2\gamma},$$

$$\begin{aligned}\mathrm{Var}(S_1) &= 4\int_0^1 \left(\int_0^t K^2(s,t)\, ds\right) dt + 8\int_0^1 \left(\int_0^t m(s)K(s,t)\, ds\right) m(t)\, dt \\ &= \frac{1}{2\gamma^2}\left(\frac{e^{4\gamma}}{4\gamma^2} + \frac{e^{2\gamma}}{\gamma^2} - \frac{2e^{2\gamma}}{\gamma} - \frac{5}{4\gamma^2} - \frac{1}{\gamma}\right) + \frac{2a^2}{\gamma^2}\left(\frac{e^{4\gamma} - 1}{4\gamma} - e^{2\gamma}\right),\end{aligned}$$

and it holds that, as $|a| \to \infty$,

$$S_1 \Rightarrow \mathrm{N}\left(\frac{a^2(e^{2\gamma}-1)}{2\gamma}, \frac{2a^2}{\gamma^2}\left(\frac{e^{4\gamma}-1}{4\gamma}-e^{2\gamma}\right)\right).$$

Note that, when $\gamma = 0$, we have

$$\begin{aligned}\mathrm{E}\left(e^{i\theta S_1}\right) &= \mathrm{E}\left[\exp\left\{i\theta\int_0^1 (W(t)+a)^2\,dt\right\}\right] \\ &= \left(\cos\sqrt{2i\theta}\right)^{-1/2}\exp\left[ia^2\theta\frac{\tan\sqrt{2i\theta}}{\sqrt{2i\theta}}\right],\end{aligned}$$

which is the same as (3.20) when $b = 0$.

The present statistic can be dealt with by the approach discussed in the next section and will be taken up again in Example 3.9.

Example 3.6 The fourth example considers

$$S_1 = \int_0^1 (Y(t) - \gamma t(1-t))^2\,dt, \tag{3.25}$$

where $\{Y(t)\}$ is the Bb with $K(s,t) = \mathrm{Cov}(Y(s), Y(t)) = \min(s,t) - st$. The c.f. of S_1 was presented in (1.88) of Section 1.5 without proof. This statistic arises in connection with goodness-of-fit tests. The null hypothesis is that the distribution follows the logistic distribution given by

$$H_0 : G(x) = F(x;\xi_0), \quad F(x;\xi) = \frac{1}{1+e^{-(x-\xi)}},$$

where ξ_0 is a specified location parameter. It was shown in Durbin (1973) that the Cramér–von Mises–Smirnov statistic constructed from a random sample of size n converges weakly to S_1 in (3.25) under the local alternative $H_1 : \xi = \xi_0 + \gamma/\sqrt{n}$.

The integral equation (3.16) with $m(s) = -\gamma s(1-s)$ is equivalent to

$$h''(t;\lambda) + \lambda h(t;\lambda) = \gamma t(1-t), \quad h(0;\lambda) = h(1;\lambda) = 0.$$

Operational calculus yields the general solution

$$\begin{aligned}h(t;\lambda) &= c_1\cos\sqrt{\lambda}t + c_2\sin\sqrt{\lambda}t + \frac{\gamma t(1-t)}{D^2+\lambda} \\ &= c_1\cos\sqrt{\lambda}t + c_2\sin\sqrt{\lambda}t + \gamma\left(\frac{2}{\lambda^2}+\frac{t(1-t)}{\lambda}\right).\end{aligned}$$

It follows from $h(0;\lambda) = h(1;\lambda) = 0$ that c_1 and c_2 are uniquely determined as

$$c_1 = -\frac{2\gamma}{\lambda^2}, \quad c_2 = \frac{2\gamma(\cos\sqrt{\lambda}-1)}{\lambda^2\sin\sqrt{\lambda}}.$$

We then obtain

$$\int_0^1 h(t;\lambda)m(t)\,dt = \gamma^2\left[\frac{8(1-\cos\sqrt{\lambda})}{\lambda^3\sqrt{\lambda}\sin\sqrt{\lambda}} - \frac{\lambda^2+10\lambda+120}{30\lambda^3}\right].$$

Thus it follows from Theorem 3.2 that

$$\mathrm{E}\left(e^{i\theta S_1}\right) = \left(\frac{\sin\sqrt{\lambda}}{\sqrt{\lambda}}\right)^{-1/2}\exp\left[\frac{4\gamma^2}{\lambda^2}\left(\frac{1-\cos\sqrt{\lambda}}{\sin\sqrt{\lambda}/\sqrt{\lambda}} - \frac{\lambda(\lambda+12)}{24}\right)\right], \tag{3.26}$$

where $\lambda = 2i\theta$. This verifies the result given in (1.88). Since $h(t;\lambda) \to \gamma t(-t^3+2t^2-1)/12$ as $\lambda \to 0$, it follows from Corollary 3.2 that

$$\mathrm{E}(S_1) = \frac{1}{6} + \gamma^2\int_0^1 t^2(1-t)^2\,dt = \frac{1}{6} + \frac{1}{30}\gamma^2,$$
$$\mathrm{Var}(S_1) = \frac{1}{45} + \frac{17}{1260}\gamma^2,$$

and it is found (Exercise 3.2.4) that, as $|\gamma| \to \infty$,

$$S_1 \Rightarrow \mathrm{N}\left(\frac{1}{30}\gamma^2, \frac{17}{1260}\gamma^2\right).$$

Example 3.7 This is a continuation of Example 3.6. We consider

$$S_1 = \int_0^1 (Y(t) - a\sin 2\pi t)^2\,dt, \tag{3.27}$$

where $\{Y(t)\}$ is the Bb. This statistic also arises from goodness-of-fit tests, and its c.f. with $a = -\gamma/2\xi_0\pi$ was presented in (1.89) of Section 1.5 without proof, which we verify here.

The integral equation (3.16) with $m(s) = -a\sin 2\pi s$ is equivalent to

$$h''(t;\lambda) + \lambda h(t;\lambda) = a\sin 2\pi t, \quad h(0;\lambda) = h(1;\lambda) = 0.$$

Operational calculus yields the general solution

$$h(t;\lambda) = c_1\cos\sqrt{\lambda}t + c_2\sin\sqrt{\lambda}t + a\frac{\sin 2\pi t}{D^2+\lambda}$$
$$= c_1\cos\sqrt{\lambda}t + c_2\sin\sqrt{\lambda}t + a\frac{\sin 2\pi t}{\lambda - 4\pi^2}.$$

It follows from $h(0;\lambda) = h(1;\lambda) = 0$ that c_1 and c_2 are uniquely determined as $c_1 = c_2 = 0$, which gives

$$h(t;\lambda) = a\frac{\sin 2\pi t}{\lambda - 4\pi^2}.$$

We then obtain

$$\frac{\lambda}{2}\int_0^1 m^2(t)\,dt + \frac{\lambda^2}{2}\int_0^1 h(t;\lambda)m(t)\,dt = \frac{a^2\pi^2\lambda}{4\pi^2-\lambda}.$$

Thus it follows from Theorem 3.2 that

$$\mathrm{E}\left(e^{i\theta S_1}\right) = \left(\frac{\sin\sqrt{\lambda}}{\sqrt{\lambda}}\right)^{-1/2} \exp\left[\frac{a^2\pi^2\lambda}{4\pi^2-\lambda}\right], \tag{3.28}$$

where $\lambda = 2i\theta$, which verifies the result given in (1.89) by putting $a = -\gamma/2\xi_0\pi$. It is found that

$$\mathrm{E}(S_1) = \left.\frac{d\mathrm{E}\left(e^{\theta S_1}\right)}{d\theta}\right|_{\theta=0} = \frac{1}{6} + \frac{1}{2}a^2,$$

$$\mathrm{Var}(S_1) = \left.\frac{d^2\mathrm{E}\left(e^{\theta S_1}\right)}{d\theta^2}\right|_{\theta=0} - \mathrm{E}^2(S_1) = \frac{1}{45} + \frac{1}{2\pi^2}a^2,$$

and it holds (Exercise 3.2.5) that, as $|a| \to \infty$,

$$S_1 \Rightarrow \mathrm{N}\left(\frac{1}{2}a^2, \frac{1}{2\pi^2}a^2\right).$$

A more general goodness-of-fit test deals with the multi-parameter case and allows some parameters to be unspecified, while the rest are specified. This general case was dealt with in Durbin (1973) and will be discussed in Chapter 4.

Exercises for Section 3.2

Exercise 3.2.1 Let the Karhunen–Loève expansion of $Y(t)$ be given by

$$Y(t) = \sum_{n=1}^{\infty} \frac{f_n(t)}{\sqrt{\lambda_n}} Z_n,$$

and define, for a continuous function $m(t)$,

$$c_n = \int_0^1 f_n(t)m(t)\,dt, \quad q(t) = \sum_{n=1}^{\infty} c_n f_n(t).$$

Then prove that

$$\int_0^1 q(t)m(t)\,dt = \int_0^1 q^2(t)\,dt.$$

Exercise 3.2.2 The c.f. of

$$S_1 = \int_0^1 (W(t) + a + bt)^2 \, dt$$

is given by

$$\mathrm{E}\left(e^{i\theta S_1}\right) = \left(\cos\sqrt{\lambda}\right)^{-1/2} \exp\left[-\frac{b(2a+b)}{2} + \frac{ab}{\cos\sqrt{\lambda}} + \frac{a^2\lambda + b^2}{2}\frac{\tan\sqrt{\lambda}}{\sqrt{\lambda}}\right],$$

where $\lambda = 2i\theta$. Prove that, as $|ab| \to \infty$,

$$S_1 \Rightarrow \mathrm{N}\left(a^2 + ab + \frac{b^2}{3}, \frac{4a^2}{3} + \frac{5ab}{3} + \frac{8b^2}{15}\right).$$

Exercise 3.2.3 Derive the c.f. of

$$S_1 = \int_0^1 (Y(t) + a + bt)^2 \, dt,$$

where $\{Y(t)\}$ is the Bb. Prove also that, as $|ab| \to \infty$,

$$S_1 \Rightarrow \mathrm{N}\left(a^2 + ab + \frac{b^2}{3}, \frac{a^2}{3} + \frac{ab}{3} + \frac{4b^2}{45}\right).$$

Exercise 3.2.4 Show that, as $|\gamma| \to \infty$,

$$S_1 = \int_0^1 (Y(t) - \gamma t(1-t))^2 \, dt \Rightarrow \mathrm{N}\left(\frac{\gamma^2}{30}, \frac{17\gamma^2}{1260}\right),$$

where $\{Y(t)\}$ is the Bb.

Exercise 3.2.5 For the Bb $\{Y(t)\}$, the c.f. of

$$S_1 = \int_0^1 (Y(t) - a\sin 2\pi t)^2 \, dt,$$

is given by

$$\mathrm{E}\left(e^{i\theta S_1}\right) = \left(\frac{\sin\sqrt{\lambda}}{\sqrt{\lambda}}\right)^{-1/2} \exp\left[\frac{a^2\pi^2\lambda}{4\pi^2 - \lambda}\right],$$

where $\lambda = 2i\theta$. Prove that, as $|a| \to \infty$,

$$S_1 \Rightarrow \mathrm{N}\left(\frac{a^2}{2}, \frac{a^2}{2\pi^2}\right).$$

3.3 Computation with the Resolvent – Case 2

In this section we deal with the statistic

$$S_2 = \int_0^1 \int_0^1 K(s,t)\, dW(s)\, dW(t) + 2a \int_0^1 K(0,t)\, dW(t) + b, \qquad (3.29)$$

where $K(s,t)$ is symmetric, continuous, and nearly definite, and a and b are constants. It follows from Mercer's theorem (Theorem 2.2) that

$$\begin{aligned} S_2 - b &= \int_0^1 \int_0^1 \sum_{n=1}^{\infty} \frac{f_n(s) f_n(t)}{\lambda_n}\, dW(s)\, dW(t) \\ &\quad + 2a \int_0^1 \sum_{n=1}^{\infty} \frac{f_n(0) f_n(t)}{\lambda_n}\, dW(t) \\ &\stackrel{\mathcal{D}}{=} \sum_{n=1}^{\infty} \frac{1}{\lambda_n} \left(Z_n^2 + 2a f_n(0) Z_n \right), \end{aligned} \qquad (3.30)$$

where $\{Z_n\} \sim \mathrm{NID}(0, 1)$, $\{\lambda_n\}$ is a sequence of eigenvalues of $K(s,t)$ repeated as many times as their multiplicities, and $\{f_n(t)\}$ is an orthonormal sequence of eigenfunctions corresponding to λ_n. Putting $g(x) = x^2 + 2a f_n(0) x$, we have

$$\begin{aligned} \mathrm{E}\left[\exp\left\{ \frac{i\theta}{\lambda_n} g(Z_n) \right\} \right] &= \frac{1}{\sqrt{2\pi}} \int_{-\infty}^{\infty} \exp\left[-\frac{1}{2} \left\{ x^2 - \frac{2i\theta}{\lambda_n} g(x) \right\} \right] dx \\ &= \left(1 - \frac{2i\theta}{\lambda_n} \right)^{-1/2} \exp\left[2a^2 (i\theta)^2 \frac{f_n^2(0)}{\lambda_n(\lambda_n - 2i\theta)} \right], \end{aligned}$$

and, using the second relation of the resolvent in (3.10) and Mercer's theorem,

$$\begin{aligned} \sum_{n=1}^{\infty} \frac{f_n^2(0)}{\lambda_n(\lambda_n - 2i\theta)} &= \frac{1}{2i\theta} \left[\sum_{n=1}^{\infty} \frac{f_n^2(0)}{\lambda_n - 2i\theta} - \sum_{n=1}^{\infty} \frac{f_n^2(0)}{\lambda_n} \right] \\ &= \frac{1}{2i\theta} \left(\Gamma(0, 0; 2i\theta) - K(0, 0) \right). \end{aligned} \qquad (3.31)$$

To obtain $\Gamma(0, 0; 2i\theta)$, we use the last relation of the resolvent in (3.10) to get

$$\Gamma(0, t; \lambda) = K(0, t) + \lambda \int_0^1 \Gamma(0, s; \lambda) K(s, t)\, ds. \qquad (3.32)$$

Solving the nonhomogeneous integral equation (3.32) for $\Gamma(0, t; \lambda)$, we obtain $\Gamma(0, 0; \lambda)$.

Thus we have the following theorem (Tanaka, 1990a).

Theorem 3.8 *Let us consider*

$$S_2 = \int_0^1 \int_0^1 K(s,t)\, dW(s)\, dW(t) + 2a \int_0^1 K(0,t)\, dW(t) + b,$$

where $K(s,t)$ is symmetric, continuous, and nearly definite, and a and b are constants. Then the c.f. of S_2 is given by

$$E\left(e^{i\theta S_2}\right) = (D(\lambda))^{-1/2} \exp\left[\frac{\lambda}{2}\left\{a^2\left(\Gamma(0,0;\lambda) - K(0,0)\right) + b\right\}\right],$$

where $\lambda = 2i\theta$, $D(\lambda)$ is the FD of $K(s,t)$, and $\Gamma(0,0;\lambda)$ is the solution $\Gamma(0,t;\lambda)$ to the equation (3.32) evaluated at $t = 0$.

The mean and variance of S_2 can be easily obtained by

Corollary 3.8 *For the statistic S_2 defined in Theorem 3.8, we have*

$$\begin{aligned}
\mathrm{E}(S_2) &= \int_0^1 K(t,t)\,dt + b = d_1 + b, \\
\mathrm{Var}(S_2) &= 4\int_0^1 \left(\int_0^t K^2(s,t)\,ds\right) dt + 4a^2 \int_0^1 K^2(0,t)\,dt \\
&= 2(d_1^2 - d_2) + 4a^2 \sum_{n=1}^{\infty} \frac{f_n^2(0)}{\lambda_n^2},
\end{aligned}$$

where d_1 and d_2 are coefficients in the series expansion for $D(\lambda)$ expressed as $1 - d_1\lambda + d_2\lambda^2/2 - \cdots$.

Examples 3.9, 3.10, and 3.11 are presented below for the computation of the c.f. of S_2 together with $\mathrm{E}(S_2)$ and $\mathrm{Var}(S_2)$.

Example 3.9 Let us define the O–U process by

$$dX(t) = \gamma X(t)\,dt + dW(t),\ X(0) = a \ \Leftrightarrow\ X(t) = ae^{\gamma t} + e^{\gamma t}\int_0^t e^{-\gamma s}\,dW(s),$$

and consider

$$\begin{aligned}
S_2 &= \int_0^1 X^2(t)\,dt = \int_0^1 \left\{e^{\gamma t}\int_0^t e^{-\gamma s}\,dW(s) + ae^{\gamma t}\right\}^2 dt \\
&= \int_0^1\int_0^1 K(s,t)\,dW(s)\,dW(t) + 2a\int_0^1 K(0,t)\,dW(t) \\
&\quad + a^2 K(0,0),
\end{aligned} \tag{3.33}$$

where

$$K(s,t) = \frac{e^{\gamma(2-s-t)} - e^{\gamma|s-t|}}{2\gamma}, \quad K(0,t) = \frac{e^{\gamma(2-t)} - e^{\gamma t}}{2\gamma}, \quad K(0,0) = \frac{e^{2\gamma} - 1}{2\gamma}.$$

It can be checked that the expression for S_2 in (3.33) is equal to S_1 in (3.23) in the sense of distribution. Note also that $\mathrm{Cov}(X(s), X(t)) = (e^{\gamma(s+t)} - e^{\gamma|s-t|})/(2\gamma)$ with $\mathrm{Cov}(X(0), X(t)) = 0$ cannot be used for the present approach. The integral equation (3.32) leads to

$$\Gamma''(0,t;\lambda)+\mu\Gamma(0,t:\lambda)=0, \quad \Gamma(0,1;\lambda)=0, \quad \Gamma'(0,0;\lambda)+\gamma\Gamma(0,0;\lambda)=-1,$$

where $\mu=\lambda-\gamma^2$, and the general solution is given by

$$\Gamma(0,t;\lambda)=c_1\cos\sqrt{\mu}\,t+c_2\sin\sqrt{\mu}\,t,$$

so that $\Gamma(0,0;\lambda)=c_1$. It follows from the two boundary conditions that

$$c_1=\frac{\sin\sqrt{\mu}}{\sqrt{\mu}\cos\sqrt{\mu}-\gamma\sin\sqrt{\mu}}, \qquad c_2=\frac{-\cos\sqrt{\mu}}{\sqrt{\mu}\cos\sqrt{\mu}-\gamma\sin\sqrt{\mu}}.$$

Thus Theorem 3.8 with the FD of $K(s,t)$ given in (2.115) leads to

$$\begin{aligned}\mathrm{E}\left(e^{i\theta S_2}\right)&=(D(2i\theta))^{-1/2}\exp\left[ia^2\theta(\Gamma(0,0;2i\theta)-K(0,0))+ia^2\theta K(0,0)\right]\\&=\left(\cos\nu-\gamma\,\frac{\sin\nu}{\nu}\right)^{-1/2}\exp\left[-\frac{\gamma}{2}+\frac{ia^2\theta\,\sin\nu/\nu}{\cos\nu-\gamma\,\sin\nu/\nu}\right],\end{aligned} \tag{3.34}$$

where $\nu=\sqrt{2i\theta-\gamma^2}$. It follows from Corollary 3.8 that

$$\begin{aligned}\mathrm{E}(S_2)&=\int_0^1 K(t,t)\,dt+a^2K(0,0)=\frac{1}{2\gamma}\left(\frac{e^{2\gamma}-1}{2\gamma}-1\right)+\frac{a^2(e^{2\gamma}-1)}{2\gamma},\\\mathrm{Var}(S_2)&=4\int_0^1\left(\int_0^t K^2(s,t)\,ds\right)dt+4a^2\int_0^1 K^2(0,t)\,dt\\&=\frac{1}{2\gamma^2}\left(\frac{e^{4\gamma}}{4\gamma^2}+\frac{e^{2\gamma}}{\gamma^2}-\frac{2e^{2\gamma}}{\gamma}-\frac{5}{4\gamma^2}-\frac{1}{\gamma}\right)+\frac{2a^2}{\gamma^2}\left(\frac{e^{4\gamma}-1}{4\gamma}-e^{2\gamma}\right).\end{aligned}$$

The present approach makes the derivation of the c.f. slightly easier than that used in Example 3.5 of the last section. This may be because of the restrictive assumption imposed on the structure of the present kernel.

Example 3.10 Let us consider

$$\begin{aligned}S_2&=\int_0^1 t^{2m}\,(W(t)+a)^2\,dt\\&=\int_0^1\int_0^1 K(s,t)\,dW(s)\,dW(t)+2a\int_0^1 K(0,t)\,dW(t)\\&\quad+a^2K(0,0),\end{aligned} \tag{3.35}$$

where $m>-1/2$ and

$$\begin{aligned}K(s,t)&=\frac{1}{2m+1}\left[1-(\max(s,t))^{2m+1}\right],\\K(0,t)&=\frac{1-t^{2m+1}}{2m+1}, \qquad K(0,0)=\frac{1}{2m+1}.\end{aligned}$$

It might be argued that the approach taken in the last section is applicable because

$$S_2 = \int_0^1 \left(Y(t) + at^m\right)^2 dt,$$

where $\mathrm{Cov}(Y(s), Y(t)) = s^m t^m \min(s,t)$. However, we find it much easier to use the present approach. Note that the FD of $K(s,t)$ was obtained in (2.133). It can be shown (Exercise 3.3.1) that the integral equation (3.32) is equivalent to

$$\Gamma''(0,t;\lambda) - \frac{2m}{t}\Gamma'(0,t;\lambda) + \lambda t^{2m}\Gamma(0,t;\lambda) = 0 \tag{3.36}$$

with the boundary conditions

$$\Gamma(0,1;\lambda) = 0, \quad \lim_{t\to 0} \frac{\Gamma'(0,t;\lambda)}{t^{2m}} = -1. \tag{3.37}$$

The differential equation (3.36) is Bessel's equation defined in (2.123) with

$$\alpha = \frac{2m+1}{2}, \quad \beta = \frac{\sqrt{\lambda}}{m+1}, \quad \gamma = m+1, \quad \nu = \frac{2m+1}{2(m+1)}.$$

The general solution is given by

$$\Gamma(0,t;\lambda) = t^{(2m+1)/2}\left\{c_1 J_\nu\left(\frac{\sqrt{\lambda}}{m+1}t^{m+1}\right) + c_2 J_{-\nu}\left(\frac{\sqrt{\lambda}}{m+1}t^{m+1}\right)\right\},$$

which yields

$$\Gamma(0,0;\lambda) = \frac{c_2}{\Gamma(-\nu+1)}\left(\frac{\sqrt{\lambda}}{2(m+1)}\right)^{-\nu}.$$

The two boundary conditions in (3.37) yield $M(\lambda)\boldsymbol{c} = (0, -1)'$, where $\boldsymbol{c} = (c_1, c_2)'$ and $M(\lambda)$ is given in (2.131). Thus c_1 and c_2 can be uniquely determined as

$$c_1 = -\frac{\Gamma(\nu+1)}{2m+1}\left(\frac{\sqrt{\lambda}}{2(m+1)}\right)^{-\nu},$$

$$c_2 = \frac{\Gamma(\nu+1)}{2m+1}\frac{J_\nu\left(\sqrt{\lambda}/(m+1)\right)}{J_{-\nu}\left(\sqrt{\lambda}/(m+1)\right)}\left(\frac{\sqrt{\lambda}}{2(m+1)}\right)^{-\nu},$$

so that we obtain

$$\Gamma(0,0;\lambda) = \frac{\Gamma(\nu+1)}{(2m+1)\Gamma(-\nu+1)}\frac{J_\nu\left(\sqrt{\lambda}/(m+1)\right)}{J_{-\nu}\left(\sqrt{\lambda}/(m+1)\right)}\left(\frac{\sqrt{\lambda}}{2(m+1)}\right)^{-2\nu} \tag{3.38}$$

Thus it follows from Theorem 3.8 and (2.133) that

$$\begin{aligned}
\mathrm{E}\left(e^{i\theta S_2}\right) &= (D(2i\theta))^{-1/2} \exp\left[ia^2\theta\, \Gamma(0,0;2i\theta)\right] \\
&= \left[\Gamma(-\nu+1)\, J_{-\nu}(\eta) \Big/ \left(\frac{\eta}{2}\right)^{-\nu}\right]^{-1/2} \\
&\quad \times \exp\left[\frac{ia^2\theta\, \Gamma(\nu+1)}{(2m+1)\,\Gamma(-\nu+1)} \frac{J_\nu(\eta)}{J_{-\nu}(\eta)} \left(\frac{\eta}{2}\right)^{-2\nu}\right], \qquad (3.39)
\end{aligned}$$

where $\nu = (2m+1)/2(m+1)$ and $\eta = \sqrt{2i\theta}/(m+1)$. We have

$$\begin{aligned}
\mathrm{E}(S_2) &= \int_0^1 K(t,t)\,dt + a^2 K(0,0) = \frac{1}{2(m+1)} + \frac{a^2}{2m+1}, \\
\mathrm{Var}(S_2) &= 4\int_0^1 \left(\int_0^t K^2(s,t)\,ds\right) dt + 4a^2 \int_0^1 K^2(0,t)\,dt \\
&= \frac{1}{(m+1)(2m+3)} + \frac{4a^2}{(m+1)(4m+3)},
\end{aligned}$$

and it can be shown (Exercise 3.3.2) from (3.39) that, as $|a| \to \infty$,

$$S_2 \Rightarrow \mathrm{N}\left(\frac{a^2}{2m+1}, \frac{4a^2}{(m+1)(4m+3)}\right).$$

Note that the case of $m = 0$ gives the same result as Example 3.3 for $b = 0$ and Example 3.5 for $\gamma = 0$ in the last section.

Example 3.11 The present approach can be effectively used to obtain the c.f. of

$$\begin{aligned}
S_2 &= \int_0^1 t^m\, (W(t) + a)\, dW(t) \\
&= \int_0^1 \int_0^1 K(s,t)\, dW(s)\, dW(t) + 2a \int_0^1 K(0,t)\, dW(t) \\
&\quad - \frac{1}{2(m+1)}, \qquad (3.40)
\end{aligned}$$

where $m > 0$ and

$$K(s,t) = \frac{1}{2}\,(\max(s,t))^m, \quad K(0,t) = \frac{1}{2}\, t^m, \quad K(0,0) = 0.$$

The expression in (3.40) is a consequence of Itô's formula (Theorem 1.4), which gives

$$d(t^m W^2(t)) = 2t^m W(t)\, dW(t) + (m t^{m-1} W^2(t) + t^m)\, dt,$$

so that

$$\begin{aligned}\int_0^1 t^m\, W(t)\, dW(t) &= \frac{1}{2}\int_0^1 \left[d\left(t^m W^2(t)\right) - \left(mt^{m-1}W^2(t) + t^m\right) dt\right]\\ &= \int_0^1\int_0^1 \frac{1}{2}\left(\max(s,t)\right)^m\, dW(s)\, dW(t) - \frac{1}{2(m+1)}.\end{aligned}$$

The FD of $K(s,t)$ was obtained in (2.143). It can be shown (Exercise 3.3.3) that the integral equation (3.32) is equivalent to

$$\Gamma''(0,t;\lambda) - \frac{m-1}{t}\Gamma'(0,t;\lambda) - \frac{\lambda m}{2} t^{m-1}\Gamma(0,t;\lambda) = 0, \tag{3.41}$$

$$\lim_{t\to 0}\frac{\Gamma'(0,t;\lambda)}{t^{m-1}} = \frac{m}{2}, \quad \Gamma'(0,1;\lambda) = m\Gamma(0,1;\lambda). \tag{3.42}$$

The differential equation (3.41) is Bessel's equation defined in (2.123) with

$$\alpha = \frac{m}{2}, \quad \beta = \frac{\sqrt{-2\lambda m}}{m+1}, \quad \gamma = \frac{m+1}{2}, \quad \nu = \frac{m}{m+1}.$$

The general solution is given by

$$\begin{aligned}\Gamma(0,t;\lambda) &= t^{m/2}\left(c_1 J_\nu\left(\beta t^\gamma\right) + c_2 J_{-\nu}\left(\beta t^\gamma\right)\right)\\ &= c_1 t^m \left(\frac{\beta}{2}\right)^\nu \left\{\frac{1}{\Gamma(\nu+1)} - \frac{\beta^2}{4\Gamma(\nu+2)} t^{m+1} + O\left(t^{2(m+1)}\right)\right\}\\ &\quad + c_2 \left(\frac{\beta}{2}\right)^{-\nu} \left\{\frac{1}{\Gamma(-\nu+1)} - \frac{\beta^2}{4\Gamma(-\nu+2)} t^{m+1} + O\left(t^{2(m+1)}\right)\right\}.\end{aligned}$$

Thus we have $\Gamma(0,0;\lambda) = c_2(\beta/2)^{-\nu}/\Gamma(-\nu+1)$. It follows from the first boundary condition in (3.42) that

$$\lim_{t\to 0}\frac{\Gamma'(0,t;\lambda)}{t^{m-1}} = c_1\left(\frac{\beta}{2}\right)^\nu \frac{m}{\Gamma(\nu+1)} = \frac{m}{2}, \quad c_1 = \frac{\Gamma(\nu+1)}{2}\left(\frac{\beta}{2}\right)^{-\nu}.$$

Using the relation $J'_\nu(z) = (J_{\nu-1}(z) - J_{\nu+1}(z))/2$, $\Gamma'(0,t;\lambda)$ is equal to

$$\begin{aligned}&c_1\left(\frac{m}{2}t^{m/2-1}J_\nu(\beta t^\gamma) + t^{m/2}J'_\nu(\beta t^\gamma)\right)\\ &\quad + c_2\left(\frac{m}{2}t^{m/2-1}J_{-\nu}(\beta t^\gamma) + t^{m/2}J'_{-\nu}(\beta t^\gamma)\right)\\ &= c_1\left(\frac{m}{2}t^{m/2-1}J_\nu(\beta t^\gamma) + t^{m-1/2}\frac{\sqrt{-2\lambda m}}{4}\left(J_{\nu-1}(\beta t^\gamma) - J_{\nu+1}(\beta t^\gamma)\right)\right)\\ &\quad + c_2\left(\frac{m}{2}t^{m/2-1}J_{-\nu}(\beta t^\gamma) + t^{m-1/2}\frac{\sqrt{-2\lambda m}}{4}\left(J_{-\nu-1}(\beta t^\gamma) - J_{-\nu+1}(\beta t^\gamma)\right)\right).\end{aligned}$$

Then it follows from the relation $J_\nu(z) = z(J_{\nu-1}(z) + J_{\nu+1}(z))/(2\nu)$ that $\Gamma'(0,1;\lambda) - m\Gamma(0,1;\lambda)$ is equal to

$$\begin{aligned}
&c_1\left(-\frac{m}{2}J_\nu(\beta) + \frac{\sqrt{-2\lambda m}}{4}(J_{\nu-1}(\beta) - J_{\nu+1}(\beta))\right)\\
&+c_2\left(-\frac{m}{2}J_{-\nu}(\beta) + \frac{\sqrt{-2\lambda m}}{4}(J_{-\nu-1}(\beta) - J_{-\nu+1}(\beta))\right)\\
&= -\frac{\sqrt{-2\lambda m}}{2}(c_1 J_{\nu+1}(\beta) - c_2 J_{-\nu-1}(\beta)).
\end{aligned}$$

Thus we obtain, from the second boundary condition in (3.42),

$$c_2 = c_1\frac{J_{\nu+1}(\beta)}{J_{-\nu-1}(\beta)} = \frac{\Gamma(\nu+1)}{2}\left(\frac{\beta}{2}\right)^{-\nu}\frac{J_{\nu+1}(\beta)}{J_{-\nu-1}(\beta)}$$

and

$$\begin{aligned}
\Gamma(0,0;\lambda) &= c_2\left(\frac{\beta}{2}\right)^{-\nu}\frac{1}{\Gamma(-\nu+1)}\\
&= \frac{\Gamma(\nu+1)}{2\Gamma(-\nu+1)}\left(\frac{\beta}{2}\right)^{-2\nu}\frac{J_{\nu+1}(\beta)}{J_{-\nu-1}(\beta)}.
\end{aligned} \tag{3.43}$$

Finally, the c.f. of S_2 in (3.40) is given from Theorem 3.8 by

$$\begin{aligned}
\mathrm{E}\left(e^{i\theta S_2}\right) &= (D(2i\theta))^{-1/2}\exp\left\{ia^2\theta\Gamma(0,0;2i\theta) - \frac{i\theta}{2(m+1)}\right\}\\
&= \left[\Gamma(-\nu)J_{-\nu-1}(\eta)\Big/\left(\frac{\eta}{2}\right)^{-\nu-1}\right]^{-1/2}\\
&\quad\times\exp\left[\frac{ia^2\theta}{2}\frac{\Gamma(\nu+1)}{\Gamma(-\nu+1)}\frac{J_{\nu+1}(\eta)}{J_{-\nu-1}(\eta)}\left(\frac{\eta}{2}\right)^{-2\nu} - \frac{i\theta}{2(m+1)}\right],
\end{aligned} \tag{3.44}$$

where $\nu = m/(m+1)$ and $\eta = \sqrt{-4i\theta m}/(m+1)$. We have

$$\begin{aligned}
\mathrm{E}(S_2) &= \int_0^1 K(t,t)\,dt - \frac{1}{2(m+1)} = 0,\\
\mathrm{Var}(S_2) &= 4\int_0^1\left(\int_0^t K^2(s,t)\,ds\right)dt + 4a^2\int_0^1 K^2(0,t)\,dt\\
&= \frac{1}{2(m+1)} + \frac{a^2}{2m+1},
\end{aligned}$$

and it holds that, as $|a| \to \infty$, $S_2 \Rightarrow \mathrm{N}(0, a^2/(2m+1))$.

Exercises for Section 3.3

Exercise 3.3.1 Show that the integral equation (3.32) with the kernel

$$K(s,t) = \frac{1}{2m+1}\left(1 - (\max(s,t))^{2m+1}\right),$$

where $m > -1/2$, is equivalent to

$$\Gamma''(0,t;\lambda) - \frac{2m}{t}\Gamma'(0,t;\lambda) + \lambda t^{2m}\Gamma(0,t;\lambda) = 0$$

with the boundary conditions

$$\Gamma(0,1;\lambda) = 0, \quad \lim_{t\to 0}\frac{\Gamma'(0,t;\lambda)}{t^{2m}} = -1.$$

Exercise 3.3.2 Suppose that the statistic S_2 has the c.f.

$$\begin{aligned} \mathrm{E}\left(e^{i\theta S_2}\right) &= \left[\Gamma(-\nu+1)\,J_{-\nu}(\eta)\Big/\left(\frac{\eta}{2}\right)^{-\nu}\right]^{-1/2} \\ &\quad\times \exp\left[\frac{ia^2\theta\,\Gamma(\nu+1)}{(2m+1)\,\Gamma(-\nu+1)}\,\frac{J_\nu(\eta)}{J_{-\nu}(\eta)}\left(\frac{\eta}{2}\right)^{-2\nu}\right], \end{aligned}$$

where $\nu = (2m+1)/2(m+1)$ and $\eta = \sqrt{2i\theta}/(m+1)$. Prove that, as $|a| \to \infty$,

$$S_2 \Rightarrow \mathrm{N}\left(\frac{a^2}{2m+1}, \frac{4a^2}{(m+1)(4m+3)}\right).$$

Exercise 3.3.3 Show that the integral equation (3.32) with the kernel

$$K(s,t) = \frac{1}{2}(\max(s,t))^m,$$

where $m > 0$, is equivalent to

$$\Gamma''(0,t;\lambda) - \frac{m-1}{t}\Gamma'(0,t;\lambda) - \frac{\lambda m}{2}t^{m-1}\Gamma(0,t;\lambda) = 0,$$

$$\lim_{t\to 0}\frac{\Gamma'(0,t;\lambda)}{t^{m-1}} = \frac{m}{2}, \quad \Gamma'(0,1;\lambda) = m\Gamma(0,1;\lambda).$$

3.4 Computation with the Resolvent – Case 3

In this section we deal with the statistic given by

$$S_3 = \int_0^1 (Y(t) + Zm(t))^2\,dt, \tag{3.45}$$

where $\{Y(t)\}$ is a zero-mean Gaussian process, $m(t)$ is a continuous and deterministic function, and Z follows $\mathrm{N}(0,1)$ and is independent of $\{Y(t)\}$.

To derive the c.f. of S_3, we could follow the approach taken in Section 3.2 by considering (Exercise 3.4.1)

$$S_3 \stackrel{\mathcal{D}}{=} \int_0^1 \left\{ \sum_{n=1}^{\infty} \frac{f_n(t)}{\sqrt{\lambda_n}} Z_n + Zm(t) \right\}^2 dt$$
$$= \sum_{n=1}^{\infty} \left(\frac{1}{\lambda_n} Z_n^2 + \frac{2c_n}{\sqrt{\lambda_n}} Z_n Z \right) + Z^2 \int_0^1 m^2(t)\, dt,$$

where

$$c_n = \int_0^1 f_n(t) m(t)\, dt.$$

However, it is more convenient to use the conditional argument. Given Z, it is recognized that S_3 is essentially the same as S_1. Then it follows from Theorem 3.2 that, given Z, the c.f. of S_3 is given by

$$\mathrm{E}\left(e^{i\theta S_3} \middle| Z \right) = (D(\lambda))^{-1/2}$$
$$\times \exp\left[Z^2 \left\{ \frac{\lambda}{2} \int_0^1 m^2(t)\, dt + \frac{\lambda^2}{2} \int_0^1 h(t;\lambda) m(t)\, dt \right\} \right],$$

where $\lambda = 2i\theta$.

Taking expectations with respect to Z on both sides, we have the following theorem.

Theorem 3.12 *Let us consider*

$$S_3 = \int_0^1 (Y(t) + Zm(t))^2\, dt,$$

where $\{Y(t)\}$ is a zero-mean Gaussian process, $m(t)$ is a continuous and deterministic function, and Z follows $N(0,1)$ and is independent of $\{Y(t)\}$. Then the c.f. of S_3 is given by

$$E\left(e^{i\theta S_3} \right) = (D(\lambda))^{-1/2} \left(1 - \lambda \int_0^1 m^2(t)\, dt - \lambda^2 \int_0^1 h(t;\lambda) m(t)\, dt \right)^{-1/2},$$

where $\lambda = 2i\theta$, $D(\lambda)$ is the FD of $K(s,t) = Cov(Y(s), Y(t))$, and $h(t;\lambda)$ is the solution of the equation in (3.16).

It should be noted that asymptotic normality of S_3 cannot be expected, unlike the statistics S_1 and S_2 dealt with in the previous sections. The mean and variance of S_3 can be computed using

Corollary 3.12 *For the statistic S_3 defined in Theorem 3.12, we have*

$$\mathrm{E}(S_3) = \int_0^1 K(t,t)\,dt + \int_0^1 m^2(t)\,dt,$$

$$\begin{aligned}\mathrm{Var}(S_3) &= 4\int_0^1\int_0^t \left(K^2(s,t) + 2m(s)m(t)K(s,t)\right)\,ds\,dt + 2\left(\int_0^1 m^2(t)\,dt\right)^2\\ &= 2(d_1^2 - d_2) + 8\int_0^1\int_0^t m(s)m(t)K(s,t)\,ds\,dt + 2\left(\int_0^1 m^2(t)\,dt\right)^2,\end{aligned}$$

where d_1 and d_2 are coefficients in the series expansion for $D(\lambda)$ expressed as $1 - d_1\lambda + d_2\lambda^2/2 - \cdots$.

We present below Examples 3.13, 3.14, and 3.15 for computing the c.f. of S_3 together with its mean and variance.

Example 3.13 Let us first deal with

$$S_3 = \int_0^1 (W(t) + Z(a+bt))^2\,dt. \tag{3.46}$$

It follows (Exercise 3.4.2) from (3.20) and Theorem 3.12 that

$$\begin{aligned}\mathrm{E}\left(e^{i\theta S_3}\right) &= \left(\cos\sqrt{\lambda}\right)^{-1/2}\left(1 + b(2a+b) - 2ab\sec\sqrt{\lambda}\right.\\ &\qquad\left. -(a^2\lambda + b^2)\frac{\tan\sqrt{\lambda}}{\sqrt{\lambda}}\right)^{-1/2}\\ &= \left((1 + b(2a+b))\cos\sqrt{\lambda} - 2ab - (a^2\lambda + b^2)\frac{\sin\sqrt{\lambda}}{\sqrt{\lambda}}\right)^{-1/2},\end{aligned}$$

where $\lambda = 2i\theta$. Putting $K(s,t) = \min(s,t)$ and $m(t) = a + bt$, we have

$$\mathrm{E}(S_3) = \int_0^1 K(t,t)\,dt + \int_0^1 m^2(t)\,dt = \frac{1}{2} + a^2 + ab + \frac{1}{3}b^2,$$

$$\mathrm{Var}(S_3) = \frac{1}{3} + \frac{1}{15}\left(20a^2 + 25ab + 8b^2\right) + 2\left(a^2 + ab + \frac{1}{3}b^2\right)^2.$$

Figure 3.2 draws probability densities of $(S_3 - \mathrm{E}(S_3))/\sqrt{\mathrm{Var}(S_3)}$ for various values of a and b. It is seen that the standardized distributions are far from standard normal.

Example 3.14 Consider

$$S_3 = \int_0^1 (Y(t) + Z(a+bt))^2\,dt, \tag{3.47}$$

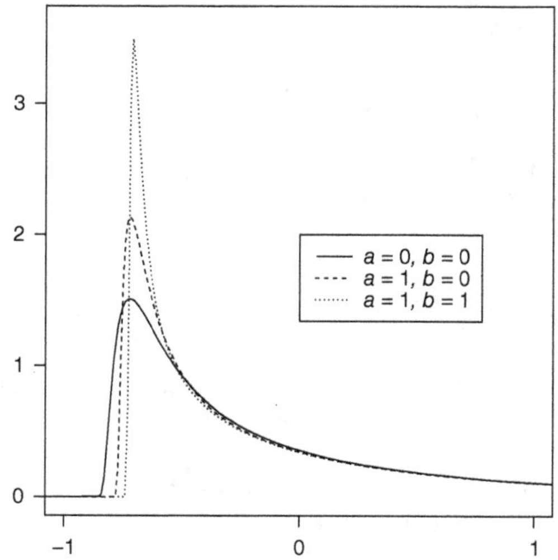

Figure 3.2 Probability densities of the normalized S_3 in (3.46).

where $\{Y(t)\}$ is the Bb. It follows from (3.22) and Theorem 3.12 that

$$\begin{aligned}
\mathrm{E}\left(e^{i\theta S_3}\right) &= \left(\frac{\sin\sqrt{\lambda}}{\sqrt{\lambda}}\right)^{-1/2}\Big(1 - b^2 - 2a(a+b)\sqrt{\lambda}\,\mathrm{cosec}\sqrt{\lambda} \\
&\qquad + \sqrt{\lambda}\left(2a(a+b)+b^2\right)\cot\sqrt{\lambda}\Big)^{-1/2} \\
&= \left((1-b^2)\frac{\sin\sqrt{\lambda}}{\sqrt{\lambda}} - 2a(a+b) + \left(2a(a+b)+b^2\right)\cos\sqrt{\lambda}\right)^{-1/2},
\end{aligned}$$

where $\lambda = 2i\theta$. Putting $K(s,t) = \min(s,t) - st$ and $m(t) = a + bt$, we have

$$\begin{aligned}
\mathrm{E}(S_3) &= \frac{1}{6} + a^2 + ab + \frac{1}{3}b^2, \\
\mathrm{Var}(S_3) &= \frac{1}{45} + \frac{1}{45}\left(15a^2 + 15ab + 4b^2\right) + 2\left(a^2 + ab + \frac{1}{3}b^2\right)^2.
\end{aligned}$$

Example 3.15 Consider

$$S_3 = \int_0^1 \left(X(t) + aZe^{\gamma t}\right)^2 dt, \tag{3.48}$$

where $\{X(t)\}$ is the O–U process defined by

$$dX(t) = \gamma X(t)\,dt + dW(t), \quad X(0) = 0 \quad \Leftrightarrow \quad X(t) = e^{\gamma t}\int_0^t e^{-\gamma s}\,dW(s).$$

Assuming that $\gamma < 0$ and $a = 1/\sqrt{-2\gamma}$,

$$Y(t) = X(t) + aZe^{\gamma t} = aZe^{\gamma t} + e^{\gamma t} \int_0^t e^{-\gamma s}\, dW(s)$$

is a zero-mean stationary O–U process with $\mathrm{Cov}(Y(s), Y(t)) = e^{\gamma|s-t|}/(-2\gamma)$. It follows (Exercise 3.4.3) from (3.24) and Theorem 3.12 that

$$\begin{aligned} \mathrm{E}\left(e^{i\theta S_3}\right) &= e^{-\gamma/2}\left(\cos\nu - \gamma\frac{\sin\nu}{\nu}\right)^{-1/2}\left(1 - \frac{2ia^2\theta\sin\nu/\nu}{\cos\nu - \gamma\sin\nu/\nu}\right)^{-1/2} \\ &= e^{-\gamma/2}\left(\cos\nu - \left(\gamma - \frac{i\theta}{\gamma}\right)\frac{\sin\nu}{\nu}\right)^{-1/2}, \end{aligned}$$

where $\nu = \sqrt{2i\theta - \gamma^2}$. Putting $K(s,t) = e^{\gamma|s-t|}/(-2\gamma)$ and $m(t) = 0$, we have

$$\begin{aligned} \mathrm{E}(S_3) &= \int_0^1 K(t,t)\, dt = -\frac{1}{2\gamma}, \\ \mathrm{Var}(S_3) &= 4\int_0^1\int_0^t K^2(s,t)\, ds\, dt = \frac{1}{4\gamma^4}\left(e^{2\gamma} - 2\gamma - 1\right). \end{aligned}$$

The result in this example can also be obtained by the approach described in the next section and will be taken up in Example 3.17.

Exercises for Section 3.4

Exercise 3.4.1 Define, for a zero-mean Gaussian process $\{Y(t)\}$,

$$\begin{aligned} S_3 &= \int_0^1 (Y(t) + Zm(t))^2\, dt \\ &\overset{\mathcal{D}}{=} \int_0^1 \left\{\sum_{n=1}^{\infty} \frac{f_n(t)}{\sqrt{\lambda_n}} Z_n + Zm(t)\right\}^2 dt \\ &= \sum_{n=1}^{\infty}\left(\frac{1}{\lambda_n} Z_n^2 + \frac{2c_n}{\sqrt{\lambda_n}} Z_n Z\right) + Z^2\int_0^1 m^2(t)\, dt, \end{aligned}$$

where Z follows N(0, 1) and is independent of $\{Y(t)\}$, $m(t)$ is a continuous and deterministic function, and

$$c_n = \int_0^1 f_n(t) m(t)\, dt.$$

Derive the c.f. of S_3 by considering the c.f. of

$$S_{N3} = \sum_{n=1}^{N}\left(\frac{1}{\lambda_n} Z_n^2 + \frac{2c_n}{\sqrt{\lambda_n}} Z_n Z\right) + Z^2\int_0^1 m^2(t)\, dt.$$

Exercise 3.4.2 Using (3.20) and Theorem 3.12, show that the c.f. of

$$S_3 = \int_0^1 (W(t) + Z(a + bt))^2 \, dt$$

is given by

$$\mathrm{E}\left(e^{i\theta S_3}\right) = \left((1 + b(2a + b))\cos\sqrt{\lambda} - 2ab - (a^2\lambda + b^2)\frac{\sin\sqrt{\lambda}}{\sqrt{\lambda}}\right)^{-1/2},$$

where $\lambda = 2i\theta$.

Exercise 3.4.3 Consider

$$S_3 = \int_0^1 \left(X(t) + aZe^{\gamma t}\right)^2 dt,$$

where $a = 1/\sqrt{-2\gamma}$ ($\gamma < 0$) and $\{X(t)\}$ is the O–U process given by $dX(t) = \gamma X(t)\, dt + dW(t)$ with $X(0) = 0$. Using (3.24) and Theorem 3.12, show that the c.f. of S_3 is given by

$$\mathrm{E}\left(e^{i\theta S_3}\right) = e^{-\gamma/2}\left(\cos\nu - \left(\gamma - \frac{i\theta}{\gamma}\right)\frac{\sin\nu}{\nu}\right)^{-1/2},$$

where $\nu = \sqrt{2i\theta - \gamma^2}$.

3.5 Computation with the Resolvent – Case 4

In this section we deal with the statistic given by

$$S_4 = \int_0^1 \int_0^1 K(s,t)\, dW(s)\, dW(t) + 2aZ \int_0^1 K(0,t)\, dW(t) + bZ^2, \quad (3.49)$$

where $K(s,t)$ is symmetric, continuous, and nearly definite; a and b are constants; and Z follows N(0, 1) and is independent of $\{W(t)\}$. The c.f. of S_4 could be obtained (Exercise 3.5.1) by the approach taken in Section 3.3 by considering

$$S_4 \stackrel{\mathcal{D}}{=} \sum_{n=1}^{\infty} \frac{1}{\lambda_n}\left(Z_n^2 + 2af_n(0)Z_nZ\right) + bZ^2.$$

However, it can be easily obtained from Theorem 3.8 by the conditional argument. In fact, we have

$$\mathrm{E}\left(e^{i\theta S_4}\middle| Z\right) = (D(\lambda))^{-1/2}\exp\left[Z^2\frac{\lambda}{2}\left\{a^2\left(\Gamma(0,0;\lambda) - K(0,0)\right) + b\right\}\right],$$

where $\lambda = 2i\theta$. Taking expectations with respect to Z on both sides, we can establish the following theorem (Tanaka, 1990a).

Theorem 3.16 *Let us consider*

$$S_4 = \int_0^1 \int_0^1 K(s,t)\, dW(s)\, dW(t) + 2aZ \int_0^1 K(0,t)\, dW(t) + bZ^2,$$

where $K(s,t)$ is symmetric, continuous, and nearly definite; a and b are constants; and Z follows $N(0,1)$ and is independent of $\{W(t)\}$. Then the c.f. of S_4 is given by

$$E\left(e^{i\theta S_4}\right) = (D(\lambda))^{-1/2} \left(1 - \lambda a^2 \left(\Gamma(0,0;\lambda) - K(0,0)\right) - \lambda b\right)^{-1/2}, \quad (3.50)$$

where $\lambda = 2i\theta$, $D(\lambda)$ is the FD of $K(s,t)$, and $\Gamma(0,0;\lambda)$ is the solution $\Gamma(0,t;\lambda)$ to the equation in (3.32) evaluated at $t = 0$.

Asymptotic normality of S_4 cannot be expected. The mean and variance of S_4 can be computed by

Corollary 3.16 *For the statistic S_4 defined in Theorem 3.16, we have*

$$\mathrm{E}(S_4) = \int_0^1 K(t,t)\, dt + b,$$

$$\mathrm{Var}(S_4) = 4 \int_0^1 \int_0^t K^2(s,t)\, ds\, dt + 4a^2 \int_0^1 K^2(0,t)\, dt + 2b^2.$$

Examples 3.17, 3.18, and 3.19 are presented below for computing the c.f. of S_4 together with $\mathrm{E}(S_4)$ and $\mathrm{Var}(S_4)$.

Example 3.17 Let $\{Y(t)\}$ be the O–U process defined by $dY(t) = \gamma Y(t)\, dt + dW(t)$ with $\gamma < 0$ and $Y(0) = Z/\sqrt{-2\gamma} \sim \mathrm{N}(0, -1/(2\gamma))$. Then $\{Y(t)\}$ becomes a zero-mean stationary O–U process. Let us consider, by putting $a = 1/\sqrt{-2\gamma}$,

$$\begin{aligned} S_4 &= \int_0^1 Y^2(t) dt = \int_0^1 \left\{ e^{\gamma t} \int_0^t e^{-\gamma s}\, dW(s) + aZe^{\gamma t} \right\}^2 dt \\ &= \int_0^1 \int_0^1 K(s,t)\, dW(s)\, dW(t) + 2aZ \int_0^1 K(0,t)\, dW(t) \\ &\quad + a^2 K(0,0) Z^2, \end{aligned} \quad (3.51)$$

where $K(s,t) = \left(e^{\gamma(2-s-t)} - e^{\gamma|s-t|}\right)/(2\gamma)$ and $K(0,t) = (e^{\gamma(2-t)} - e^{\gamma t})/(2\gamma)$. It follows from (3.34) and Theorem 3.16 that, by putting $b = a^2 K(0,0) = (1 - e^{2\gamma})/(4\gamma^2)$,

$$
\begin{aligned}
\mathrm{E}\left(e^{i\theta S_4}\right) &= (D(2i\theta))^{-1/2}\left(1-2ia^2\theta\Gamma(0,0;2i\theta)\right)^{-1/2} \\
&= e^{-\gamma/2}\left[\left(\cos\nu-\gamma\,\frac{\sin\nu}{\nu}\right)\left(1-\frac{2ia^2\theta\,(\sin\nu/\nu)}{\cos\nu-\gamma\,(\sin\nu/\nu)}\right)\right]^{-1/2} \\
&= e^{-\gamma/2}\left(\cos\nu-\left(\gamma-\frac{i\theta}{\gamma}\right)\frac{\sin\nu}{\nu}\right)^{-1/2},
\end{aligned}
$$

where $\nu = \sqrt{2i\theta-\gamma^2}$. This result, of course, is the same as that in Example 3.15, and it can also be derived from (3.34) by the conditional argument (Exercise 3.5.2).

We also have

$$
\begin{aligned}
\mathrm{E}(S_4) &= \int_0^1 \frac{e^{\gamma(2-2t)}-1}{2\gamma}\,dt + \frac{1-e^{2\gamma}}{4\gamma^2} = -\frac{1}{2\gamma}, \\
\mathrm{Var}(S_4) &= 4\int_0^1\int_0^t K^2(s,t)\,ds\,dt - \frac{2}{\gamma}\int_0^1 K^2(0,t)\,dt + \frac{1}{2\gamma^2}K^2(0,0) \\
&= \frac{1}{4\gamma^4}\left(e^{2\gamma}-2\gamma-1\right).
\end{aligned}
$$

Example 3.18 Let us consider

$$
\begin{aligned}
S_4 &= \int_0^1 t^{2m}\,(W(t)+aZ)^2\,dt \\
&= \int_0^1\int_0^1 \frac{1}{2m+1}\left[1-(\max(s,t))^{2m+1}\right]dW(s)\,dW(t) \\
&\quad + 2aZ\int_0^1 \frac{1-t^{2m+1}}{2m+1}\,dW(t) + \frac{a^2}{2m+1}Z^2, \qquad (3.52)
\end{aligned}
$$

where $m > -1/2$. This statistic is an extended version of S_2 in (3.35) of Example 3.10, where the constant a is replaced by the random variable aZ. Thus we have, using (3.39) and the conditional argument,

$$
\begin{aligned}
\mathrm{E}\left(e^{i\theta S_4}\middle| Z\right) &= \left[\Gamma(-\nu+1)\,J_{-\nu}(\eta)\Big/\left(\frac{\eta}{2}\right)^{-\nu}\right]^{-1/2} \\
&\quad\times\exp\left[\frac{ia^2Z^2\theta\,\Gamma(\nu+1)}{(2m+1)\,\Gamma(-\nu+1)}\,\frac{J_\nu(\eta)}{J_{-\nu}(\eta)}\left(\frac{\eta}{2}\right)^{-2\nu}\right],
\end{aligned}
$$

where $\nu = (2m+1)/2(m+1)$ and $\eta = \sqrt{2i\theta}/(m+1)$. Then, taking expectations with respect to Z on both sides, we have

$$
\begin{aligned}
\mathrm{E}\left(e^{i\theta S_4}\right) &= \left[\Gamma(-\nu+1)\, J_{-\nu}(\eta) \Big/ \left(\frac{\eta}{2}\right)^{-\nu}\right]^{-1/2} \\
&\quad \times \left(1 - \frac{2ia^2\theta\Gamma(\nu+1)}{(2m+1)\Gamma(-\nu+1)} \frac{J_\nu(\eta)}{J_{-\nu}(\eta)} \left(\frac{\eta}{2}\right)^{-2\nu}\right)^{-1/2} \\
&= \left(\frac{\Gamma(-\nu+1)\, J_{-\nu}(\eta)}{(\eta/2)^{-\nu}} - \frac{2ia^2\theta\Gamma(\nu+1) J_\nu(\eta)}{(2m+1)(\eta/2)^{\nu}}\right)^{-1/2}. \qquad (3.53)
\end{aligned}
$$

This result can also be obtained from Theorem 3.16 (Exercise 3.5.3).

Putting $K(s,t) = (1 - (\max(s,t))^{2m+1})/(2m+1)$ and $b = a^2/(2m+1)$, it follows from Corollary 3.16 that

$$
\begin{aligned}
\mathrm{E}(S_4) &= \int_0^1 K(t,t)\, dt + b = \frac{1}{2(m+1)} + \frac{a^2}{2m+1}, \\
\mathrm{Var}(S_4) &= 4\int_0^1 \int_0^t K^2(s,t)\, ds\, dt + 4a^2 \int_0^1 K^2(0,t)\, dt + 2b^2 \\
&= \frac{1}{(m+1)(2m+3)} + \frac{4a^2}{(m+1)(4m+3)} + \frac{2a^4}{(2m+1)^2}.
\end{aligned}
$$

Example 3.19 Consider

$$
\begin{aligned}
S_4 &= \int_0^1 t^m (W(t) + aZ)\, dW(t) \\
&= \int_0^1 \int_0^1 K(s,t)\, dW(s)\, dW(t) + 2aZ \int_0^1 K(0,t)\, dW(t) \\
&\quad - \frac{1}{2(m+1)}, \qquad (3.54)
\end{aligned}
$$

where $m > 0$ and $K(s,t) = (\max(s,t))^m/2$. This statistic contains the Itô integral and is an extended version of S_2 in (3.40) of Example 3.11. Using the conditional argument, we obtain $\mathrm{E}\left(e^{i\theta S_4} \middle| Z\right)$ from (3.44) by replacing a by aZ. Taking expectations with respect to Z, we have (Exercise 3.5.4)

$$
\begin{aligned}
\mathrm{E}\left(e^{i\theta S_4}\right) &= \mathrm{E}\left[\mathrm{E}\left(e^{i\theta S_4} \middle| Z\right)\right] \qquad (3.55) \\
&= \left[\Gamma(-\nu) J_{-\nu-1}(\eta) \Big/ \left(\frac{\eta}{2}\right)^{-\nu-1}\right]^{-1/2} \exp\left\{-\frac{i\theta}{2(m+1)}\right\} \\
&\quad \times \left(1 - ia^2\theta\, \frac{\Gamma(\nu+1)}{\Gamma(-\nu+1)} \frac{J_{\nu+1}(\eta)}{J_{-\nu-1}(\eta)} \left(\frac{\eta}{2}\right)^{-2\nu}\right)^{-1/2} \\
&= \left[\left(\frac{\Gamma(-\nu) J_{-\nu-1}(\eta)}{(\eta/2)^{-\nu-1}} + \frac{ia^2\theta\Gamma(\nu+1) J_{\nu+1}(\eta)}{\nu(\eta/2)^{\nu-1}}\right) e^{i\theta/(m+1)}\right]^{-1/2},
\end{aligned}
$$

where $\nu = m/(m+1)$ and $\eta = \sqrt{-4i\theta m}/(m+1)$.

It holds that

$$\mathrm{E}(S_4) = \int_0^1 K(t,t)\,dt - \frac{1}{2(m+1)} = 0,$$

$$\begin{aligned}\mathrm{Var}(S_4) &= 4\int_0^1\int_0^t K^2(s,t)\,ds\,dt + 4a^2\int_0^1 K^2(0,t)\,dt\\ &= \frac{1}{2(m+1)} + \frac{a^2}{2m+1},\end{aligned}$$

and it is seen that the first two moments of S_2 in (3.40) and S_4 in (3.54) coincide, although their distributions are different from each other.

Exercises for Section 3.5

Exercise 3.5.1 Define

$$\begin{aligned}S_4 &= \int_0^1\int_0^1 K(s,t)\,dW(s)\,dW(t) + 2aZ\int_0^1 K(0,t)\,dW(t) + bZ^2\\ &\stackrel{\mathcal{D}}{=} \sum_{n=1}^{\infty}\frac{1}{\lambda_n}\left(Z_n^2 + 2af_n(0)Z_nZ\right) + bZ^2,\end{aligned}$$

where $K(s,t)$ is symmetric, continuous, and nearly definite; a and b are constants; and Z follows N(0, 1) and is independent of $\{W(t)\}$. Derive the c.f. of S_4 by considering the c.f. of

$$S_{N4} = \sum_{n=1}^{N}\frac{1}{\lambda_n}\left(Z_n^2 + 2af_n(0)Z_nZ\right) + bZ^2.$$

Exercise 3.5.2 Let $\{Y(t)\}$ be a stationary O–U process whose differential is given by $dY(t) = \gamma Y(t)\,dt + dW(t)$ with $\gamma < 0$ and $Y(0) = Z/\sqrt{-2\gamma}$, where Z follows N(0, 1) and is independent of $\{W(t)\}$. Using the result in (3.34) and the conditional argument, derive the c.f. of

$$S_4 = \int_0^1 Y^2(t)\,dt.$$

Exercise 3.5.3 Using Theorem 3.16 and the result in (3.38), derive the c.f. of

$$S_4 = \int_0^1 t^{2m}\left(W(t) + aZ\right)^2\,dt,$$

where $m > -1/2$, and Z follows N(0, 1) and is independent of $\{W(t)\}$.

Exercise 3.5.4 Using Theorem 3.16 and the result in (3.44), derive the c.f. of

$$S_4 = \int_0^1 t^m \, (W(t) + aZ) \, dW(t),$$

where $m > 0$, and Z follows N(0, 1) and is independent of $\{W(t)\}$.

Part II

Applications

Part II is composed of Chapters 4 through 8 and attempts to apply our theory discussed in Part I to various statistical problems. Four topics are presented, concentrating on one particular topic in each chapter. The selected topics are concerned with testing or estimation problems in various models ranging from the i.i.d. population to the time series situation.

Chapter 4 considers, under the i.i.d. population, goodness-of-fit tests for testing if the underlying distribution $G(x)$ follows a specified distribution $F(x;\boldsymbol{\xi})$ parameterized by a parameter vector $\boldsymbol{\xi}$. As such a test, we consider

$$H_0 : G(x) = F(x;\boldsymbol{\xi}_0) \quad \text{versus} \quad H_1 : G(x) = F(x;\boldsymbol{\xi}), \quad \boldsymbol{\xi} = \boldsymbol{\xi}_0 + \frac{\boldsymbol{\gamma}}{\sqrt{n}}, \tag{II.1}$$

where $\boldsymbol{\xi}_0$ is a specified vector of nuisance parameters, $\boldsymbol{\gamma}$ is any constant vector, and n is the sample size of observations taken from the i.i.d. population. It is noted here that H_1 is a local alternative close to the null H_0 on the order of $1/\sqrt{n}$. To test (II.1), we use the Cramér–von Mises–Smirnov statistic

$$W_n^2(\boldsymbol{\xi}_0) = n\int_{-\infty}^{\infty}\Big(F_n(x) - F(x;\boldsymbol{\xi}_0)\Big)^2 dF(x;\boldsymbol{\xi}_0), \tag{II.2}$$

where $F_n(x)$ is the empirical distribution function. Then it follows from Durbin (1973) that, as $n \to \infty$ under H_1,

$$W_n^2(\boldsymbol{\xi}_0) \Rightarrow W^2(\boldsymbol{\xi}_0) = \int_0^1 \big(Y(t) + \boldsymbol{\gamma}'\boldsymbol{g}_{\xi_0}(t)\big)^2\, dt, \tag{II.3}$$

where $Y(t)$ is a zero-mean Gaussian process with $\mathrm{Cov}(Y(s), Y(t)) = \min(s,t) - st$, and $\boldsymbol{g}_{\xi_0}(t)$ is a vector-valued continuous and deterministic function. The c.f. of $W^2(\boldsymbol{\xi}_0)$ can be obtained from the computational method associated with the resolvent discussed in Chapter 3. Other hypotheses are also discussed where a subset of nuisance parameters is estimated from the sample and the remaining parameters are specified at some values.

Chapter 5 deals with state space models where one of the parameters in the measurement equation is suspected to change over time. To check this we consider a model

$$y_j = \beta_j x_j + \boldsymbol{\gamma}' \boldsymbol{z}_j + \varepsilon_j, \quad \beta_j = \beta_{j-1} + \eta_j \qquad (j = 1, \ldots, n), \qquad \text{(II.4)}$$

where

(i) regressor variables $\{x_j\}$ and $\{\boldsymbol{z}_j\}$ are scalar and $p \times 1$ nonstochastic sequences, respectively;

(ii) $\begin{pmatrix} \varepsilon_j \\ \eta_j \end{pmatrix} \sim \text{NID}\left(\mathbf{0}, \begin{pmatrix} \sigma_\varepsilon^2 & 0 \\ 0 & \sigma_\eta^2 \end{pmatrix}\right)$;

(iii) $\boldsymbol{\gamma}$ is a $p \times 1$ unknown constant vector, and $\{\beta_j\}$ follows a random walk starting from an unknown constant β_0.

Our concern is to test if β_j is a constant over time, that is, $\beta_j = \beta_0$ for all j. This is equivalent to testing $\rho = \sigma_\eta^2/\sigma_\varepsilon^2 = 0$, which leads to the testing problem

$$H_0 : \rho = \frac{\sigma_\eta^2}{\sigma_\varepsilon^2} = 0 \quad \text{versus} \quad H_1 : \rho = \frac{c}{n^\kappa}, \qquad \text{(II.5)}$$

where c and κ are positive constants. There is no uniformly best test for the present problem. We use the locally best invariant (LBI) test, which rejects H_0 when $R_n = \partial L(\boldsymbol{\delta}) / \partial\rho\big|_{H_0}$ takes large values, where $L(\boldsymbol{\delta})$ is the log-likelihood for $\boldsymbol{y} = (y_1, \ldots, y_n)'$ in (II.4) with $\boldsymbol{\delta} = (\rho, \beta_0, \boldsymbol{\gamma}', \sigma_\varepsilon^2)'$, and the derivative is evaluated at the MLE under H_0. To examine the asymptotic distribution of R_n, we specify the regressors x_j and $\boldsymbol{z}_j$ in (II.4). For $x_j = 1$ and $\boldsymbol{z}_j = j$, it is shown that, as $n \to \infty$ under $\rho = c/n^2$,

$$\frac{1}{n^2} R_n \Rightarrow R(c) = \int_0^1 \int_0^1 \left(K(s,t) + cK^{(2)}(s,t)\right) dW(s)\, dW(t), \qquad \text{(II.6)}$$

where

$$K(s,t) = \min(s,t) - st - 3st(1-s)(1-t), \quad K^{(2)}(s,t) = \int_0^1 K(s,u)K(u,t)\, du. \qquad \text{(II.7)}$$

Moreover the c.f. of $R(c)$ in (II.6) is expressed as

$$\mathrm{E}\left(e^{i\theta R(c)}\right) = \left[D\left(i\theta + \sqrt{-\theta^2 + 2ic\theta}\right) D\left(i\theta - \sqrt{-\theta^2 + 2ic\theta}\right)\right]^{-1/2}, \qquad \text{(II.8)}$$

where $D(\lambda)$ is the FD of $K(s,t)$ in (II.7) given by

$$D(\lambda) = \frac{12}{\lambda^2}\left(2 - \sqrt{\lambda}\sin\sqrt{\lambda} - 2\cos\sqrt{\lambda}\right). \tag{II.9}$$

The other cases of regressors are also considered, and the limiting local powers of the LBI tests under H_1 in (II.5) are computed, where powers are compared with the conditional case where the initial value β_0 is set at 0.

Chapter 6 deals with the moving average (MA) model defined by

$$y_j = \boldsymbol{\beta}'\boldsymbol{x}_j + \varepsilon_j - \alpha\varepsilon_{j-1} \quad (j = 1, \ldots, n), \tag{II.10}$$

where $\boldsymbol{\beta}$ and $\boldsymbol{x}_j$ are coefficient and regression vectors, respectively whereas $\varepsilon_j\ (j = 0, 1, \ldots, n) \sim \mathrm{NID}(0, \sigma_\varepsilon^2)$. Our concern is to check the possibility of overdifferencing the series, which leads to testing

$$H_0 : \alpha = 1 \quad \text{versus} \quad H_1 : \alpha = 1 - \frac{c}{n^\kappa}, \tag{II.11}$$

where c and κ are positive constants. We employ a test that rejects H_0 when $S_n = \partial^2 L(\boldsymbol{\delta}) / \partial\alpha^2 \big|_{H_0}$ takes large values, where $\boldsymbol{\delta} = (\alpha, \boldsymbol{\beta}', \sigma_\varepsilon^2)'$, $L(\boldsymbol{\delta})$ is the log-likelihood for $\boldsymbol{y} = (y_1, \ldots, y_n)'$ in (II.10), and the second derivative is evaluated at the MLE under H_0. The test is shown to be LBI and unbiased (LBIU).

For $\boldsymbol{x}_j = 1$ in (II.10) it holds that, as $n \to \infty$ under $\alpha = 1 - c/n$,

$$\frac{1}{n^2} S_n \Rightarrow S(c) = \int_0^1\int_0^1 \left(K(s,t) + c^2 K^{(2)}(s,t)\right) dW(s)\, dW(t), \tag{II.12}$$

where $K(s,t)$ and $K^{(2)}(s,t)$ are the same as those defined in (II.7), and the c.f. of $S(c)$ is expressed as

$$\mathrm{E}\left(e^{i\theta S(c)}\right) = \left[D\left(i\theta + \sqrt{-\theta^2 + 2ic^2\theta}\right) D\left(i\theta - \sqrt{-\theta^2 + 2ic^2\theta}\right)\right]^{-1/2},$$

where $D(\lambda)$ is the FD of $K(s,t)$. It is recognized that the present LBIU test has a close relationship with the LBI test in the state space model (see (II.6), (II.7), (II.8), and (II.9)). The reason is described in the text. Other MA models with different regressors are also discussed, and limiting local powers are computed and compared with the conditional model where the initial value ε_0 that drives the error process is set to 0.

Chapter 7 deals with the autoregressive (AR) model defined by

$$y_j = \boldsymbol{\beta}'\boldsymbol{x}_j + u_j, \quad u_j = \rho u_{j-1} + \varepsilon_j \quad (j = 1, \ldots, n), \tag{II.13}$$

where $y_0 = 0$ and $\{\varepsilon_j\} \sim$ i.i.d.$(0, \sigma_\varepsilon^2)$. Our concern is to examine if the error term u_j exhibits nonstationary behavior after the regressor $\boldsymbol{x}_j$ is fitted to y_j, which leads to the unit root test

$$H_0 : \rho = 1 \quad \text{versus} \quad H_1 : \rho = 1 - \frac{c}{n^\kappa}, \tag{II.14}$$

where c and κ are positive constants. The most frequently used test is the LSE-based test, which employs the LSE $\hat{\rho}$ of ρ. For the simplest model where $\boldsymbol{x}_j = 0$, it holds that, as $n \to \infty$ under $\rho = 1 - c/n$,

$$n(\hat{\rho} - 1) \Rightarrow Q(c) = \int_0^1 Y(t)\, dY(t) \Bigg/ \int_0^1 Y^2(t)\, dt, \tag{II.15}$$

where $Y(t)$ is the Ornstein–Uhlenbeck (O–U) process given by $dY(t) = -cY(t)\, dt + dW(t)$ with $Y(0) = 0$. The null hypothesis is rejected when $Q(c)$ takes small values. The statistic $Q(c)$ takes the ratio form $Q(c) = U(c)/V(c)$, unlike the test statistics dealt with before. The same is true for other models with different regressor $\boldsymbol{x}_j$ in (II.13).

We can also consider the LBI and LBIU tests for the unit root. As an example, for $\boldsymbol{x}_j = (1, j)$, the LBIU test rejects H_0 when

$$T(c) = \int_0^1 (Y(t) - tY(1))^2\, dt \tag{II.16}$$

takes small values; its c.f. can be derived from our approach. Limiting local powers are compared between the LSE-based, LBI, and LBIU tests, which reveals that the LBI test behaves very badly, and the LBIU test is superior to the LSE-based test.

Chapter 7 also introduces another approach based on Girsanov's theorem to derive the c.f. associated with the AR unit root statistics and cointegration distributions.

Chapter 8 first extends the AR unit root distribution to the multiple unit root distribution given by

$$R(g) = \frac{1}{2} F_g^2(1) \Bigg/ \int_0^1 F_g^2(t)\, dt \quad (g = 1, 2, \ldots), \tag{II.17}$$

where $F_g(t)$ is g-fold integrated Bm defined by

$$F_g(t) = \int_0^t F_{g-1}(s)\, ds = \int_0^t \frac{(t-s)^g}{g!}\, dW(s) \quad (g = 1, 2, \ldots), \tag{II.18}$$

and the c.f. associated with $R(g)$ is derived for $g = 1, 2$.

The multiple unit root distribution is a prelude to the fractional unit root distribution, which replaces $F_g(t)$ in (II.17) by $B_H(t)$, fractional Bm (fBm). fBm $B_H(t)$ is a zero-mean Gaussian process with

$$\begin{aligned} K_H(s,t) &= \mathrm{Cov}(B_H(s), B_H(t)) \\ &= \frac{1}{2}\left[s^{2H} + t^{2H} - |s-t|^{2H}\right] \quad (1/2 < H < 1), \end{aligned} \tag{II.19}$$

where $B_H(t)$ reduces to Bm when $H = 1/2$. A striking feature of fBm is that increments are not independent, unlike Bm. The derivation of the associated c.f. of the fractional unit root distribution given by

$$Q_H = \frac{1}{2}\,B_H^2(1)\Big/\int_0^1 B_H^2(t)\,dt = \frac{1}{2}\,B_H^2(1)\Big/S_H \tag{II.20}$$

remains to be done, where the c.f. of $S_H = \int_0^1 B_H^2(t)\,dt$ might be expected to be easily derived, but this is also still unsolved.

An approximation to $B_H(t)$ is suggested using the fundamental martingale introduced by Norros, Valkeila, and Virtamo (1999), which leads to approximating $B_H(t)$ by

$$C_H(t) = \sqrt{2(1-H)}\,t^{2H-1}\int_0^t u^{1/2-H}\,dW(u). \tag{II.21}$$

Accordingly S_H is approximated by

$$\begin{aligned} T_H &= \int_0^1 C_H^2(t)\,dt \\ &= \frac{2(1-H)}{4H-1}\int_0^1\int_0^1 [1-(\max(s,t))]\,(st)^{1/2-H}\,dW(s)\,dW(t), \end{aligned}$$

and the statistic Q_H in (II.20) is approximated by

$$R_H = \frac{1}{2}\,C_H^2(1)\Big/\int_0^1 C_H^2(t)\,dt.$$

The c.f. of T_H and the c.f. associated with R_H can be obtained by our approach, where an interesting moment property is found to hold for R_H.

Note: Numerical integration is carried out in each chapter of Part II to compute distribution functions, percentage points, powers of tests, and so on, where the statistics take the form of purely quadratic functionals of Bm, quadratic plus linear or bilinear functionals of Bm, or a ratio of those functionals. For each type of statistic, we need an appropriate formula for numerical integration; these are summarized in Appendix B.

4

Fredholm Determinants for Goodness-of-Fit Tests

Goodness-of-fit tests were introduced in Section 1.5.1 to test if the underlying distribution follows the hypothesized distribution parameterized by a single parameter. In particular, two cases were discussed. One is the unspecified case where the parameter is a nuisance parameter to be estimated, and the other is the specified case where the parameter is set at some value under the null. In this chapter, we extend the discussions to the multi-parameter case where the hypothesized distribution depends on a vector of parameters. Then, following Durbin (1973), we deal with the situation where a subset of nuisance parameters is estimated from the sample and where the remaining parameters are specified at some values. Our main purpose here is to derive the c.f.s of asymptotic distributions of test statistics under this situation, for which our approach discussed in Chapters 2 and 3 is effectively used. At the same time we note the difficulty in deriving the c.f.s explicitly for the present problem.

4.1 Introduction

In Section 1.5.1 we dealt with goodness-of-fit tests, where we tested if the common continuous distribution follows a distribution parameterized by a single parameter. Two cases were discussed. One is the case where the parameter is a nuisance parameter to be estimated, and the other the case where the parameter is specified at some value. For both cases the modified Cramér–von Mises–Smirnov statistics were used, whose c.f.s were presented without proof, whereas how to derive these c.f.s was discussed in Chapters 2 and 3.

In this chapter we extend the discussions to the multi-parameter case where the hypothesized distribution depends on a vector of parameters. Let $X_1, \ldots, X_n$ be a random sample from the common distribution $G(x)$, which is assumed to be absolutely continuous. The testing problem considered here is to test if $G(x) = F(x;\boldsymbol{\xi})$, where $F(x;\boldsymbol{\xi})$ is a distribution function parameterized by a $p \times 1$ vector $\boldsymbol{\xi}$ of parameters. According to Durbin (1973), let $\boldsymbol{\xi} = (\boldsymbol{\xi}_1', \boldsymbol{\xi}_2')$ where $\boldsymbol{\xi}_1$ is a vector of p_1 parameters to be specified under the null and where $\boldsymbol{\xi}_2$ is a vector of p_2 parameters whose value is unknown and is to be estimated from the data.

Then we naturally deal with three cases. The first is the unspecified case (Case 4.1) where $p = p_2$ so that $\boldsymbol{\xi} = \boldsymbol{\xi}_2$ is a vector of nuisance parameters, the second the specified case (Case 4.2) where $p = p_1$ so that $\boldsymbol{\xi} = \boldsymbol{\xi}_1$ is a specified vector under the null, and the third the mixed case (Case 4.3) where $p = p_1 + p_2$ with $p_1 > 0$ and $p_2 > 0$, a subset of nuisance parameters is estimated from the sample, and the remaining parameters are specified at some values.

In the next section we discuss Cases 4.1 and 4.2, whereas Case 4.3 is dealt with in Section 4.3. In Section 4.4 we remark that the possibility of the derivation of the FD associated with our statistics is quite restrictive, for which the reason is also stated.

4.2 Cases with Nuisance Parameters Unspecified and Specified

We first deal with Case 4.1, where the parameters are all unspecified and the asymptotic distribution under the null is considered, which is followed by Case 4.2 where the parameters are all specified and the asymptotic distribution under a local alternative is discussed.

Case 4.1 The testing problem considered here is

$$H_0 : G(x) = F(x;\boldsymbol{\xi}),$$

where $\boldsymbol{\xi}$ is a $p \times 1$ vector of nuisance parameters. Proceeding in the same way as in the single-parameter case discussed in Section 1.5.1, we use the modified Cramér–von Mises–Smirnov statistic

$$W_n^2(\hat{\boldsymbol{\xi}}) = n \int_{-\infty}^{\infty} \left(F_n(x) - F(x;\hat{\boldsymbol{\xi}})\right)^2 dF(x;\hat{\boldsymbol{\xi}}), \tag{4.1}$$

where $\hat{\boldsymbol{\xi}}$ is a suitable estimator of $\boldsymbol{\xi}$, and $F_n(x)$ is the empirical distribution function defined by

$$F_n(x) = \frac{1}{n} \sum_{j=1}^{n} \epsilon\left(x - X_j\right), \qquad \epsilon(x) = \begin{cases} 1 & (x \geq 0), \\ 0 & (x < 0). \end{cases}$$

Let us consider the asymptotic distribution of $W_n^2(\hat{\boldsymbol{\xi}})$ under H_0. Suppose that $\hat{\boldsymbol{\xi}}$ is the MLE of $\boldsymbol{\xi}$ that satisfies some regularity conditions. Putting

$$\boldsymbol{g}_\xi(t) = \left.\frac{\partial F(x;\boldsymbol{\xi})}{\partial \boldsymbol{\xi}}\right|_{x=x(t;\xi)}, \qquad x(t;\boldsymbol{\xi}) = \inf\{x : F(x;\boldsymbol{\xi}) = t\},$$

it follows from Durbin (1973) that

$$\begin{aligned} W_n^2(\hat{\boldsymbol{\xi}}) &= \int_{-\infty}^{\infty} \left(\sqrt{n}\,(F_n(x) - F(x;\boldsymbol{\xi})) - \sqrt{n}\,(\hat{\boldsymbol{\xi}} - \boldsymbol{\xi})' \frac{\partial F(x;\boldsymbol{\xi}^*)}{\partial \boldsymbol{\xi}}\right)^2 dF(x;\hat{\boldsymbol{\xi}}) \\ &= \int_0^1 \left(Y_n(t) - \sqrt{n}\,(\hat{\boldsymbol{\xi}} - \boldsymbol{\xi})'\,\boldsymbol{g}_\xi(t)\right)^2 dt + \delta_n \\ &= \int_0^1 Z_n^2(t)\,dt + \delta_n, \end{aligned}$$

where $\boldsymbol{\xi}^*$ lies between $\boldsymbol{\xi}$ and $\hat{\boldsymbol{\xi}}$ and $\delta_n \to 0$ in probability, whereas

$$Y_n(t) = \sqrt{n}\,(H_n(t) - t), \quad H_n(t) = \frac{1}{n}\sum_{j=1}^{n} \epsilon\Big(t - F(X_j;\boldsymbol{\xi})\Big),$$

$$Z_n(t) = Y_n(t) - \sqrt{n}\,(\hat{\boldsymbol{\xi}} - \boldsymbol{\xi})'\,\boldsymbol{g}_\xi(t).$$

It holds that $\{Y_n(t)\} \Rightarrow \{Y(t)\}$ with $\mathrm{Cov}(Y(s), Y(t)) = \min(s,t) - st$ and

$$\boldsymbol{\zeta}_n = \sqrt{n}(\hat{\boldsymbol{\xi}} - \boldsymbol{\xi}) \Rightarrow \boldsymbol{\zeta} \sim \mathrm{N}\left(\boldsymbol{0},\, J^{-1}(\boldsymbol{\xi})\right),$$

where

$$J(\boldsymbol{\xi}) = \mathrm{E}\left[\left(\frac{\partial \log f(x;\boldsymbol{\xi})}{\partial \boldsymbol{\xi}}\right)\left(\frac{\partial \log f(x;\boldsymbol{\xi})}{\partial \boldsymbol{\xi}}\right)'\right], \quad f(x;\boldsymbol{\xi}) = \frac{\partial F(x;\boldsymbol{\xi})}{\partial x}.$$

Then we obtain

$$\begin{aligned} W_n^2(\hat{\boldsymbol{\xi}}) &\Rightarrow \int_0^1 \left(Y(t) - \boldsymbol{\zeta}'\boldsymbol{g}_\xi(t)\right)^2 dt = \int_0^1 Z^2(t)\,dt \\ &= W^2(\boldsymbol{\xi}) \stackrel{\mathcal{D}}{=} \int_0^1\int_0^1 K(s,t)\,dW(s)\,dW(t), \end{aligned} \tag{4.2}$$

where $\{Z(t)\} = \{Y(t) - \boldsymbol{\zeta}'\boldsymbol{g}_\xi(t)\}$ is a zero-mean Gaussian process and

$$K(s,t) = \mathrm{Cov}(Z(s),\, Z(t)) = \min(s,t) - st - \boldsymbol{g}_\xi'(s)J^{-1}(\boldsymbol{\xi})\boldsymbol{g}_\xi(t). \tag{4.3}$$

It needs to be ensured that the distribution of $W^2(\boldsymbol{\xi})$ under H_0 does not depend on $\boldsymbol{\xi}$. This is equivalent to saying that $\boldsymbol{g}_\xi'(s)J^{-1}(\boldsymbol{\xi})\boldsymbol{g}_\xi(t)$ does not depend on $\boldsymbol{\xi}$. In many important cases it does not depend on $\boldsymbol{\xi}$. These include cases where $F(x;\boldsymbol{\xi})$ is transformable into a form in which $\boldsymbol{\xi}$ is a vector of location and scale parameters (Durbin, 1973).

As an example dealt with in Nabeya (1992), let us take up the Cauchy distribution

$$F(x;\boldsymbol{\xi}) = \frac{1}{2} + \frac{1}{\pi}\tan^{-1}\frac{x-\alpha}{\beta}, \quad f(x;\boldsymbol{\xi}) = \frac{1}{\beta\pi}\frac{1}{1+(x-\alpha)^2/\beta^2},$$

where $\boldsymbol{\xi} = (\alpha, \beta)'$. Since

$$F(x;\boldsymbol{\xi}) = t \quad \Leftrightarrow \quad (x-\alpha)/\beta = -\cot \pi t$$

and

$$\frac{\partial F(x;\boldsymbol{\xi})}{\partial \alpha} = \frac{-1}{\beta\pi}\frac{1}{1+(x-\alpha)^2/\beta^2},$$
$$\frac{\partial F(x;\boldsymbol{\xi})}{\partial \beta} = \frac{-(x-\alpha)}{\beta^2\pi}\frac{1}{1+(x-\alpha)^2/\beta^2},$$

we have $\boldsymbol{g}_{\xi}(t) = (g_\alpha(t), g_\beta(t))'$, where (Exercise 4.2.1)

$$g_\alpha(t) = \frac{-1}{\beta\pi}\frac{1}{1+\cot^2\pi t} = \frac{-1}{\beta\pi}\sin^2 \pi t, \tag{4.4}$$
$$g_\beta(t) = \frac{\cot \pi t}{\beta\pi}\frac{1}{1+\cot^2 \pi} = \frac{1}{2\beta\pi}\sin 2\pi t. \tag{4.5}$$

We also have

$$\frac{\partial \log f(x;\boldsymbol{\xi})}{\partial \alpha} = \frac{2(x-\alpha)/\beta^2}{1+(x-\alpha)^2/\beta^2},$$
$$\frac{\partial \log f(x;\boldsymbol{\xi})}{\partial \beta} = -\frac{1}{\beta} + \frac{2(x-\alpha)^2/\beta^3}{1+(x-\alpha)^2/\beta^2},$$

which yields (Exercise 4.2.2)

$$J(\boldsymbol{\xi}) = \mathrm{E}\begin{pmatrix} \left(\dfrac{\partial \log f(X;\boldsymbol{\xi})}{\partial \alpha}\right)^2 & \dfrac{\partial \log f(X;\boldsymbol{\xi})}{\partial \alpha}\dfrac{\partial \log f(X;\boldsymbol{\xi})}{\partial \beta} \\ \dfrac{\partial \log f(X;\boldsymbol{\xi})}{\partial \beta}\dfrac{\partial \log f(X;\boldsymbol{\xi})}{\partial \alpha} & \left(\dfrac{\partial \log f(X;\boldsymbol{\xi})}{\partial \beta}\right)^2 \end{pmatrix}$$
$$= \begin{pmatrix} \dfrac{1}{2\beta^2} & 0 \\ 0 & \dfrac{1}{2\beta^2} \end{pmatrix}. \tag{4.6}$$

Thus we have

$$\boldsymbol{g}_{\xi}'(s)J^{-1}(\boldsymbol{\xi})\boldsymbol{g}_{\xi}(t) = \frac{2}{\pi^2}\sin^2 \pi s \sin^2 \pi t + \frac{1}{2\pi^2}\sin 2\pi s \sin 2\pi t,$$

which does not depend on $\boldsymbol{\xi}$, and the kernel $K(s,t)$ associated with the limiting statistic $W^2(\boldsymbol{\xi})$ in (4.2) is given by

$$K(s,t) = \min(s,t) - st - \frac{2}{\pi^2}\sin^2\pi s\,\sin^2\pi t - \frac{1}{2\pi^2}\sin 2\pi s\,\sin 2\pi t.$$

The c.f. of $W^2(\boldsymbol{\xi})$ is $(D(2i\theta))^{-1/2}$, where $D(\lambda)$ is the FD of $K(s,t)$ already obtained in (2.188) as

$$D(\lambda) = \left(\frac{\sin\sqrt{\lambda}}{\sqrt{\lambda}} + \frac{1-\cos\sqrt{\lambda}}{\pi^2\left(1-\frac{\lambda}{4\pi^2}\right)}\right)\Big/\left(1-\frac{\lambda}{4\pi^2}\right)^2$$
$$= 1 - \left(\frac{1}{6} - \frac{1}{\pi^2}\right)\lambda + \left(\frac{1}{120} - \frac{1}{8\pi^2} + \frac{9}{16\pi^4}\right)\lambda^2 - \cdots.$$

We also have

$$\mathrm{E}\left(W^2(\boldsymbol{\xi})\right) = \frac{1}{6} - \frac{1}{\pi^2},\quad \mathrm{Var}\left(W^2(\boldsymbol{\xi})\right) = \frac{1}{45} - \frac{1}{6\pi^2} - \frac{1}{4\pi^4}.$$

Case 4.2 Here we deal with the testing problem

$$H_0: G(x) = F(x;\boldsymbol{\xi}_0) \quad \text{versus} \quad H_1: G(x) = F(x;\boldsymbol{\xi}),\quad \boldsymbol{\xi} = \boldsymbol{\xi}_0 + \frac{\boldsymbol{\gamma}}{\sqrt{n}}, \tag{4.7}$$

where $\boldsymbol{\xi}_0$ is a specified vector of $\boldsymbol{\xi}$ and $\boldsymbol{\gamma}$ is any constant vector. This case was also discussed in Durbin (1973) as a special case. We consider here the situation under the local alternative H_1 as well. As the test statistic, we use the Cramér–von Mises–Smirnov statistic

$$W_n^2(\boldsymbol{\xi}_0) = n\int_{-\infty}^{\infty}\left(F_n(x) - F(x;\boldsymbol{\xi}_0)\right)^2 dF(x;\boldsymbol{\xi}_0), \tag{4.8}$$

where $F_n(x)$ is the empirical distribution function. The null hypothesis H_0 is rejected if $W_n^2(\boldsymbol{\xi}_0)$ is sufficiently large.

Let us consider the limiting distribution of $W_n^2(\boldsymbol{\xi}_0)$ as $n\to\infty$. We first deal with the distribution under H_0. Putting $G(x) = F(x;\boldsymbol{\xi}_0) = t$, we have

$$W_n^2(\boldsymbol{\xi}_0) = n\int_{-\infty}^{\infty}\left(F_n(x) - F(x;\boldsymbol{\xi}_0)\right)^2 dF(x;\boldsymbol{\xi}_0)$$
$$= n\int_0^1 (H_n(t)-t)^2\,dt = \int_0^1 Y_n^2(t)\,dt,$$

where $Y_n(t) = \sqrt{n}\,(H_n(t)-t)$ and $H_n(t) = \frac{1}{n}\sum_{j=1}^n \epsilon\left(t - F(X_j;\boldsymbol{\xi}_0)\right)$. It follows that $\{Y_n(t)\} \Rightarrow \{Y(t)\}$, where $\{Y(t)\}$ is the Bb with $\mathrm{Cov}(Y(s),\,Y(t)) = \min(s,t) - st$. Thus it follows that

$$W_n^2(\xi_0) \Rightarrow W^2(\xi_0) = \int_0^1 Y^2(t)\,dt, \tag{4.9}$$

and the c.f. of $W^2(\xi_0)$ is given by

$$\mathrm{E}\left(e^{i\theta W^2(\xi_0)}\right) = (D(2i\theta))^{-1/2} = \left(\frac{\sin\sqrt{2i\theta}}{\sqrt{2i\theta}}\right)^{-1/2},$$

where $D(\lambda)$ is the FD of $K(s,t) = \min(s,t) - st$. It holds that $\mathrm{E}\left(W^2(\xi_0)\right) = 1/6$ and $\mathrm{Var}\left(W^2(\xi_0)\right) = 1/45$ under H_0.

We next consider the limiting distribution of $W_n^2(\xi_0)$ under $H_1 : \xi = \xi_0 + \gamma/\sqrt{n}$. Expanding $F(x;\xi_0)$ around $\xi = \xi_0 + \gamma/\sqrt{n}$, we consider, in place of $W_n^2(\xi_0)$,

$$\begin{aligned} R_n^2(\xi_0) &= n\int_{-\infty}^{\infty}\left(F_n(x) - F(x;\xi) + \frac{\gamma'}{\sqrt{n}}\left.\frac{\partial F(x;\xi)}{\partial \xi}\right|_{\xi=\xi_0}\right)^2 dF(x;\xi) \\ &= \int_0^1 \left(Y_n(t) + \gamma' g_{\xi_0}(t)\right)^2 dt, \end{aligned} \tag{4.10}$$

where $Y_n(t) = \sqrt{n}\,(H_n(t) - t)$, $H_n(t) = \frac{1}{n}\sum_{j=1}^{n} \epsilon\left(t - F(X_j;\xi)\right)$, and

$$g_{\xi_0}(t) = \left.\frac{\partial F(x;\xi)}{\partial \xi}\right|_{x=x(t;\xi_0)}, \qquad x(t;\xi_0) = \inf\{x : F(x;\xi_0) = t\}.$$

It follows from Durbin (1973) that, under H_1 and some regularity conditions, the limiting distributions of $W_n^2(\xi_0)$ and $R_n^2(\xi_0)$ are the same and

$$R_n^2(\xi_0) \Rightarrow R^2(\xi_0) = \int_0^1 \left(Y(t) + \gamma' g_{\xi_0}(t)\right)^2 dt, \tag{4.11}$$

where $\{Y(t)\}$ is the Bb. The statistic $R^2(\xi_0)$ is the quadratic functional of a nonzero-mean Gaussian process $\{Y(t)+\gamma' g_{\xi_0}(t)\}$. To obtain the c.f. of $R^2(\xi_0)$, we need not only the FD $D(\lambda)$ of $K(s,t) = \min(s,t) - st$, but also the resolvent $\Gamma(s,t;\lambda)$ of $K(s,t)$. For this purpose we can use the approach discussed in Section 3.2, where it was indicated that we can avoid the computation of $\Gamma(s,t;\lambda)$ itself. In fact, putting $\gamma' g_{\xi_0} = m(t)$, we have

$$\begin{aligned} &\mathrm{E}\left(e^{i\theta R^2(\xi_0)}\right) \\ &= (D(\lambda))^{-1/2}\exp\left[\frac{\lambda}{2}\int_0^1 m^2(t)\,dt + \frac{\lambda^2}{2}\int_0^1\int_0^1 \Gamma(s,t;\lambda)m(s)m(t)\,ds\,dt\right] \\ &= (D(\lambda))^{-1/2}\exp\left[\frac{\lambda}{2}\int_0^1 m^2(t)\,dt + \frac{\lambda^2}{2}\int_0^1 h(t;\lambda)m(t)\,dt\right], \end{aligned} \tag{4.12}$$

where $\lambda = 2i\theta$ and $h(t;\lambda)$ satisfies the nonhomogeneous integral equation given by

$$h(t;\lambda) = \int_0^1 K(s,t)m(s)\,ds + \lambda \int_0^1 K(s,t)h(s;\lambda)\,ds. \tag{4.13}$$

As an example, let us take up the Cauchy distribution dealt with in Case 4.1. Putting $\boldsymbol{\gamma} = (\gamma_1, \gamma_2)'$ and $\boldsymbol{\xi}_0 = (\alpha_0, \beta_0)'$, we have, from (4.4) and (4.5),

$$\begin{aligned} m(t) = \boldsymbol{\gamma}' \boldsymbol{g}_{\xi_0}(t) &= (\gamma_1, \gamma_2) \left(\frac{-1}{\beta_0 \pi} \sin^2 \pi t, \ \frac{1}{2\beta_0 \pi} \sin 2\pi t \right)' \\ &= \frac{\gamma_1}{2\beta_0 \pi} (\cos 2\pi t - 1) + \frac{\gamma_2}{2\beta_0 \pi} \sin 2\pi t. \end{aligned}$$

It can be shown (Exercise 4.2.3) that the integral equation (4.13) is equivalent to

$$h''(t;\lambda) + \lambda h(t;\lambda) = -m(t), \quad h(0;\lambda) = 0, \quad h(1;\lambda) = 0.$$

Operational calculus with the differential operator D yields the general solution given by

$$\begin{aligned} h(t;\lambda) &= c_1 \cos \sqrt{\lambda} t + c_2 \sin \sqrt{\lambda} t + \frac{\gamma_1}{2\beta_0 \pi} \frac{1 - \cos 2\pi t}{D^2 + \lambda} - \frac{\gamma_2}{2\beta_0 \pi} \frac{\sin 2\pi t}{D^2 + \lambda} \\ &= c_1 \cos \sqrt{\lambda} t + c_2 \sin \sqrt{\lambda} t + \frac{\gamma_1}{2\lambda \beta_0 \pi} - \frac{\gamma_1 \cos 2\pi t + \gamma_2 \sin 2\pi t}{2\beta_0 \pi (\lambda - 4\pi^2)}. \end{aligned}$$

Then the boundary conditions $h(0;\lambda) = h(1;\lambda) = 0$ determine c_1 and c_2 uniquely as

$$c_1 = \frac{2\pi \gamma_1}{\lambda \beta_0 (\lambda - 4\pi^2)}, \qquad c_2 = c_1 \frac{1 - \cos \sqrt{\lambda}}{\sin \sqrt{\lambda}},$$

and we obtain, using computerized algebra,

$$\begin{aligned} A(\lambda; \boldsymbol{\xi}_0, \boldsymbol{\gamma}) &= \frac{\lambda}{2} \int_0^1 m^2(t)\,dt + \frac{\lambda^2}{2} \int_0^1 h(t;\lambda)m(t)\,dt \\ &= \left(\frac{\gamma_1}{\beta_0} \right)^2 \left[\frac{\lambda}{4(4\pi^2 - \lambda)} + \frac{4\pi^2}{(4\pi^2 - \lambda)^2} \frac{1 - \cos \sqrt{\lambda}}{\sin \sqrt{\lambda}/\sqrt{\lambda}} \right] \\ &\quad + \left(\frac{\gamma_2}{\beta_0} \right)^2 \frac{\lambda}{4(4\pi^2 - \lambda)}. \end{aligned}$$

Thus it follows from (4.12) that

$$\phi(\theta) = \mathrm{E}\left(e^{i\theta R^2(\xi_0)} \right) = \left(\frac{\sin \sqrt{\lambda}}{\sqrt{\lambda}} \right)^{-1/2} \exp\left[A(\lambda; \boldsymbol{\xi}_0, \boldsymbol{\gamma}) \right], \tag{4.14}$$

where $\lambda = 2i\theta$. The distribution of $R^2(\boldsymbol{\xi}_0)$ depends on the scale parameter β_0 but not on the location parameter α_0. We obtain, from Corollary 3.2,

$$\mathrm{E}\left(R^2(\xi_0)\right) = \frac{1}{6} + \frac{3(\gamma_1/\beta_0)^2 + (\gamma_2/\beta_0)^2}{8\pi^2},$$

$$\mathrm{Var}\left(R^2(\xi_0)\right) = \frac{1}{45} + \left(\frac{1}{12\pi^2} + \frac{5}{8\pi^4}\right)\left(\frac{\gamma_1}{\beta_0}\right)^2 + \frac{1}{8\pi^4}\left(\frac{\gamma_2}{\beta_0}\right)^2,$$

and it holds (Exercise 4.2.4) that, as $|\gamma_1| \to \infty$ and $|\gamma_2| \to \infty$,

$$R^2(\xi_0) \Rightarrow \mathrm{N}\left(\frac{3(\gamma_1/\beta_0)^2 + (\gamma_2/\beta_0)^2}{8\pi^2}, \left(\frac{1}{12\pi^2} + \frac{5}{8\pi^4}\right)\left(\frac{\gamma_1}{\beta_0}\right)^2 + \frac{1}{8\pi^4}\left(\frac{\gamma_2}{\beta_0}\right)^2\right). \tag{4.15}$$

Using Lévy's inversion formula, the limiting power at the $100\kappa\%$ significance level under H_1 in (4.7) can be computed as

$$P\left(R^2(\xi_0) \geq x(\kappa)\right) = 1 - \frac{1}{\pi}\int_0^\infty \mathrm{Re}\left[\frac{1 - e^{-i\theta x(\kappa)}}{i\theta}\,\phi(\theta)\right] d\theta, \tag{4.16}$$

where $\phi(\theta)$ is the c.f. of $R^2(\xi_0)$ given in (4.14) and $x(\kappa)$ is the upper $100\kappa\%$ point of the null distribution of $R^2(\xi_0)$. Numerical integration for computing distribution functions and powers can be done by Simpson's rule, although care must be taken in computing $\phi(\theta)$, which involves the square root of complex values. See Tanaka (1996, 2017) for computational details.

Figure 4.1 plots the powers of the test based on $R^2(\xi_0)$ at the $1\%, 5\%$, and 10% significance levels against the horizontal axis $a = (\gamma_1/\beta_0)^2 = (\gamma_2/\beta_0)^2$,

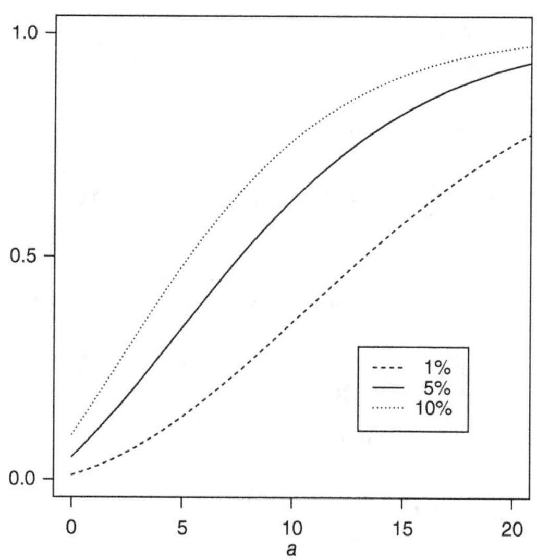

Figure 4.1 Powers of the test based on $R^2(\xi_0)$ against $a = (\gamma_1/\beta_0)^2 = (\gamma_2/\beta_0)^2$.

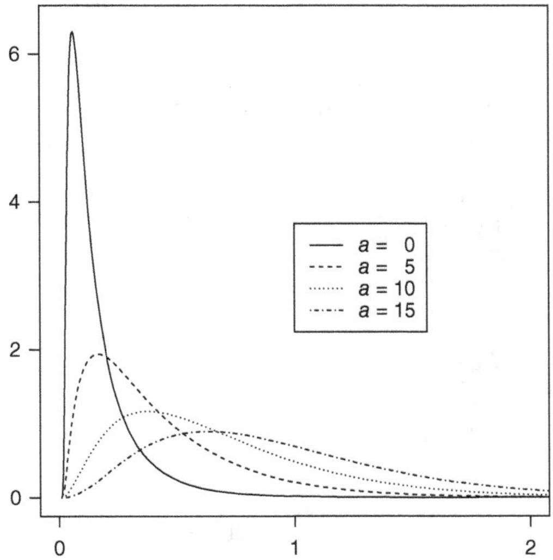

Figure 4.2 Probability densities of $R^2(\boldsymbol{\xi}_0)$ for $a = (\gamma_1/\beta_0)^2 = (\gamma_2/\beta_0)^2 = 0, 5, 10, 15$.

where $x(0.01) = 0.74346, x(0.05) = 0.46136$, and $x(0.1) = 0.34730$. For reference Figure 4.2 presents the probability densities of $R^2(\boldsymbol{\xi}_0)$ for $a = 0, 5, 10$, and 15. Note that the case of $a = 0$ corresponds to the density under H_0. It is seen that the distribution tends to normality as a becomes large, as is anticipated.

Exercises for Section 4.2

Exercise 4.2.1 For the Cauchy distribution

$$F(x;\boldsymbol{\xi}) = \frac{1}{2} + \frac{1}{\pi}\tan^{-1}\frac{x-\alpha}{\beta}, \quad \boldsymbol{\xi} = (\alpha,\beta)',$$

prove that

$$\left.\frac{\partial F(x;\boldsymbol{\xi})}{\partial \boldsymbol{\xi}}\right|_{x=x(t;\xi)} = \begin{pmatrix} \dfrac{-1}{\beta\pi}\sin^2 \pi t \\ \dfrac{1}{2\beta\pi}\sin 2\pi t \end{pmatrix}, \quad x(t;\boldsymbol{\xi}) = \inf\{x : F(x;\boldsymbol{\xi}) = t\}.$$

Exercise 4.2.2 For the density of the Cauchy distribution

$$f(x;\boldsymbol{\xi}) = \frac{1}{\beta\pi}\frac{1}{1+(x-\alpha)^2/\beta^2}, \quad \boldsymbol{\xi} = (\alpha,\beta)',$$

prove that

$$\mathrm{E}\left[\left(\frac{\partial \log f(X;\boldsymbol{\xi})}{\partial \alpha}\right)^2\right] = \mathrm{E}\left[\left(\frac{\partial \log f(X;\boldsymbol{\xi})}{\partial \beta}\right)^2\right] = \frac{1}{2\beta^2},$$

$$\mathrm{E}\left[\left(\frac{\partial \log f(X;\boldsymbol{\xi})}{\partial \alpha}\,\frac{\partial \log f(X;\boldsymbol{\xi})}{\partial \beta}\right)\right] = 0,$$

using the formula

$$\int_0^\infty \frac{u^{a-1}}{(1+u^c)^b}\,du = \frac{1}{c}\,B\left(b-\frac{a}{c},\frac{a}{c}\right) \quad \left(a>0,\, b>\frac{a}{c},\, c>0\right),$$

where $B(p,q)$ is the beta function defined by

$$B(p,q) = \int_0^1 t^{p-1}(1-t)^{q-1}\,dt = \frac{\Gamma(p)\Gamma(q)}{\Gamma(p+q)} \quad (p>0,\, q>0).$$

Exercise 4.2.3 Show that the integral equation

$$h(t;\lambda) = \int_0^1 K(s,t)m(s)\,ds + \lambda\int_0^1 K(s,t)h(s;\lambda)\,ds,$$

where $K(s,t)=\min(s,t)-st$, is equivalent to

$$h''(t;\lambda)+\lambda h(t;\lambda) = -m(t), \quad h(0;\lambda)=0, \quad h(1;\lambda)=0.$$

Exercise 4.2.4 Prove that, for the c.f. of $R^2(\boldsymbol{\xi}_0)$ given by (4.14), it holds that, as $|\gamma_1|\to\infty$ and $|\gamma_2|\to\infty$, $R^2(\boldsymbol{\xi}_0)$ converges in distribution to

$$\mathrm{N}\left(\frac{3(\gamma_1/\beta_0)^2+(\gamma_2/\beta_0)^2}{8\pi^2},\, \left(\frac{1}{12\pi^2}+\frac{5}{8\pi^4}\right)\left(\frac{\gamma_1}{\beta_0}\right)^2+\frac{1}{8\pi^4}\left(\frac{\gamma_2}{\beta_0}\right)^2\right).$$

4.3 Mixed Case with Nuisance Parameters

In this section we deal with Case 4.3, where the parameter $\boldsymbol{\xi}$ is a $p=p_1+p_2$ dimensional vector with $p_1>0$ and $p_2>0$. We estimate a subset of nuisance parameters from the sample, whereas the remaining parameters are specified at some values. Two examples are given as Cases 4.3.1 and 4.3.2, employing the Cauchy distribution.

Case 4.3 Let $\boldsymbol{\xi}$ be a $p\times 1$ vector of parameters, and put $\boldsymbol{\xi}=(\boldsymbol{\xi}_1',\boldsymbol{\xi}_2')'$, where $\boldsymbol{\xi}_1$ and $\boldsymbol{\xi}_2$ are $p_1\times 1$ and $p_2\times 1$ vectors, respectively. Then we consider the testing problem

$$H_0: G(x)=F(x;\boldsymbol{\xi}_{10},\boldsymbol{\xi}_2) \quad \text{versus} \quad H_1: \boldsymbol{\xi}_1=\boldsymbol{\xi}_{10}+\frac{\boldsymbol{\gamma}}{\sqrt{n}},$$

where $\boldsymbol{\xi}_{10}$ is a specified vector of $\boldsymbol{\xi}$, and $\boldsymbol{\xi}_2$ is a vector of unspecified nuisance parameters to be estimated.

Durbin (1973) considered a modified Cramér–von Mises–Smirnov statistic

$$W_n^2 = n \int_{-\infty}^{\infty} \left(F_n(x) - F\left(x; \boldsymbol{\xi}_{10}, \hat{\boldsymbol{\xi}}_2\right)\right)^2 dF\left(x; \boldsymbol{\xi}_{10}, \hat{\boldsymbol{\xi}}_2\right),$$

where $F_n(x)$ is the empirical distribution function. Suppose that $\hat{\boldsymbol{\xi}}_2$ is the MLE of $\boldsymbol{\xi}_2$ under the condition that $\boldsymbol{\xi}_1 = \boldsymbol{\xi}_{10}$. Then, under some regularity conditions, Durbin (1973) showed that, under H_1,

$$W_n^2 = \int_0^1 \left(Y_n(t) + \boldsymbol{\gamma}'\boldsymbol{g}_1(t) - \sqrt{n}\left(\hat{\boldsymbol{\xi}}_2 - \boldsymbol{\xi}_2\right)'\boldsymbol{g}_2(t)\right)^2 dt + o_p(1),$$

where $o_p(a_n)$ stands for a term that has stochastic order less than a_n, $Y_n(t) = \sqrt{n}(H_n(t) - t)$, $H_n(t) = \frac{1}{n}\sum_{j=1}^{n} \epsilon\left(t - F(X_j; \boldsymbol{\xi})\right)$, and

$$\boldsymbol{g}_{\xi}(t) = \begin{pmatrix} \boldsymbol{g}_1(t) \\ \boldsymbol{g}_2(t) \end{pmatrix} = \begin{pmatrix} \partial F(x;\boldsymbol{\xi})/\partial \boldsymbol{\xi}_1 \\ \partial F(x;\boldsymbol{\xi})/\partial \boldsymbol{\xi}_2 \end{pmatrix}\Bigg|_{x=x(t;\xi)}, \quad x(t;\boldsymbol{\xi}) = \inf\{x : F(x;\boldsymbol{\xi}) = t\}.$$

Moreover it holds that, under H_1,

$$\boldsymbol{\zeta}_n = \sqrt{n}(\hat{\boldsymbol{\xi}}_2 - \boldsymbol{\xi}_2) \Rightarrow \boldsymbol{\zeta} \sim \mathrm{N}\left(J_{22}^{-1} J_{21}\boldsymbol{\gamma},\ J_{22}^{-1}\right),$$

where, for $j, k = 1, 2$,

$$J_{jk} = \mathrm{E}\left[\left(\frac{\partial \log f(x;\boldsymbol{\xi})}{\partial \boldsymbol{\xi}_j}\right)\left(\frac{\partial \log f(x;\boldsymbol{\xi})}{\partial \boldsymbol{\xi}_k}\right)'\right], \quad f(x;\boldsymbol{\xi}) = \frac{\partial F(x;\boldsymbol{\xi})}{\partial x}.$$

Then we obtain, as $n \to \infty$ under H_1,

$$\begin{aligned} W_n^2 &\Rightarrow \int_0^1 \left[Y(t) - \left(\boldsymbol{\zeta} - J_{22}^{-1}J_{21}\boldsymbol{\gamma}\right)'\boldsymbol{g}_2(t) + \boldsymbol{\gamma}'\left(\boldsymbol{g}_1(t) - J_{12}J_{22}^{-1}\boldsymbol{g}_2(t)\right)\right]^2 dt \\ &= W^2 = \int_0^1 (Z(t) + m(t))^2\, dt, \qquad (4.17) \end{aligned}$$

where $Z(t) = Y(t) - \left(\boldsymbol{\zeta} - J_{22}^{-1}J_{21}\boldsymbol{\gamma}\right)'\boldsymbol{g}_2(t)$ is a zero-mean Gaussian process and

$$K(s,t) = \mathrm{Cov}(Z(s), Z(t)) = \min(s,t) - st - \boldsymbol{g}_2'(s)J_{22}^{-1}\boldsymbol{g}_2(t), \qquad (4.18)$$

$$m(t) = \boldsymbol{\gamma}'\left(\boldsymbol{g}_1(t) - J_{12}J_{22}^{-1}\boldsymbol{g}_2(t)\right). \qquad (4.19)$$

The c.f. of W^2 in (4.17) can be derived by the approach taken in Section 3.2, which was also used in Case 4.2 of the last section.

As an example let us consider again the Cauchy distribution

$$F(x;\xi_1,\xi_2) = \frac{1}{2} + \frac{1}{\pi}\tan^{-1}\frac{x-\alpha}{\beta},$$

and we consider the two situations concerning the testing problem. One is the case where α is specified with β unspecified so that $\xi_{10} = \alpha$ and $\xi_2 = \beta$. The other is the case where β is specified with α unspecified so that $\xi_{10} = \beta$ and $\xi_2 = \alpha$.

Case 4.3.1. We first deal with the former case, which is

$$H_0 : G(x) = F(x;\alpha_0,\beta) \quad \text{versus} \quad H_1 : G(x) = F(x;\alpha,\beta),\ \alpha = \alpha_0 + \frac{\gamma}{\sqrt{n}}. \tag{4.20}$$

It follows from (4.4), (4.5), and (4.6) that $g_1(t) = (\cos 2\pi t - 1)/2\beta\pi$, $g_2(t) = \sin 2\pi t/2\beta\pi$ and $J_{11} = J_{22} = 1/2\beta^2$, $J_{12} = J_{21} = 0$. Thus we have, from (4.18) and (4.19),

$$\begin{aligned} K(s,t) &= \min(s,t) - st - \frac{g_2(s)g_2(t)}{J_{22}} \\ &= \min(s,t) - st - \frac{1}{2\pi^2}\sin 2\pi s \sin 2\pi t, \qquad (4.21) \\ m(t) &= \gamma g_1(t) \\ &= \frac{\gamma}{2\beta\pi}(\cos 2\pi t - 1). \qquad (4.22) \end{aligned}$$

The FD of $K(s,t)$ in (4.21) can be derived (Exercise 4.3.1) from Darling's formula in (2.159) or from the relationship between the eigenvalues of $\min(s,t) - st$ and the corresponding eigenfunctions as

$$D(\lambda) = \frac{\sin\sqrt{\lambda}}{\sqrt{\lambda}} \Big/ \left(1 - \frac{\lambda}{4\pi^2}\right). \tag{4.23}$$

To derive the c.f. of W^2 in (4.17), we consider the integral equation

$$h(t;\lambda) = \int_0^1 K(s,t)m(s)\,ds + \lambda\int_0^1 K(s,t)h(s;\lambda)\,ds, \tag{4.24}$$

which leads to

$$h''(t;\lambda) + \lambda h(t;\lambda) = -m(t) + 2\lambda a \sin 2\pi t \tag{4.25}$$

with three conditions

$$h(0;\lambda) = 0, \quad h(1;\lambda) = 0, \quad a = \int_0^1 h(s;\lambda)\sin 2\pi s\,ds. \tag{4.26}$$

It can be shown (Exercise 4.3.2) that the integral equation (4.24) is equivalent to the differential equation (4.25) with two boundary conditions and $a = 0$ in (4.26).

Using operational calculus, we obtain, as the general solution to (4.25),

$$h(t;\lambda) = c_1 \cos\sqrt{\lambda}t + c_2 \sin\sqrt{\lambda}t + \frac{\gamma}{2\lambda\beta\pi} + \frac{\gamma}{2\beta\pi}\frac{\cos 2\pi t}{4\pi^2 - \lambda}.$$

From the two boundary conditions we have

$$c_1 = \frac{-2\gamma\pi}{\lambda\beta(4\pi^2 - \lambda)}, \quad c_2 = \frac{c_1(1 - \cos\sqrt{\lambda})}{\sin\sqrt{\lambda}},$$

and computerized algebra gives

$$\begin{aligned}\frac{\lambda}{2}\int_0^1 m^2(s)\,ds &+ \frac{\lambda^2}{2}\int_0^1 h(s;\lambda)m(s)\,ds \\ &= \frac{\gamma^2}{\beta^2}\left(\frac{\lambda}{4(4\pi^2-\lambda)} + \frac{4\pi^2\sqrt{\lambda}}{(4\pi^2-\lambda)^2}\frac{1-\cos\sqrt{\lambda}}{\sin\sqrt{\lambda}}\right).\end{aligned}$$

Thus it follows from Theorem 3.2 that

$$\begin{aligned}\mathrm{E}\left(e^{i\theta W^2}\right) = \left[\frac{\sin\sqrt{\lambda}}{\sqrt{\lambda}}\Big/\left(1 - \frac{\lambda}{4\pi^2}\right)\right]^{-1/2} \exp\left[\left(\frac{\gamma}{\beta}\right)^2\left\{\frac{\lambda}{4(4\pi^2-\lambda)}\right.\right. \\ \left.\left. + \frac{4\pi^2}{(4\pi^2-\lambda)^2}\frac{1-\cos\sqrt{\lambda}}{\sin\sqrt{\lambda}/\sqrt{\lambda}}\right\}\right], \qquad (4.27)\end{aligned}$$

where $\lambda = 2i\theta$. It is seen that the limiting distribution does not depend on α_0, although it does depend on β. We obtain, from Corollary 3.2,

$$\mathrm{E}\left(W^2\right) = \frac{1}{6} - \frac{1}{4\pi^2} + \frac{3}{8\pi^2}\left(\frac{\gamma}{\beta}\right)^2,$$

$$\mathrm{Var}\left(W^2\right) = \frac{1}{45} - \frac{1}{8\pi^4} + \left(\frac{1}{12\pi^2} + \frac{5}{8\pi^4}\right)\left(\frac{\gamma}{\beta}\right)^2,$$

and it holds that, as $|\gamma| \to \infty$,

$$W^2 \Rightarrow \mathrm{N}\left(\frac{3}{8\pi^2}\left(\frac{\gamma}{\beta}\right)^2, \left(\frac{1}{12\pi^2} + \frac{5}{8\pi^4}\right)\left(\frac{\gamma}{\beta}\right)^2\right).$$

Using the inversion formula described in (4.16), Figure 4.3 plots the power $P\left(W^2 > x(\kappa)\right)$ under H_1 in (4.20), where the powers are computed at the $100\kappa\%$ significance level ($\kappa = 0.01, 0.05, 0.1$) against the horizontal axis $a = (\gamma/\beta)^2$ with $x(0.01) = 0.71351, x(0.05) = 0.43062$, and $x(0.1) = 0.31567$. Figure 4.4 presents the probability densities of W^2 for $a = 0, 5, 10$, and 15. Note that the case of $a = 0$ corresponds to the density of W^2 under H_0.

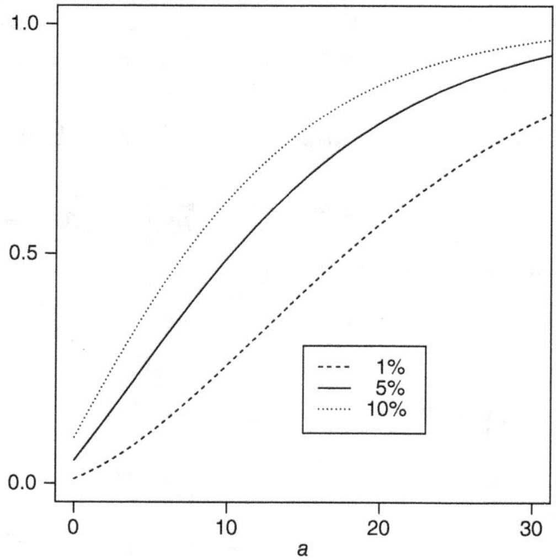

Figure 4.3 Powers of the test based on W^2: Case 4.3.1.

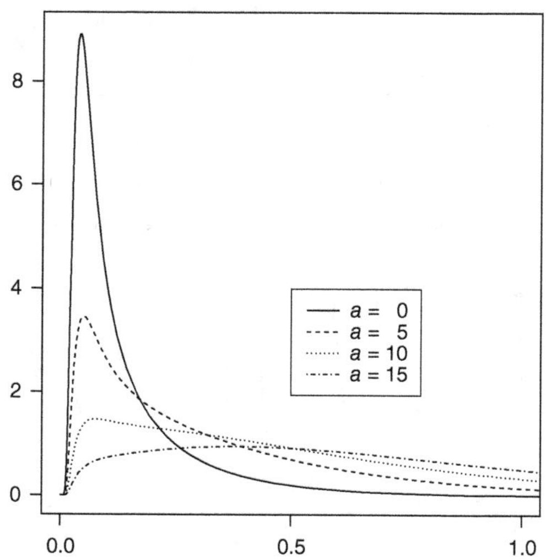

Figure 4.4 Probability densities of W^2: Case 4.3.1.

Case 4.3.2. We next consider

$$H_0 : G(x) = F(x;\alpha,\beta_0) \quad \text{versus} \quad H_1 : G(x) = F(x;\alpha,\beta),\ \beta = \beta_0 + \frac{\gamma}{\sqrt{n}}. \tag{4.28}$$

It follows from (4.4), (4.5), and (4.6) that $g_1(t) = \sin 2\pi t/2\beta_0\pi$, $g_2(t) = (\cos 2\pi t - 1)/2\beta_0\pi$, and $J_{11} = J_{22} = 1/2\beta_0^2$, $J_{12} = J_{21} = 0$. We also have, from (4.18) and (4.19),

$$K(s,t) = \min(s,t) - st - \frac{1}{2\pi^2}(\cos 2\pi s - 1)(\cos 2\pi t - 1), \tag{4.29}$$

$$m(t) = \frac{\gamma}{2\beta_0\pi} \sin 2\pi t. \tag{4.30}$$

The FD of $K(s,t)$ in (4.29) was earlier obtained in (2.165) from Darling's formula as

$$D(\lambda) = \left(\frac{\sin\sqrt{\lambda}}{\sqrt{\lambda}} + \frac{1-\cos\sqrt{\lambda}}{\pi^2\left(1-\dfrac{\lambda}{4\pi^2}\right)} \right) \Bigg/ \left(1 - \frac{\lambda}{4\pi^2}\right).$$

To derive the c.f. of W^2 in (4.17), we consider the integral equation (4.24) with $K(s,t)$ in (4.29) and $m(t)$ in (4.30), which leads to

$$h''(t;\lambda) + \lambda h(t;\lambda) = -m(t) - 4\lambda a \cos 2\pi t \tag{4.31}$$

with three conditions

$$h(0;\lambda) = 0, \quad h(1;\lambda) = 0, \quad a = \int_0^1 h(s;\lambda)\sin^2 \pi s\, ds. \tag{4.32}$$

It can be shown (Exercise 4.3.3) that the integral equation (4.24) is equivalent to the differential equation (4.31) with the three conditions in (4.32) by verifying that

$$\int_0^1 h(s;\lambda) \cos 2\pi s\, ds = 0.$$

Using operational calculus we obtain, as the general solution to (4.31),

$$h(t;\lambda) = c_1 \cos\sqrt{\lambda}t + c_2 \sin\sqrt{\lambda}t + \frac{\gamma}{2\beta_0\pi}\frac{\sin 2\pi t}{4\pi^2-\lambda} + 4\lambda a \frac{\cos 2\pi t}{4\pi^2-\lambda},$$

and the two boundary conditions in (4.32) give

$$c_1 + \frac{4\lambda a}{4\pi^2-\lambda} = 0, \quad c_1\cos\sqrt{\lambda} + c_2 \sin\sqrt{\lambda} + \frac{4\lambda a}{4\pi^2-\lambda} = 0.$$

Then it holds (Exercise 4.3.4) that

$$\int_0^1 h(s;\lambda)m(s)\,ds = \frac{1}{4\pi^2-\lambda}\,\frac{\gamma^2}{8\,\beta_0^2\,\pi^2}$$

so that

$$\frac{\lambda}{2}\int_0^1 m^2(s)\,ds + \frac{\lambda^2}{2}\int_0^1 h(s;\lambda)m(s)\,ds = \frac{\lambda}{4(4\pi^2-\lambda)}\left(\frac{\gamma}{\beta_0}\right)^2.$$

Thus it follows from Theorem 3.2 that

$$\begin{aligned} \mathrm{E}\left(e^{i\theta W^2}\right) &= \left[\left(\frac{\sin\sqrt{\lambda}}{\sqrt{\lambda}} + \frac{1-\cos\sqrt{\lambda}}{\pi^2\left(1-\frac{\lambda}{4\pi^2}\right)}\right)\Big/\left(1-\frac{\lambda}{4\pi^2}\right)\right]^{-1/2} \\ &\quad\times \exp\left[\frac{\lambda}{4(4\pi^2-\lambda)}\left(\frac{\gamma}{\beta_0}\right)^2\right], \end{aligned} \tag{4.33}$$

where $\lambda = 2i\theta$. The limiting distribution does not depend on α, although it does depend on β_0. We obtain, from Corollary 3.2,

$$\mathrm{E}\left(W^2\right) = \frac{1}{6} - \frac{3}{4\pi^2} + \frac{1}{8\pi^2}\left(\frac{\gamma}{\beta_0}\right)^2,$$

$$\mathrm{Var}\left(W^2\right) = \frac{1}{45} - \frac{1}{6\pi^2} - \frac{1}{8\pi^4} + \frac{1}{8\pi^4}\left(\frac{\gamma}{\beta_0}\right)^2,$$

and it holds that, as $|\gamma| \to \infty$,

$$W^2 \Rightarrow \mathrm{N}\left(\frac{1}{8\pi^2}\left(\frac{\gamma}{\beta_0}\right)^2, \frac{1}{8\pi^4}\left(\frac{\gamma}{\beta_0}\right)^2\right).$$

Figure 4.5 plots the power $P\left(W^2 > x(\kappa)\right)$ under H_1 in (4.28), where the powers are computed at the $100\kappa\%$ significance level ($\kappa = 0.01, 0.05, 0.1$) against the horizontal axis $a = (\gamma/\beta_0)^2$ with $x(0.01) = 0.31926, x(0.05) = 0.21619$, and $x(0.1) = 0.17252$. Figure 4.6 presents the probability densities of W^2 for $a = 0, 5, 10$, and 15. Note that the case of $a = 0$ corresponds to the density of W^2 under H_0. The distribution of Case 4.3.1 is seen to spread out more than that of Case 4.3.2, and the power performance of the former is slightly better than that of the latter.

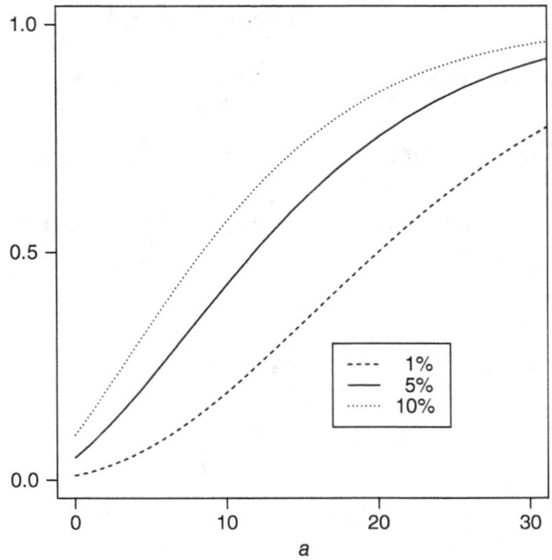

Figure 4.5 Powers of the test based on W^2: Case 4.3.2.

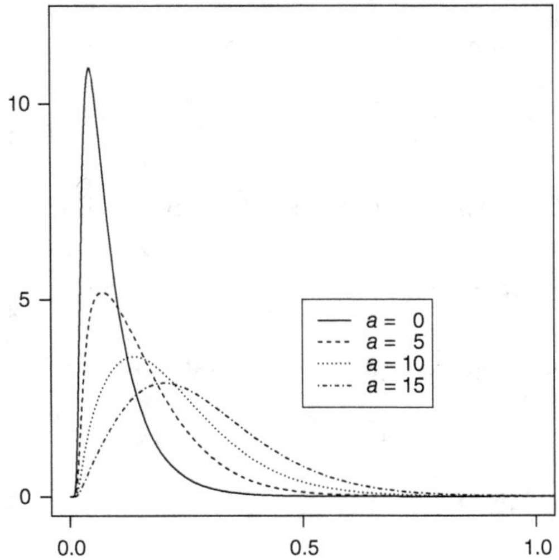

Figure 4.6 Probability densities of W^2: Case 4.3.2.

Exercises for Section 4.3

Exercise 4.3.1 Derive the FD of

$$K(s,t) = \min(s,t) - st - \frac{1}{2\pi^2}\sin 2\pi s \sin 2\pi t$$

in two ways: one is by Darling's formula in (2.159) and the other using the relationship between the eigenvalues of $\min(s,t) - st$ and the corresponding eigenfunctions.

Exercise 4.3.2 Consider the integral equation

$$h(t) = \int_0^1 K(s,t)m(s)\,ds + \lambda \int_0^1 K(s,t)h(s)\,ds,$$
$$m(s) = \frac{\gamma}{2\beta\pi}(\cos 2\pi s - 1),$$

where $K(s,t)$ is the kernel given in Exercise 4.3.1. Show that the above integral equation is equivalent to

$$h''(t) + \lambda h(t) = -m(t) + 2\lambda a \sin 2\pi t$$

with the three conditions

$$h(0) = 0, \quad h(1) = 0, \quad a = \int_0^1 h(s)\sin 2\pi s\,ds = 0.$$

Exercise 4.3.3 Consider the integral equation given in Exercise 4.3.2, where

$$K(s,t) = \min(s,t) - st - \frac{2}{\pi^2}\sin^2 \pi s \sin^2 \pi t, \quad m(t) = \frac{\gamma}{2\beta\pi}\sin 2\pi t.$$

Show that the above integral equation is equivalent to

$$h''(t) + \lambda h(t) = -m(t) - 4\lambda a \cos 2\pi t$$

with the three conditions

$$h(0) = 0, \quad h(1) = 0, \quad a = \int_0^1 h(s)\sin^2 \pi s\,ds.$$

Exercise 4.3.4 For the functions $m(t)$ and $h(t)$ with the three conditions given in Exercise 4.3.3, show that

$$\int_0^1 h(t)m(t)\,dt = \frac{1}{4\pi^2 - \lambda}\frac{\gamma^2}{8\beta^2\pi^2}.$$

4.4 Limitation of the Present Approach

In this chapter we have taken up the Cauchy distribution as a hypothesized distribution to derive explicitly the limiting c.f. of the modified Cramér–von Mises–Smirnov statistic. In Chapters 1 and 3 we also dealt with the logistic distribution parameterized by a single parameter. These, however, are exceptionally successful cases for deriving the c.f. of test statistics. Most distributions frequently used as hypothesized distributions are not amenable to the explicit derivation of the c.f., the reason for which we describe here.

Let us consider the testing problem

$$H_0 : G(x) = F(x;\xi),$$

where ξ is a scalar nuisance parameter to be estimated. Then it follows from the discussions in Section 4.2 that

$$W_n^2(\hat{\xi}) \Rightarrow W^2(\xi) \stackrel{\mathcal{D}}{=} \int_0^1 \int_0^1 K(s,t)\, dW(s)\, dW(t), \tag{4.34}$$

where $\hat{\xi}$ is the MLE of ξ and

$$K(s,t) = \min(s,t) - st - \sigma^2(\xi) g_\xi(s) g_\xi(t), \tag{4.35}$$

$$g_\xi(t) = \left.\frac{\partial F(x;\xi)}{\partial \xi}\right|_{x=x(t;\xi)}, \quad x(t;\xi) = \inf\{x : F(x;\xi) = t\}, \tag{4.36}$$

$$\sigma^{-2}(\xi) = \mathrm{E}\left[\left(\frac{\partial \log f(X;\xi)}{\partial \xi}\right)^2\right], \quad f(x;\xi) = \frac{\partial F(x;\xi)}{\partial x}. \tag{4.37}$$

The first difficulty with the present problem is whether we can compute $g_\xi(t)$ in (4.36) explicitly. Suppose that

$$F(x;\xi) = \int_{-\infty}^{x} f(u;\xi)\, du = \Phi(x-\xi), \quad f(x;\xi) = \frac{1}{\sqrt{2\pi}} \exp\left\{-\frac{(x-\xi)^2}{2}\right\},$$

where $\Phi(x)$ is the standard normal distribution function. Thus we have

$$F(x;\xi) = \Phi(x-\xi) = t \quad \Leftrightarrow \quad x = \xi + \Phi^{-1}(t)$$

so that

$$\begin{aligned} g_\xi(t) &= \left.\frac{\partial F(x;\xi)}{\partial \xi}\right|_{x=x(t;\xi)} = \left.\frac{\partial \Phi(x-\xi)}{\partial \xi}\right|_{x=x(t;\xi)} \\ &= -\frac{1}{\sqrt{2\pi}} \exp\left\{-\frac{(x-\xi)^2}{2}\right\}\Bigg|_{x=x(t;\xi)} \\ &= -\frac{1}{\sqrt{2\pi}} \exp\left\{-\frac{\left(\Phi^{-1}(t)\right)^2}{2}\right\}. \end{aligned}$$

It seems impossible to obtain $g_\xi(t)$ explicitly. The situation becomes more complicated if we deal with $F(x;\boldsymbol{\xi}) = \Phi((x-\mu)/\sigma)$, where $\boldsymbol{\xi} = (\mu, \sigma^2)'$.

The second difficulty is whether we can derive the FD of the kernel $K(s,t)$, even if $g_\xi(t)$ and $K(s,t)$ can be obtained explicitly. As an example of this, consider the exponential distribution with a scale parameter ξ (> 0) given by

$$F(x;\xi) = 1 - e^{-\xi x}, \quad f(x;\xi) = \xi e^{-\xi x}.$$

It follows from $F(x;\xi) = t$ that $e^{-\xi x} = 1 - t$ and $\xi x = -\log(1-t)$, so that

$$\begin{aligned} g_\xi(t) &= \left.\frac{\partial F(x;\xi)}{\partial \xi}\right|_{x=x(t;\xi)} = \left. x e^{-\xi x}\right|_{x=x(t;\xi)} \\ &= -\frac{1}{\xi}(1-t)\log(1-t). \end{aligned}$$

We also have $\sigma^{-2}(\xi) = \mathrm{E}\left[(X - 1/\xi)^2\right] = \mathrm{Var}(X) = 1/\xi^2$. Thus we obtain the kernel

$$K(s,t) = \min(s,t) - st - (1-s)(1-t)\log(1-s)\,\log(1-t). \tag{4.38}$$

Darling's formula in (2.83) gives the FD of $K(s,t)$ in (4.38) as

$$\begin{aligned} D(\lambda) = \frac{\sin\sqrt{\lambda}}{\sqrt{\lambda}} - 2\int_0^1 \left(\int_0^t (1+\log(1-s))\cos\sqrt{\lambda}s\, ds\right) \\ \times (1+\log(1-t))\,\cos\sqrt{\lambda}(1-t)\,dt. \end{aligned}$$

It is difficult to execute the above integrals explicitly.

In Section 1.5.1, we dealt with the logistic distribution with a location parameter. The associated kernel was $K(s,t) = \min(s,t) - st - 3st(1-s)(1-t)$. The FD of this kernel can be easily derived from Darling's formula and is given in (1.79). The situation is quite different if we include a scale parameter. Consider the logistic distribution with a scale parameter ξ (> 0) given by

$$F(x;\xi) = \frac{1}{1+e^{-x/\xi}}, \quad f(x;\xi) = \frac{e^{-x/\xi}}{\xi(1+e^{-x/\xi})^2}.$$

It follows from $F(x;\xi) = t$ that $1 + e^{-x/\xi} = 1/t$, so that

$$\begin{aligned} g_\xi(t) &= \left.\frac{\partial F(x;\xi)}{\partial \xi}\right|_{x=x(t;\xi)} = \left.\frac{-xe^{-x/\xi}}{\xi^2(1+e^{-x/\xi})^2}\right|_{x=x(t;\xi)} \\ &= \frac{t(1-t)}{\xi}\log\frac{1-t}{t}. \end{aligned}$$

Putting $r(x) = f(x; 1)$, we also have

$$\begin{aligned}
\sigma^{-2}(\xi) &= \mathrm{E}\left[\left(\frac{\partial \log f(X;\xi)}{\partial \xi}\right)^2\right] \\
&= \frac{1}{\xi^2}\int_{-\infty}^{\infty}\left(1 + x\frac{r'(x)}{r(x)}\right)^2 r(x)\,dx \\
&= \frac{1}{\xi^2}\left[\frac{1}{2}\int_0^{\infty}\frac{x^2\tanh^2(x/2)}{\cosh^2(x/2)}\,dx - 1\right] \\
&= \frac{A}{\xi^2} \qquad (A = 1.42996).
\end{aligned}$$

Thus we obtain

$$K(s,t) = \min(s,t) - st - \frac{1}{A}st(1-s)(1-t)\log\frac{(1-s)}{s}\log\frac{(1-t)}{t},$$

which is more involved than $K(s,t)$ in (4.38).

In conclusion, the computation of the FD associated with the Cramér–von Mises–Smirnov statistic is difficult if the underlying distributions are normal or exponential. The computation is also difficult if the distribution is logistic and depends on a nuisance scale parameter. The difficulty also applies to most other common distributions.

5

Fredholm Determinants in the State Space Model

Using a simple state space model, an LBI test for parameter constancy was introduced in Section 1.5.2, where the state equation follows a random walk that describes the parameter variation over time. The c.f. of the limiting distribution of the LBI statistic under the null distribution was also described. This chapter extends the measurement equation to consider various tests for parameter constancy. We also discuss how to compute the limiting powers under a sequence of local alternatives, and we compare the powers under the two situations where the initial value of the random walk is known or unknown.

5.1 Weak Convergence of Quadratic Forms

In Section 1.5.2 we presented Theorem 1.11 concerning the weak convergence of quadratic forms in connection with the FD. For convenience, an extended version of Theorem 1.11 is presented here together with its proof (Nabeya and Tanaka, 1988, 1990a).

Theorem 5.1 *Consider*

$$S_n = \frac{1}{n}\boldsymbol{\varepsilon}' B_n \boldsymbol{\varepsilon} = \frac{1}{n}\sum_{j=1}^{n}\sum_{k=1}^{n} B_n(j,k)\varepsilon_j\varepsilon_k,$$

where $\{\varepsilon_j\} \sim$ *i.i.d.*$(0,1)$ *and* B_n *is an* $n \times n$ *symmetric matrix. Assume that, for the* (j,k)*th element* $B_n(j,k)$ *of* B_n*, there exists a symmetric, continuous, and nearly definite function* $K(s,t)$ *satisfying*

$$\lim_{n\to\infty}\max_{1\le j,k\le n}\left|B_n(j,k) - K\left(\frac{j}{n},\frac{k}{n}\right)\right| = 0. \tag{5.1}$$

Then it holds that, as $n \to \infty$,

$$S_n \Rightarrow S = \sum_{j=1}^{\infty} \frac{1}{\lambda_j} Z_j^2 \stackrel{\mathcal{D}}{=} \int_0^1 \int_0^1 K(s,t)\, dW(s)\, dW(t), \tag{5.2}$$

where $\{\lambda_j\}$ is a sequence of eigenvalues of $K(s,t)$ repeated as many times as their multiplicities, and $\{Z_j\} \sim NID(0,1)$. Moreover, the c.f. of S is given by

$$E\left(e^{i\theta S}\right) = \prod_{j=1}^{\infty} \left(1 - \frac{2i\theta}{\lambda_j}\right)^{-1/2} = (D(2i\theta))^{-1/2}, \tag{5.3}$$

where $D(\lambda)$ is the FD of $K(s,t)$.

Proof Let K_n be the $n \times n$ matrix whose (j,k)th element is $K(j/n, k/n)$. Then, because of (5.1), it holds that

$$\frac{1}{n}\, \boldsymbol{\varepsilon}'(B_n - K_n)\boldsymbol{\varepsilon} = \frac{1}{n} \sum_{j=1}^{n} \sum_{k=1}^{n} \left(B_n(j,k) - K\left(\frac{j}{n}, \frac{k}{n}\right)\right) \varepsilon_j \varepsilon_k$$

converges to 0 in probability. Thus we can assume that $B_n(j,k) = K(j/n, k/n)$, so that $S_n = \boldsymbol{\varepsilon}' B_n \boldsymbol{\varepsilon}/n = \boldsymbol{\varepsilon}' K_n \boldsymbol{\varepsilon}/n$. It follows from Mercer's theorem (Theorem 2.2) that $K(s,t)$ can be expanded as

$$K(s,t) = \sum_{j=1}^{\infty} \frac{1}{\lambda_j} f_j(s) f_j(t)$$

with absolute and uniform convergence for $(s,t) \in [0,1] \times [0,1]$, where $\{f_j(t)\}$ is an orthonormal sequence of eigenfunctions corresponding to the eigenvalues λ_j $(j = 1, 2, \ldots)$. Since $f_j(t)$ is bounded for $t \in [0,1]$, the Lindeberg condition is satisfied, so that, as $n \to \infty$,

$$\frac{1}{\sqrt{n}} \sum_{j=1}^{n} f_\ell\left(\frac{j}{n}\right) \varepsilon_j \Rightarrow \mathrm{N}(0,1) \quad \text{for each } \ell.$$

Putting $f(t) = c_1 f_1(t) + c_2 f_2(t) + \cdots + c_m f_m(t)$ for any real $c_1, \ldots, c_m$, it follows that, as $n \to \infty$ with m fixed,

$$\frac{1}{\sqrt{n}} \sum_{j=1}^{n} f\left(\frac{j}{n}\right) \varepsilon_j = \sum_{\ell=1}^{m} c_\ell \frac{1}{\sqrt{n}} \sum_{j=1}^{n} f_\ell\left(\frac{j}{n}\right) \varepsilon_j \Rightarrow \mathrm{N}\left(0, \sum_{\ell=1}^{m} c_\ell^2\right), \tag{5.4}$$

which implies that the joint distribution of $\sum_{j=1}^{n} f_\ell(j/n)\varepsilon_j/\sqrt{n}$ $(\ell = 1, \ldots, m)$ tends to $\mathrm{N}(\mathbf{0}, I_m)$.

Let us consider

$$S_n = \frac{1}{n}\, \boldsymbol{\varepsilon}' K_n \boldsymbol{\varepsilon} = \frac{1}{n}\, \boldsymbol{\varepsilon}' \tilde{K}_{n,m} \boldsymbol{\varepsilon} + \frac{1}{n}\, \boldsymbol{\varepsilon}'(K_n - \tilde{K}_{n,m})\boldsymbol{\varepsilon}, \tag{5.5}$$

where $\tilde{K}_{n,m}$ is the $n \times n$ matrix whose (j,k)th element is given by

$$\tilde{K}_{n,m}(j,k) = \tilde{K}_m\left(\frac{j}{n}, \frac{k}{n}\right), \quad \tilde{K}_m(s,t) = \sum_{\ell=1}^{m} \frac{1}{\lambda_\ell} f_\ell(s) f_\ell(t).$$

Because of the properties of $K(s,t)$ and $\tilde{K}_m(s,t)$, for every $\delta > 0$, there exists a positive integer m_1 such that, for all $m > m_1$,

$$|\tilde{K}_m(s,t) - K(s,t)| < \delta, \quad P\left(\frac{1}{n}|\boldsymbol{\varepsilon}'(\tilde{K}_{n,m} - K_n)\boldsymbol{\varepsilon}| > \delta\right) < \delta$$

for all n. Furthermore, it follows from (5.4) that, as $n \to \infty$ with m fixed,

$$\frac{1}{n}\boldsymbol{\varepsilon}'\tilde{K}_{n,m}\boldsymbol{\varepsilon} = \sum_{\ell=1}^{m} \frac{1}{\lambda_\ell}\left(\frac{1}{\sqrt{n}}\sum_{j=1}^{n} f_\ell\left(\frac{j}{n}\right)\varepsilon_j\right)^2 \Rightarrow V_m = \sum_{\ell=1}^{m} \frac{1}{\lambda_\ell} Z_\ell^2,$$

where $\{Z_\ell\} \sim \mathrm{NID}(0,1)$ and $V_m \Rightarrow \sum_{\ell=1}^{\infty} Z_\ell^2/\lambda_\ell$ as $m \to \infty$. Thus it follows from Anderson (1971, theorem 7.7.1) that S_n in (5.5) satisfies (5.2). The c.f. of S given in (5.3) follows from the Anderson–Darling theorem (Theorem 2.3), and the theorem is established. □

The above theorem is useful for obtaining the limiting null distributions of various test statistics discussed subsequently. However, it needs to be extended further to deal with distributions under a sequence of local alternatives. For this purpose we have the following theorem (Nabeya, 1989; Tanaka, 1990b).

Theorem 5.2 *Assume the conditions described in Theorem 5.1, and suppose that the statistic T_n is given by*

$$T_n = \frac{1}{n}\boldsymbol{\varepsilon}' B_n \boldsymbol{\varepsilon} + \frac{\gamma}{n^2}\boldsymbol{\varepsilon}' B_n^2 \boldsymbol{\varepsilon}, \tag{5.6}$$

where γ is any constant. Then it holds that $T_n \Rightarrow T$ as $n \to \infty$, where

$$T = \sum_{j=1}^{\infty}\left(\frac{1}{\lambda_j} + \frac{\gamma}{\lambda_j^2}\right) Z_j^2 \tag{5.7}$$

$$\stackrel{\mathcal{D}}{=} \int_0^1\int_0^1 \left(K(s,t) + \gamma K^{(2)}(s,t)\right) dW(s)\, dW(t),$$

with $K^{(2)}(s,t)$ being the iterated kernel of $K(s,t)$ defined in (2.26). Moreover, the c.f. of T is given by

$$E\left(e^{i\theta T}\right) = \prod_{j=1}^{\infty}\left[1 - 2i\theta\left(\frac{1}{\lambda_j} + \frac{\gamma}{\lambda_j^2}\right)\right]^{-1/2}$$

$$= \left[D\left(i\theta + \sqrt{-\theta^2 + 2i\gamma\theta}\right) D\left(i\theta - \sqrt{-\theta^2 + 2i\gamma\theta}\right)\right]^{-1/2}, \tag{5.8}$$

where $D(\lambda)$ is the FD of $K(s,t)$.

Proof It follows from condition (5.1) that

$$\lim_{n\to\infty} \max_{1\le j,k\le n} \left| \frac{1}{n} B_n^{(2)}(j,k) - K^{(2)}\left(\frac{j}{n}, \frac{k}{n}\right) \right| = 0,$$

where $B_n^{(2)}(j,k)$ is the (j,k)th element of B_n^2. Thus it holds that

$$\lim_{n\to\infty} \max_{1\le j,k\le n} \left| B_n(j,k) + \frac{\gamma}{n} B_n^{(2)}(j,k) - \left\{ K\left(\frac{j}{n}, \frac{k}{n}\right) + \gamma K^{(2)}\left(\frac{j}{n}, \frac{k}{n}\right) \right\} \right| = 0.$$

We can now consider

$$T_n = \frac{1}{n} \boldsymbol{\varepsilon}' K_n \boldsymbol{\varepsilon} + \frac{\gamma}{n^2} \boldsymbol{\varepsilon}' K_n^{(2)} \boldsymbol{\varepsilon},$$

where $K_n^{(2)}$ is the $n \times n$ matrix whose (j,k)th element is $K^{(2)}(j/n, k/n)$. Noting that $K^{(2)}(s,t)$ can be expanded (see (2.28)) as

$$K^{(2)}(s,t) = \sum_{j=1}^{\infty} \frac{1}{\lambda_j^2} f_j(s) f_j(t)$$

and using Theorem 5.1, we can prove (5.7). The result in (5.8) was earlier presented in Theorem 2.10, which establishes the theorem. □

The mean and variance of T in (5.7) can be computed by differentiating $\mathrm{E}\left(e^{i\theta T}\right)$. Alternatively, we obtain

Corollary 5.2 *For the statistic T in (5.7), it holds that*

$$\mathrm{E}(T) = \int_0^1 \left(K(t,t) + \gamma K^{(2)}(t,t) \right) dt \;=\; d_1 + \gamma\left(d_1^2 - d_2\right),$$

$$\begin{aligned}
\mathrm{Var}(T) &= 2 \int_0^1 \int_0^1 \left(K(s,t) + \gamma K^{(2)}(s,t) \right)^2 ds\, dt \\
&= 2 \int_0^1 \left(K^{(2)}(t,t) + 2\gamma K^{(3)}(t,t) + \gamma^2 K^{(4)}(t,t) \right) dt \\
&= 2 \sum_{j=1}^{\infty} \left(\frac{1}{\lambda_j^2} + \frac{2\gamma}{\lambda_j^3} + \frac{\gamma^2}{\lambda_j^4} \right) = 2 \sum_{j=1}^{\infty} \frac{(\lambda_j + \gamma)^2}{\lambda_j^4},
\end{aligned}$$

where d_1 and d_2 are coefficients in the series expansion for $D(\lambda)$ expressed as $1 - d_1\lambda + d_2\lambda^2/2 - \cdots$, and $K^{(i)}(s,t)$ is the ith iterated kernel defined by

$$K^{(i)}(s,t) = \int_0^1 K^{(i-1)}(s,u) K(u,t)\, du, \quad K^{(1)}(s,t) = K(s,t).$$

5.2 LBI Tests for Parameter Constancy

Let us consider the following state space model:

$$y_j = \beta_j x_j + \boldsymbol{\gamma}' \boldsymbol{z}_j + \varepsilon_j, \quad \beta_j = \beta_{j-1} + \eta_j \qquad (j = 1, \ldots, n), \tag{5.9}$$

where

(i) regressor variables $\{x_j\}$ and $\{\boldsymbol{z}_j\}$ are scalar and $p \times 1$ nonstochastic sequences, respectively;

(ii) $\begin{pmatrix} \varepsilon_j \\ \eta_j \end{pmatrix} \sim \mathrm{NID}\left(\mathbf{0}, \begin{pmatrix} \sigma_\varepsilon^2 & 0 \\ 0 & \sigma_\eta^2 \end{pmatrix}\right)$;

(iii) $\boldsymbol{\gamma}$ is a $p \times 1$ unknown constant vector, and the state variable $\{\beta_j\}$ follows a random walk starting from an unknown constant β_0.

The above model is an extension of the simple state space model dealt with in Section 1.5.2. Our concern here is to test if β_j is a constant over time, that is, $\beta_j = \beta_0$ for all j. This is equivalent to testing $\rho = \sigma_\eta^2/\sigma_\varepsilon^2 = 0$. In particular, we consider the testing problem

$$H_0 : \rho = \frac{\sigma_\eta^2}{\sigma_\varepsilon^2} = 0 \qquad \text{versus} \qquad H_1(c, \kappa) : \rho = \frac{c}{n^\kappa}, \tag{5.10}$$

where c and κ are positive constants. The testing problem in the present model was discussed in Nabeya and Tanaka (1988) and Nabeya (1989), which we explain below.

Noting that $y_j = \beta_0 x_j + \boldsymbol{\gamma}' \boldsymbol{z}_j + (\eta_1 + \cdots + \eta_j) x_j + \varepsilon_j$, we rewrite (5.9) as

$$\boldsymbol{y} = \beta_0 \boldsymbol{x} + Z\boldsymbol{\gamma} + D_x C \boldsymbol{\eta} + \boldsymbol{\varepsilon} \sim \mathrm{N}\left(\beta_0 \boldsymbol{x} + Z\boldsymbol{\gamma}, \, \sigma_\varepsilon^2 \left(I_n + \rho D_x C C' D_x\right)\right), \tag{5.11}$$

where $\boldsymbol{y}, \boldsymbol{x}, \boldsymbol{\eta}$, and $\boldsymbol{\varepsilon}$ are $n \times 1$ vectors; $Z = (\boldsymbol{z}_1, \ldots, \boldsymbol{z}_n)' : n \times p$; $D_x = \mathrm{diag}(x_1, \ldots, x_n)$; and C is the $n \times n$ random walk-generating matrix whose (j,k)th element is 0 for $j < k$ and 1 for $j \geq k$. Thus we have the the log-likelihood for $\boldsymbol{y}$ given by

$$\begin{aligned} L(\boldsymbol{\delta}) = & -\frac{n}{2} \log 2\pi\sigma_\varepsilon^2 - \frac{1}{2} \log |I_n + \rho D_x C C' D_x| - \frac{1}{2\sigma_\varepsilon^2} \\ & \times (\boldsymbol{y} - \beta_0 \boldsymbol{x} - Z\boldsymbol{\gamma})' (I_n + \rho D_x C C' D_x)^{-1} (\boldsymbol{y} - \beta_0 \boldsymbol{x} - Z\boldsymbol{\gamma}), \end{aligned} \tag{5.12}$$

where $\boldsymbol{\delta} = (\rho, \beta_0, \boldsymbol{\gamma}', \sigma_\varepsilon^2)'$. Since it was shown in Section 1.5.2 that the LM principle is much more convenient for deriving the LBI test, we evaluate $\partial L(\boldsymbol{\delta})/\partial\rho$ at the MLE $\tilde{\boldsymbol{\delta}}$ of $\boldsymbol{\delta}$ under H_0, which yields (Exercise 5.2.1)

$$\left.\frac{\partial L(\boldsymbol{\delta})}{\partial \rho}\right|_{H_0} = -\frac{1}{2}\mathrm{tr}(D_x C C' D_x) + \frac{n}{2}\,\frac{\boldsymbol{y}' M D_x C C' D_x M \boldsymbol{y}}{\boldsymbol{y}' M \boldsymbol{y}},$$

where

$$M = I_n - (\boldsymbol{x}, Z)\left((\boldsymbol{x}, Z)'(\boldsymbol{x}, Z)\right)^{-1}(\boldsymbol{x}, Z)'.$$

Then the LBI test rejects H_0 when

$$R_n = \frac{\boldsymbol{y}' M D_x C C' D_x M \boldsymbol{y}}{\boldsymbol{y}' M \boldsymbol{y}} \tag{5.13}$$

takes large values.

To derive the limiting c.f. associated with R_n, we need to specify regressor variables $\{x_j\}$ and $\{\boldsymbol{z}_j\}$. Here we consider three cases:

$$\text{(a)}\quad x_j = j^m,\ \boldsymbol{z}_j = \boldsymbol{0},\quad \text{(b)}\quad x_j = 1,\ \boldsymbol{z}_j = j^q,\quad \text{(c)}\quad x_j = 1,\ \boldsymbol{z}_j = (j, j^2)', \tag{5.14}$$

where $m > -1/2$ and $q = 1, 2, 3, 4$. For these three cases, it is shown (Exercise 5.2.2) that, as $n \to \infty$ under $\rho = c/n^{\kappa}$, where $\kappa = 2m + 2$ for Case (a) and $\kappa = 2$ for Cases (b) and (c),

$$\frac{1}{n}\,\boldsymbol{y}' M \boldsymbol{y} \to \sigma_\varepsilon^2 \quad \text{in probability.} \tag{5.15}$$

Thus we can concentrate on $\boldsymbol{y}' M D_x C C' D_x M \boldsymbol{y}$ to derive the asymptotic distribution of R_n in (5.13).

In the next section (Section 5.3) we derive, under the sequence of local alternatives $H_1(c, \kappa)$ in (5.10), the limiting c.f.s of the suitably normalized R_n for the three cases in (5.14). In Section 5.4 we deal with the LBI tests under the situation where the initial value β_0 is known and, in Section 5.5, compare the asymptotic distributions of test statistics between the two situations.

Exercises for Section 5.2

Exercise 5.2.1 For $\boldsymbol{y} = \beta_0 \boldsymbol{x} + Z\boldsymbol{\gamma} + D_x C \boldsymbol{\eta} + \boldsymbol{\varepsilon}$ in (5.11), prove that

$$\left.\frac{\partial L(\boldsymbol{\delta})}{\partial \rho}\right|_{H_0} = -\frac{1}{2}\mathrm{tr}(D_x C C' D_x) + \frac{n}{2}\,\frac{\boldsymbol{y}' M D_x C C' D_x M \boldsymbol{y}}{\boldsymbol{y}' M \boldsymbol{y}},$$

where $L(\boldsymbol{\delta})$ is the log-likelihood for $\boldsymbol{y}$.

Exercise 5.2.2 For Cases (a), (b), and (c) of x_j and $\boldsymbol{z}_j$ given in (5.14), prove that, as $n \to \infty$ under $\rho = c/n^{\kappa}$, where $\kappa = 2m + 2$ for Case (a) and $\kappa = 2$ for Cases (b) and (c),

$$\frac{1}{n}\,\boldsymbol{y}' M \boldsymbol{y} \to \sigma_\varepsilon^2 \quad \text{in probability.}$$

5.3 Asymptotic Distributions of the LBI Statistics – 1

Putting $\Omega(\rho; x) = I_n + \rho D_x CC' D_x$, we first note that

$$My = M(D_x C\boldsymbol{\eta} + \boldsymbol{\varepsilon}) \stackrel{\mathcal{D}}{=} M\Omega^{1/2}(\rho; x)\boldsymbol{\zeta} \sim \mathrm{N}\left(\mathbf{0}, \sigma_\varepsilon^2 M\Omega(\rho; x)M\right),$$

where $\boldsymbol{\zeta} \sim \mathrm{N}(\mathbf{0}, \sigma_\varepsilon^2 I_n)$. Then the following distributional equivalence holds:

$$\begin{aligned} y'MD_xCC'D_xMy &\stackrel{\mathcal{D}}{=} \boldsymbol{\zeta}'\Omega^{1/2}(\rho; x)MD_xCC'D_xM\Omega^{1/2}(\rho; x)\boldsymbol{\zeta} \\ &\stackrel{\mathcal{D}}{=} \boldsymbol{\zeta}'C'D_xM(I_n + \rho D_xCC'D_x)MD_xC\boldsymbol{\zeta} \\ &= \boldsymbol{\zeta}'\left(C'D_xMD_xC + \rho(C'D_xMD_xC)^2\right)\boldsymbol{\zeta} \\ &= \boldsymbol{\zeta}'A_n\boldsymbol{\zeta} + \rho\boldsymbol{\zeta}'A_n^2\boldsymbol{\zeta}, \end{aligned} \tag{5.16}$$

where $A_n = C'D_xMD_xC$. To be more specific, let us consider the three cases of regressor variables x_j and z_j given in (5.14).

Case (a): $x_j = j^m$, $z_j = \mathbf{0}$ $\quad (m > -1/2)$.

Since $M = I_n - xx'/x'x$, the (j,k)th element of A_n is given (Exercise 5.3.1) by

$$A_n(j,k) = \sum_{\ell=\max(j,k)}^{n} \ell^{2m} - \sum_{\ell=j}^{n} \ell^{2m} \sum_{\ell=k}^{n} \ell^{2m} \Bigg/ \sum_{\ell=1}^{n} \ell^{2m},$$

so that $B_n(j,k) = A_n(j,k)/n^{2m+1}$ converges uniformly to

$$\begin{aligned} K(s,t;m) &= \frac{1}{2m+1}\left[1 - (\max(s,t))^{2m+1} - (1 - s^{2m+1})(1 - t^{2m+1})\right] \\ &= \frac{1}{2m+1}\left[(\min(s,t))^{2m+1} - (st)^{2m+1}\right]. \end{aligned} \tag{5.17}$$

The FD of $K(s,t;m)$ was earlier obtained in (2.141). Putting $\rho = c/n^{2m+2}$, it follows from (5.16) that

$$\begin{aligned} \frac{1}{n^{2m+2}} y'MD_xCC'D_xMy &\stackrel{\mathcal{D}}{=} \frac{1}{n^{2m+2}}\left[\boldsymbol{\zeta}'A_n\boldsymbol{\zeta} + \frac{c}{n^{2m+2}}\boldsymbol{\zeta}'A_n^2\boldsymbol{\zeta}\right] \\ &= \frac{1}{n}\boldsymbol{\zeta}'B_n\boldsymbol{\zeta} + \frac{c}{n^2}\boldsymbol{\zeta}'B_n^2\boldsymbol{\zeta}. \end{aligned}$$

Then we obtain the following theorem using Theorem 5.2 and (5.15).

Theorem 5.3 *Consider the state space model*

$$y_j = \beta_j j^m + \varepsilon_j, \quad \beta_j = \beta_{j-1} + \eta_j \quad (m > -1/2),$$

where $\{\beta_j\}$ starts from an unknown constant β_0, and $\{\varepsilon_j\}$ and $\{\eta_j\}$ satisfy the conditions described in (5.9). Then it holds that, as $n \to \infty$ under $\rho = c/n^{2m+2}$,

$$\frac{1}{n^{2m+1}} R_n = \frac{y'MD_xCC'D_xMy/(n^{2m+2}\sigma_\varepsilon^2)}{y'My/(n\sigma_\varepsilon^2)} \Rightarrow R(m;c),$$

$$R(m;c) = \int_0^1 \int_0^1 \left(K(s,t;m) + cK^{(2)}(s,t;m) \right) dW(s)\,dW(t), \quad (5.18)$$

$$E(R(m;c)) = \frac{1}{2(m+1)(4m+3)} + \frac{c}{2(m+1)(4m+3)^2(6m+5)},$$

where $K(s,t;m)$ is given in (5.17) and $K^{(2)}(s,t;m)$ is the iterated kernel of $K(s,t;m)$. Moreover, the c.f. of $R(m;c)$ is given by (5.8) with γ replaced by c, and the FD of $K(s,t;m)$ is given by

$$D(\lambda) = \Gamma(\nu+1) J_\nu \left(\frac{\sqrt{\lambda}}{m+1} \right) \Big/ \left(\frac{\sqrt{\lambda}}{2(m+1)} \right)^\nu, \quad (5.19)$$

where $\nu = (2m+1)/2(m+1)$ and $J_\nu(z)$ is the Bessel function of the first kind defined in (2.58).

Note that, when $m = 0$,

$$J_\nu(\sqrt{\lambda}) = J_{1/2}(\sqrt{\lambda}) = \sqrt{\frac{2}{\pi\sqrt{\lambda}}} \sin\sqrt{\lambda},$$

$$D(\lambda) = \Gamma\left(\frac{3}{2}\right) J_{1/2}(\sqrt{\lambda}) \Big/ \sqrt{\frac{\sqrt{\lambda}}{2}} = \frac{\sin\sqrt{\lambda}}{\sqrt{\lambda}}.$$

It may also be noted from (5.19) that

$$\lim_{m\to\infty} D\left((m+1)^2\lambda\right) = 2J_1(\sqrt{\lambda})/\sqrt{\lambda},$$

from which we obtain the following corollary (Exercise 5.3.2).

Corollary 5.3 *For $R(m;c)$ defined in (5.18), it holds that*

$$\lim_{m\to\infty} \mathrm{E}\left(e^{i\theta(m+1)^2 R(m;c)}\right) = \left(\frac{2J_1\left(\sqrt{2i\theta}\right)}{\sqrt{2i\theta}} \right)^{-1/2}. \quad (5.20)$$

It can be seen that, as m gets large, the distribution of $(m+1)^2 R(m;c)$ becomes independent of the local alternative and closer to the null distribution.

Figure 5.1 plots the probability densities of $R(m;c)$ for $m = 0, 0.5, 0.8, 1$ under $H_0 : c = 0$. It can be seen that the distributions are unimodal. They are shifted to the left and do not tend to normality as m gets large. Figure 5.2 plots probability densities of $R(m;c)$ for $m = 1$ and $c = 0, 20, 50, 100$. The distributions are shifted to the right as c gets large, that is, as the alternative gets away from the null distribution. Power functions will be presented in

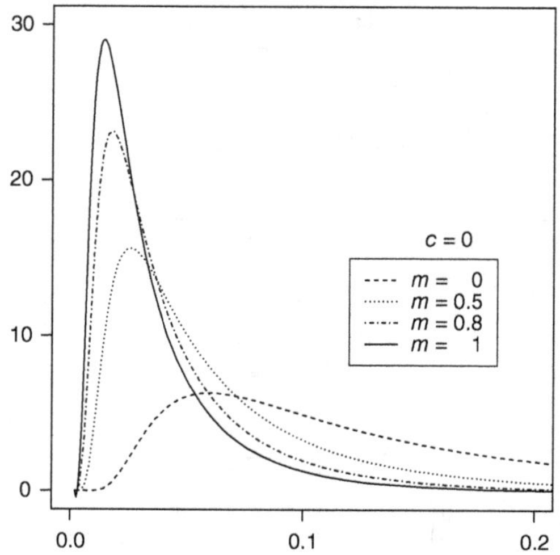

Figure 5.1 Probability densities of $R(m; c)$ for $c = 0$.

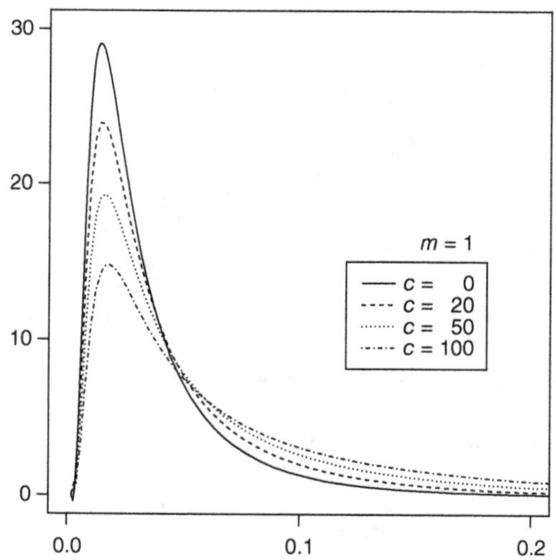

Figure 5.2 Probability densities of $R(m; c)$ for $m = 1$.

Section 5.5 to compare with these under a different initial condition of the state equation.

Case (b): $x_j = 1,\ z_j = j^q \quad (q = 1, 2, 3, 4)$.

Putting $\boldsymbol{x} = \boldsymbol{e} = (1, \ldots, 1)' : n \times 1$, $\boldsymbol{z} = (1^q, 2^q, \ldots, n^q)'$, and noting that $D_x = I_n$, we have

$$A_n = C'MC, \quad M = I_n - (\boldsymbol{e}, \boldsymbol{z}) \left((\boldsymbol{e}, \boldsymbol{z})'(\boldsymbol{e}, \boldsymbol{z}) \right)^{-1} (\boldsymbol{e}, \boldsymbol{z})'.$$

Then it can be shown (Exercise 5.3.3) that $A_n(j,k)/n$ converges uniformly to

$$K(s,t;q) = \min(s,t) - st - \frac{2q+1}{q^2} st(1-s^q)(1-t^q). \tag{5.21}$$

The FD of $K(s,t;q)$ for $q = 4$ was presented earlier in (2.163). The FDs for $q = 1, 2, 3$ can also easily be obtained by Darling's formula given in (2.83) as

$$D(\lambda) = \frac{\sin\sqrt{\lambda}}{\sqrt{\lambda}} - 2\int_0^1 \left(\int_0^t \psi'(s) \cos\sqrt{\lambda}s \, ds \right) \psi'(t) \cos\sqrt{\lambda}(1-t) \, dt,$$

where $\psi(t) = \sqrt{2q+1}\, t(1-t^q)/q$.

We now have the following theorem.

Theorem 5.4 *Consider the state space model*

$$y_j = \beta_j + \gamma\, j^q + \varepsilon_j, \quad \beta_j = \beta_{j-1} + \eta_j \quad (q = 1, 2, 3, 4),$$

where $\{\beta_j\}$, $\{\varepsilon_j\}$, and $\{\eta_j\}$ satisfy the conditions described in Theorem 5.3. Then it holds that, as $n \to \infty$ under $\rho = c/n^2$,

$$\frac{1}{n} R_n = \frac{\boldsymbol{y}'MCC'M\boldsymbol{y}/\left(n^2\sigma_\varepsilon^2\right)}{\boldsymbol{y}'M\boldsymbol{y}/\left(n\sigma_\varepsilon^2\right)} \Rightarrow R(q;c),$$

$$R(q;c) = \int_0^1 \int_0^1 \left(K(s,t;q) + cK^{(2)}(s,t;q) \right) dW(s)\, dW(t), \tag{5.22}$$

where $K(s,t;q)$ is defined in (5.21) and $K^{(2)}(s,t;q)$ is the iterated kernel of $K(s,t;q)$. Moreover, the c.f. of $R(q;c)$ is given by (5.8) with γ replaced by c, and the FD $D(\lambda;q)$ of $K(s,t;q)$ for $q = 1, 2, 3, 4$ is given by

$$\begin{aligned}
D(\lambda;1) &= \frac{12}{\lambda^2} \left(2 - \sqrt{\lambda}\sin\sqrt{\lambda} - 2\cos\sqrt{\lambda} \right), \\
D(\lambda;2) &= \frac{15}{\lambda^3} \left(\sqrt{\lambda}(3-\lambda)\sin\sqrt{\lambda} - 3\lambda\cos\sqrt{\lambda} \right), \\
D(\lambda;3) &= \frac{224}{\lambda^4} \left(4 - 2\lambda - \sqrt{\lambda}\left(2 - \frac{4\lambda}{3} + \frac{\lambda^2}{10} \right) \sin\sqrt{\lambda} \right. \\
&\qquad \left. - \left(4 - 2\lambda + \frac{\lambda^2}{2} \right) \cos\sqrt{\lambda} \right), \\
D(\lambda;4) &= \frac{1350}{\lambda^5} \left(\sqrt{\lambda}\left(6 - 4\lambda + \frac{7\lambda^2}{10} - \frac{\lambda^3}{42} \right) \sin\sqrt{\lambda} \right. \\
&\qquad \left. - \lambda\left(6 - 2\lambda + \frac{\lambda^2}{6} \right) \cos\sqrt{\lambda} \right).
\end{aligned}$$

We have

$$\mathrm{E}(R(1;c)) = \frac{1}{15} + \frac{11}{12{,}600}\,c, \quad \mathrm{E}(R(2;c)) = \frac{1}{14} + \frac{1}{882}\,c,$$
$$\mathrm{E}(R(3;c)) = \frac{13}{162} + \frac{2437}{1{,}443{,}420}\,c, \quad \mathrm{E}(R(4;c)) = \frac{41}{462} + \frac{3221}{1{,}387{,}386}\,c.$$

Figure 5.3 plots the probability densities of $R(q;c)$ for $q = 1, 2, 3, 4$ under $H_0 : c = 0$, whereas Figure 5.4 shows those for $q = 1$ and $c = 0, 20, 50, 100$. It can be seen from Figure 5.3 the that distributions are unimodal, and the mode does not change much as q becomes large. It can be seen from Figure 5.4 that the mode dose not change much as c becomes large, although the peak decreases and the right-hand tail becomes thick as c gets large. Power functions will be graphically presented in Section 5.5.

Case (c): $x_j = 1,\ z_j = (j, j^2)'$.

Putting $\boldsymbol{x} = \boldsymbol{e} = (1, \ldots, 1)$, $Z = (\boldsymbol{d}, \boldsymbol{f})$, where $\boldsymbol{d} = (1, 2, \ldots, n)'$, $\boldsymbol{f} = (1^2, 2^2, \ldots, n^2)'$, we have

$$A_n = C'MC, \quad M = I_n - (\boldsymbol{e}, \boldsymbol{d}, \boldsymbol{f})\left((\boldsymbol{e}, \boldsymbol{d}, \boldsymbol{f})'(\boldsymbol{e}, \boldsymbol{d}, \boldsymbol{f})\right)^{-1}(\boldsymbol{e}, \boldsymbol{d}, \boldsymbol{f}).$$

Then it can be shown (Exercise 5.3.4) that $A_n(j,k)/n$ converges uniformly to

$$K(s,t) = \min(s,t) - st - 2st(1-s)(1-t)(4 - 5s - 5t + 10st). \quad (5.23)$$

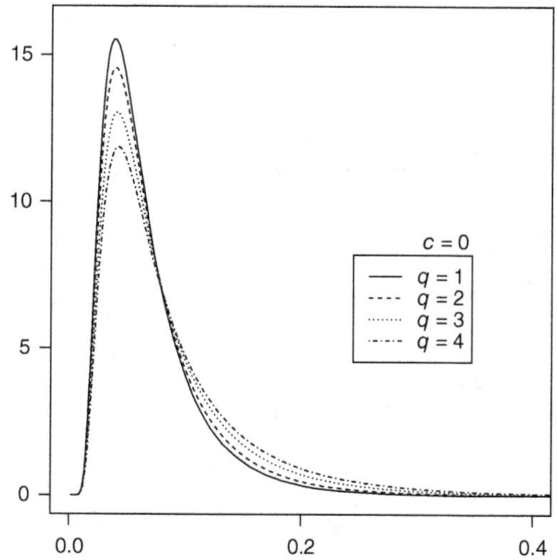

Figure 5.3 Probability densities of $R(q;c)$ for $q = 1, 2, 3, 4$ and $c = 0$.

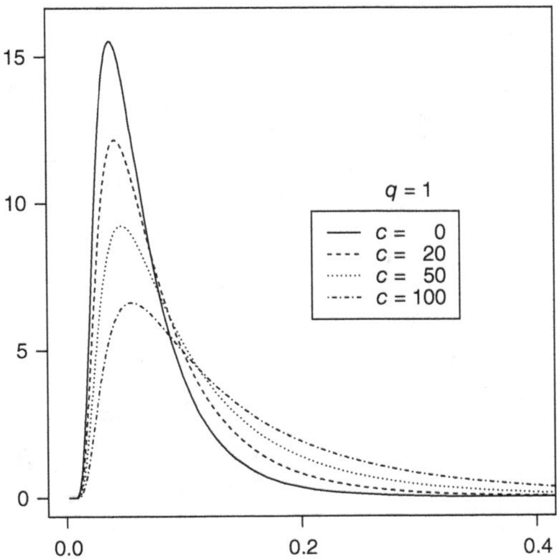

Figure 5.4 Probability densities of $R(q;c)$ for $q = 1$.

The FD of $K(s,t)$ was presented earlier in (2.179), which was efficiently obtained from Theorem 2.20 by putting

$$K(s,t) = \min(s,t) - st - \sum_{j=1}^{3} \psi_j(s)\psi_j(t),$$

where

$$\psi_1(t) = \sqrt{2}it(1-t), \quad \psi_2(t) = \sqrt{10}t^2(1-t), \quad \psi_3(t) = \sqrt{10}t(1-t)^2.$$

We now have the following theorem.

Theorem 5.5 *Consider the state space model*

$$y_j = \beta_j + \gamma_1 j + \gamma_2 j^2 + \varepsilon_j, \quad \beta_j = \beta_{j-1} + \eta_j,$$

where $\{\beta_j\}$*,* $\{\varepsilon_j\}$*, and* $\{\eta_j\}$ *satisfy the conditions described in Theorem 5.3. Then it holds that, as* $n \to \infty$ *under* $\rho = c/n^2$*,*

$$\frac{1}{n} R_n = \frac{y'MCC'My/\left(n^2\sigma_\varepsilon^2\right)}{y'My/\left(n\sigma_\varepsilon^2\right)} \Rightarrow R(c),$$

$$R(c) = \int_0^1 \int_0^1 \left(K(s,t) + cK^{(2)}(s,t)\right) dW(s)\, dW(t), \tag{5.24}$$

$$E(R(c)) = d_1 + c\left(d_1^2 - d_2\right) = \frac{3}{70} + \frac{11}{44{,}100}\, c,$$

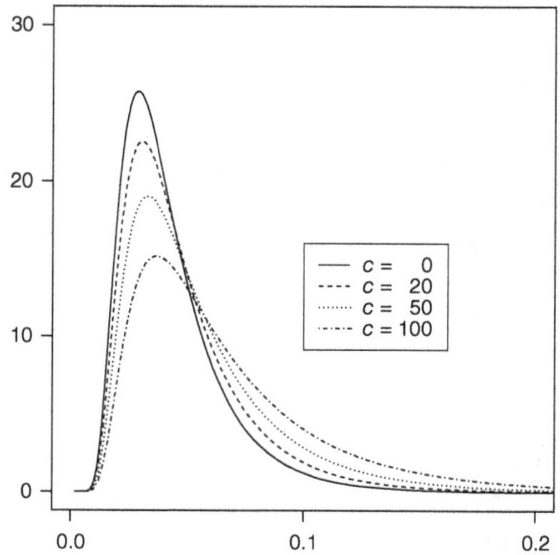

Figure 5.5 Probability densities of $R(c)$ for $c = 0$, 20, 50, 100.

where $K(s,t)$ is defined in (5.23) and $K^{(2)}(s,t)$ is the iterated kernel of $K(s,t)$. Moreover, the c.f. of $R(c)$ is given by (5.8) with γ replaced by c, and the FD $D(\lambda)$ of $K(s,t)$ is given by

$$D(\lambda) = \frac{720}{\lambda^4}\left(4(6+\lambda) - \sqrt{\lambda}(24-\lambda)\sin\sqrt{\lambda} - 8(3-\lambda)\cos\sqrt{\lambda}\right). \quad (5.25)$$

Figure 5.5 plots the probability densities of $R(c)$ for $c = 0, 20, 50, 100$. The distributions are unimodal and have a similar nature to those in Figure 5.4. Powers of the test based on $R(c)$ will be graphically presented in Section 5.5, comparing them with these under a different initial condition on the state equation.

In the above discussions we have concentrated on the local alternative of the form $\rho = c/n^\kappa$, where $\kappa = 2m+2$ for Case (a), whereas $\kappa = 2$ for Cases (b) and (c). However, it is possible to think of other values of κ.

For example, let us take up Case (c). For τ (≥ 0) to be determined later, consider (Exercise 5.3.5)

$$\begin{aligned} P\left(\frac{1}{n^{1+\tau}}R_n \leq x\right) &= P\left(\frac{1}{n^{1+\tau}}\frac{\mathbf{y}'MCC'M\mathbf{y}}{\mathbf{y}'M\mathbf{y}} \leq x\right) \\ &= P\left((\boldsymbol{\varepsilon}+C\boldsymbol{\eta})'M\left(\frac{xM}{n} - \frac{MCC'M}{n^{2+\tau}}\right)M(\boldsymbol{\varepsilon}+C\boldsymbol{\eta}) \geq 0\right) \\ &= P\left(\boldsymbol{\zeta}'Q_n(\tau)\boldsymbol{\zeta} \geq 0\right), \end{aligned} \quad (5.26)$$

where $\boldsymbol{\zeta} \sim \mathrm{N}(\mathbf{0}, \sigma_\varepsilon^2 I_n)$ and

$$Q_n(\tau) = \left(\frac{x}{n}M - \frac{1}{n^{2+\tau}}MCC'M\right)M(I_n + \rho CC')M$$
$$= \frac{xM}{n} + \frac{cxMCC'M}{n^{\kappa+1}} - \frac{MCC'M}{n^{2+\tau}} - \frac{c(MCC'M)^2}{n^{\kappa+2+\tau}}. \qquad (5.27)$$

It follows from (5.26) and (5.27) that

$$P\left(\frac{1}{n^{1+\tau}}R_n \le x\right) = P\left(\frac{\dfrac{\boldsymbol{\zeta}'MCC'M\boldsymbol{\zeta}}{n^{2+\tau}} + \dfrac{c\boldsymbol{\zeta}'(MCC'M)^2\boldsymbol{\zeta}}{n^{\kappa+2+\tau}}}{\dfrac{\boldsymbol{\zeta}'M\boldsymbol{\zeta}}{n} + \dfrac{c\boldsymbol{\zeta}'MCC'M\boldsymbol{\zeta}}{n^{\kappa+1}}} \le x\right). \qquad (5.28)$$

Note here that

$$\boldsymbol{\zeta}'M\boldsymbol{\zeta} = O_p(n), \quad \boldsymbol{\zeta}'MCC'M\boldsymbol{\zeta} = O_p(n^2), \quad \boldsymbol{\zeta}'(MCC'M)^2\boldsymbol{\zeta} = O_p(n^4),$$

where $O_p(a_n)$ stands for a term that has stochastic order a_n at most, and

$$S_n = \frac{1}{n^2\sigma_\varepsilon^2}\boldsymbol{\zeta}'\left(MCC'M + \frac{c}{n^2}(MCC'M)^2\right)\boldsymbol{\zeta}$$
$$\Rightarrow \int_0^1\int_0^1 \left(K(s,t) + cK^{(2)}(s,t)\right) dW(s)\,dW(t).$$

Taking into account the above facts, we can now summarize the limiting distribution of $R_n/n^{1+\tau}$ as $n \to \infty$ under $\rho = c/n^\kappa$ as follows.

$$\frac{1}{n^{1+\tau}}R_n \Rightarrow \begin{cases} \displaystyle\int_0^1\int_0^1 K(s,t)\,dW(s)\,dW(t) & (\kappa > 2, \tau = 0), \\[2ex] \displaystyle\int_0^1\int_0^1 [K(s,t) + cK^{(2)}(s,t)]\,dW(s)\,dW(t) & (\kappa = 2, \tau = 0), \\[2ex] \displaystyle c\int_0^1\int_0^1 K^{(2)}(s,t)\,dW(s)\,dW(t) & (1 < \kappa < 2, \tau = 2-\kappa), \\[2ex] \dfrac{\displaystyle c\int_0^1\int_0^1 K^{(2)}(s,t)\,dW(s)\,dW(t)}{\displaystyle 1 + c\int_0^1\int_0^1 K(s,t)\,dW(s)\,dW(t)} & (\kappa = 1, \tau = 1), \\[2ex] \dfrac{\displaystyle \int_0^1\int_0^1 K^{(2)}(s,t)\,dW(s)\,dW(t)}{\displaystyle \int_0^1\int_0^1 K(s,t)\,dW(s)\,dW(t)} & (0 \le \kappa < 1, \tau = 1). \end{cases}$$

It may be concluded from the above results that

(1) When $\kappa > 2$, the distribution of the LBI statistic R_n/n is asymptotically the same as the null distribution, which means that the LBI test has no asymptotic power when ρ is within an n^{-2} neighborhood of the origin.
(2) When $\kappa = 2$, that is, when ρ is away from the origin on the order of n^{-2}, R_n/n has a nondegenerate limiting distribution, which diverges to ∞ as $c \to \infty$, so that the test is consistent. The c.f. of the limiting distribution is given in Theorem 5.5 for the present value of κ.
(3) When $\kappa < 2$, that is, when ρ is away from the origin by more than the order of n^{-2}, R_n/n diverges. In particular, R_n/n is of stochastic order n when $\kappa \le 1$ including the fixed alternative ($\kappa = 0$).

Exercises for Section 5.3

Exercise 5.3.1 Let C be an $n \times n$ matrix whose (j,k)th element is 0 for $j < k$ and 1 for $j \ge k$. Define also

$$\boldsymbol{x} = (1^m, 2^m, \ldots, n^m)', \quad D_x = \mathrm{diag}(x_1, \ldots, x_n), \quad M = I_n - \boldsymbol{x}\boldsymbol{x}'/\boldsymbol{x}'\boldsymbol{x}.$$

Show that the (j,k)th element of $C'D_xMD_xC$ is given by

$$A_n(j,k) = \sum_{\ell=\max(j,k)}^{n} \ell^{2m} - \sum_{\ell=j}^{n} \ell^{2m} \sum_{\ell=k}^{n} \ell^{2m} \Big/ \sum_{\ell=1}^{n} \ell^{2m}.$$

Exercise 5.3.2 Consider the c.f. of $R(m;c)$ given by

$$\mathrm{E}\left(e^{i\theta R(m;c)}\right) = \left[D\left(i\theta + \sqrt{-\theta^2 + 2ic\theta}\right) D\left(i\theta - \sqrt{-\theta^2 + 2ic\theta}\right)\right]^{-1/2},$$

where

$$D(\lambda) = \Gamma\left(\frac{4m+3}{2(m+1)}\right) J_\nu\left(\frac{\sqrt{\lambda}}{m+1}\right) \Big/ \left(\frac{\sqrt{\lambda}}{2(m+1)}\right)^\nu, \quad \nu = \frac{2m+1}{2(m+1)},$$

and $J_\nu(z)$ is the Bessel function of the first kind. Then prove that

$$\lim_{m \to \infty} \mathrm{E}\left(e^{i\theta(m+1)^2 R(m;c)}\right) = \left(\frac{2J_1(\sqrt{2i\theta})}{\sqrt{2i\theta}}\right)^{-1/2}.$$

Exercise 5.3.3 Let the matrix C be the same as that defined in Exercise 5.3.1. Define also

$$\boldsymbol{e} = (1, \ldots, 1)', \quad \boldsymbol{z} = (1^q, 2^q, \ldots, n^q)', \quad M = I_n - X(X'X)^{-1}X',$$

where q is a positive integer and $X = (\boldsymbol{e}, \boldsymbol{z})$. Denoting by $A_n(j,k)$ the (j,k)th element of $C'MC$, show that

$$\lim_{n\to\infty} \max_{1\le j,k\le n} \left| \frac{1}{n} A_n(j,k) - K\left(\frac{j}{n}, \frac{k}{n}; q\right) \right| = 0,$$

where

$$K(s,t;q) = \min(s,t) - st - \frac{2q+1}{q^2} st(1-s^q)(1-t^q).$$

Exercise 5.3.4 Let the matrix C be the same as that defined in Exercise 5.3.1. Define also

$$\boldsymbol{e} = (1, \dots, 1)', \quad \boldsymbol{d} = (1, \dots, n)', \quad \boldsymbol{f} = (1^2, \dots, n^2)',$$
$$M = I_n - X(X'X)^{-1}X',$$

where $X = (\boldsymbol{e}, \boldsymbol{d}, \boldsymbol{f})$. Denoting by $A_n(j,k)$ the (j,k)th element of $C'MC$, show that

$$\lim_{n\to\infty} \max_{1\le j,k\le n} \left| \frac{1}{n} A_n(j,k) - K\left(\frac{j}{n}, \frac{k}{n}\right) \right| = 0,$$

where

$$K(s,t) = \min(s,t) - st - 2st(1-s)(1-t)(4-5s-5t+10st).$$

Exercise 5.3.5 Consider

$$\boldsymbol{y} = X\boldsymbol{\delta} + \boldsymbol{\varepsilon} + C\boldsymbol{\eta} \sim \mathrm{N}\left(X\boldsymbol{\xi}, \sigma_\varepsilon^2(I_n + \rho CC')\right),$$

where $\rho = c/n^\kappa$. Then prove that

$$P\left(\frac{1}{n^{1+\tau}} \frac{\boldsymbol{y}'MCC'M\boldsymbol{y}}{\boldsymbol{y}'M\boldsymbol{y}} \le x\right) = P\left(\boldsymbol{\zeta}'Q_n(\tau)\boldsymbol{\zeta} \ge 0\right),$$

where $M = I_n - X(X'X)^{-1}X'$, $\boldsymbol{\zeta} \sim \mathrm{N}(\boldsymbol{0}, \sigma_\varepsilon^2 I_n)$ and

$$Q_n(\tau) = \frac{xM}{n} + \frac{cxMCC'M}{n^{\kappa+1}} - \frac{MCC'M}{n^{2+\tau}} - \frac{c(MCC'M)^2}{n^{\kappa+2+\tau}}.$$

5.4 Asymptotic Distributions of the LBI Statistics – 2

In this section we derive the asymptotic distribution of the LBI statistic, assuming that the initial value β_0 that generates the random walk $\{\beta_j\}$ is known. Without loss of generality we put $\beta_0 = 0$. Then the present model is given by

$$\boldsymbol{y} = Z\boldsymbol{\gamma} + D_x C\boldsymbol{\eta} + \boldsymbol{\varepsilon} \sim \mathrm{N}\left(Z\boldsymbol{\gamma}, \sigma_\varepsilon^2(I_n + \rho D_x CC'D_x)\right), \tag{5.29}$$

where the definitions of Z, D_x, and C are the same as in (5.11).

The log-likelihood for $\boldsymbol{y}$ in the present model is given by

$$\begin{aligned} L(\boldsymbol{\delta}) &= -\frac{n}{2}\log 2\pi\sigma_\varepsilon^2 - \frac{1}{2}\log|I_n + \rho D_x CC' D_x| \\ &\quad - \frac{1}{2\sigma_\varepsilon^2}(\boldsymbol{y} - Z\boldsymbol{\gamma})'(I_n + \rho D_x CC' D_x)^{-1}(\boldsymbol{y} - Z\boldsymbol{\gamma}), \end{aligned} \qquad (5.30)$$

where $\boldsymbol{\delta} = (\rho, \boldsymbol{\gamma}', \sigma_\varepsilon^2)'$. Evaluating $\partial L(\boldsymbol{\delta})/\partial\rho$ at the MLE $\tilde{\boldsymbol{\delta}}$ of $\boldsymbol{\delta}$ under H_0, we have

$$\left.\frac{\partial L(\boldsymbol{\delta})}{\partial \rho}\right|_{H_0} = -\frac{1}{2}\text{tr}(D_x CC' D_x) + \frac{n}{2}\,\frac{\boldsymbol{y}' M D_x CC' D_x M \boldsymbol{y}}{\boldsymbol{y}' M \boldsymbol{y}},$$

where

$$M = I_n - Z(Z'Z)^{-1}Z'.$$

Then the LBI test in the present model rejects H_0 when

$$S_n = \frac{\boldsymbol{y}' M D_x CC' D_x M \boldsymbol{y}}{\boldsymbol{y}' M \boldsymbol{y}} \qquad (5.31)$$

takes large values.

Let us consider Cases (a), (b), and (c) of the regressor variables x_j and $\boldsymbol{z}_j$ given in (5.14). For the same reason as before, we can concentrate on $\boldsymbol{y}' M D_x CC' D_x M \boldsymbol{y}$ to derive the asymptotic distribution of S_n. Note here that

$$\boldsymbol{y}' M D_x CC' D_x M \boldsymbol{y} \stackrel{\mathcal{D}}{=} \boldsymbol{\zeta}' A_n \boldsymbol{\zeta} + \rho \boldsymbol{\zeta}' A_n^2 \boldsymbol{\zeta},$$

where $\boldsymbol{\zeta} \sim \mathrm{N}(\boldsymbol{0},\, \sigma_\varepsilon^2 I_n)$ and $A_n = C' D_x M D_x C$.

Case (a): $x_j = j^m$, $\boldsymbol{z}_j = \boldsymbol{0}$ $\quad (m > -1/2)$.

We put $\rho = c/n^{2m+2}$. Since $M = I_n$, the (j,k)th element of A_n is given by

$$A_n(j,k) = \sum_{\ell=\max(j,k)}^{n} \ell^{2m},$$

so that $B_n(j,k) = A_n(j,k)/n^{2m+1}$ converges uniformly to

$$K(s,t;m) = \frac{1}{2m+1}\left[1 - (\max(s,t))^{2m+1}\right]. \qquad (5.32)$$

The FD of $K(s,t;m)$ was obtained earlier in (2.133).

Then we obtain the following theorem parallel to Theorem 5.3.

Theorem 5.6 *Consider the state space model*

$$y_j = \beta_j\, j^m + \varepsilon_j, \quad \beta_j = \beta_{j-1} + \eta_j, \quad \beta_0 = 0 \quad (m > -1/2),$$

where $\{\varepsilon_j\}$ and $\{\eta_j\}$ satisfy the same conditions as in Theorem 5.3. Then it holds that, as $n \to \infty$ under $\rho = c/n^{2m+2}$,

$$\frac{1}{n^{2m+1}} S_n = \frac{y' D_x C C' D_x y / \left(n^{2m+2}\sigma_\varepsilon^2\right)}{y' M y / \left(n\sigma_\varepsilon^2\right)} \Rightarrow S(m;c),$$

$$S(m;c) = \int_0^1 \int_0^1 \left(K(s,t;m) + cK^{(2)}(s,t;m)\right) dW(s)\, dW(t), \qquad (5.33)$$

$$E(S(m;c)) = \frac{1}{2(m+1)} + \frac{c}{2(m+1)(2m+3)}, \qquad (5.34)$$

where $K(s,t;m)$ is given in (5.32) and $K^{(2)}(s,t;m)$ is the iterated kernel of $K(s,t;m)$. Moreover, the c.f. of $S(m;c)$ is given by (5.8) with γ replaced by c, and the FD of $K(s,t;m)$ is given by

$$D(\lambda) = \Gamma(-\nu+1)\, J_{-\nu}\left(\frac{\sqrt{\lambda}}{m+1}\right) \Big/ \left(\frac{\sqrt{\lambda}}{2(m+1)}\right)^{-\nu}, \qquad (5.35)$$

where $\nu = (2m+1)/2(m+1)$ and $J_{-\nu}(z)$ is the Bessel function of the first kind defined in (2.58).

The distribution of $S(m;c)$ contrasts with that of $R(m;c)$ when m is large. We have (Exercise 5.4.1)

Corollary 5.6 *For $S(m;c)$ defined in (5.33), it holds that*

$$\lim_{m\to\infty} E\left(e^{i\theta(m+1)S(m;c)}\right) = (1 - i\theta)^{-1/2}, \qquad (5.36)$$

which is the c.f. of $Z^2/2$ with $Z \sim N(0,1)$.

It evidently holds that $S(m;c) = O_p(m^{-1})$, whereas $R(m;c) = O_p(m^{-2})$. In fact, under $H_0 : \rho = 0$ $(c = 0)$, we have (Exercise 5.4.2)

$$\mathrm{E}(S(m;c)) = \frac{1}{2(m+1)}, \quad \mathrm{Var}(S(m;c)) = \frac{1}{(m+1)(2m+3)},$$

$$\mathrm{E}(R(m;c)) = \frac{1}{2(m+1)(4m+3)},$$

$$\mathrm{Var}(R(m;c)) = \frac{1}{(m+1)(6m+5)(4m+3)^2}.$$

Figure 5.6 plots the probability densities of $S(m;c)$ for $m = 0, 0.5, 0.8, 1$ under $H_0 : c = 0$, whereas Figure 5.7 shows those for $m = 1$ and $c = 0, 20, 50, 100$. These two figures are parallel to Figures 5.1 and 5.2, where the densities of $R(m;c)$ are presented using the same m and c. It is seen that the present distributions have longer right-hand tails with smaller means.

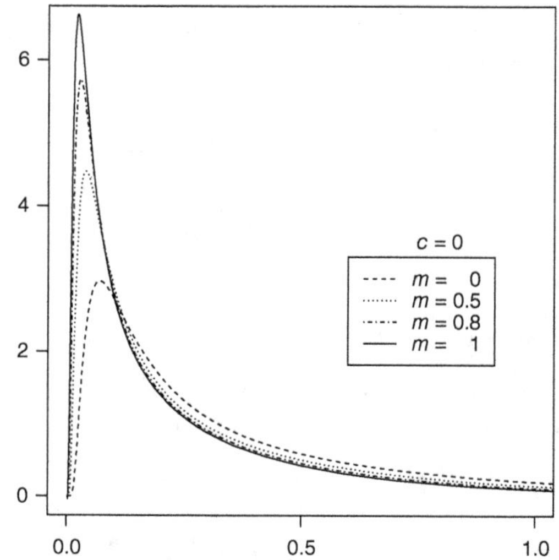

Figure 5.6 Probability densities of $S(m; c)$ for $c = 0$.

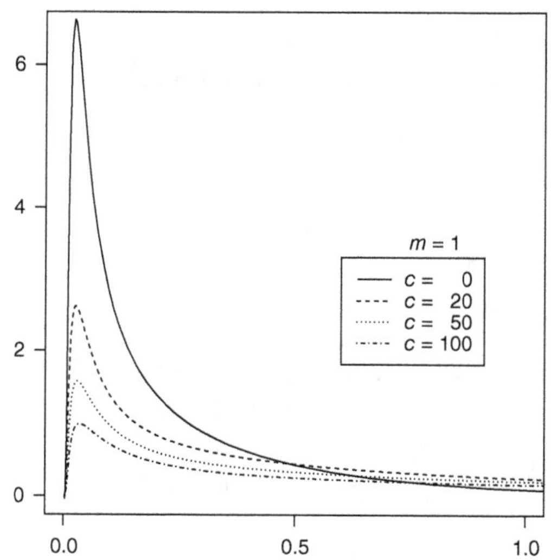

Figure 5.7 Probability densities of $S(m; c)$ for $m = 1$.

Case (b): $x_j = 1,\ z_j = j^q \quad (q = 1, 2, 3, 4)$.

Putting $z = (1^q, 2^q, \ldots, n^q)'$ and noting that $D_x = I_n$, we have

$$A_n = C'MC, \quad M = I_n - zz'/z'z.$$

Then it can be shown (Exercise 5.4.3) that $A_n(j,k)/n$ converges uniformly to

$$K(s,t;q) = 1 - \max(s,t) - \frac{2q+1}{(q+1)^2}\left(1 - s^{q+1}\right)\left(1 - t^{q+1}\right). \tag{5.37}$$

The FD of $K(s,t;q)$ for $q = 1$ was presented earlier in (2.166). The FDs for $q = 2, 3, 4$ can also be obtained by the formula established in (2.87) as

$$D(\lambda) = \cos\sqrt{\lambda} + 2\sqrt{\lambda}\int_0^1\left(\int_0^t \psi'(s)\sin\sqrt{\lambda}s\,ds\right)\psi'(t)\cos\sqrt{\lambda}(1-t)\,dt,$$

where $\psi(t) = \sqrt{2q+1}\,(1 - t^{q+1})/(q+1)$.

We now have the following theorem parallel to Theorem 5.4.

Theorem 5.7 *Consider the state space model*

$$y_j = \beta_j + \gamma\, j^q + \varepsilon_j, \quad \beta_j = \beta_{j-1} + \eta_j, \quad \beta_0 = 0 \quad (q = 1, 2, 3, 4),$$

where $\{\varepsilon_j\}$ and $\{\eta_j\}$ satisfy the same conditions as in Theorem 5.4. Then it holds that, as $n \to \infty$ under $\rho = c/n^2$,

$$\frac{1}{n}S_n = \frac{\mathbf{y}'MCC'M\mathbf{y}/\left(n^2\sigma_\varepsilon^2\right)}{\mathbf{y}'M\mathbf{y}/\left(n\sigma_\varepsilon^2\right)} \Rightarrow S(q;c),$$

$$S(q;c) = \int_0^1\int_0^1\left(K(s,t;q) + cK^{(2)}(s,t;q)\right)dW(s)\,dW(t), \tag{5.38}$$

where $K(s,t;q)$ is defined in (5.37) and $K^{(2)}(s,t;q)$ is the iterated kernel of $K(s,t;q)$. Moreover, the c.f. of $S(q;c)$ is given by (5.8) with γ replaced by c, and the FD $D(\lambda;q)$ of $K(s,t;q)$ for $q = 1, 2, 3, 4$ is given by

$$\begin{aligned}
D(\lambda;1) &= \frac{3}{\lambda^2}\left(\sqrt{\lambda}\sin\sqrt{\lambda} - \lambda\cos\sqrt{\lambda}\right),\\
D(\lambda;2) &= \frac{20}{\lambda^3}\left(-2\lambda + \sqrt{\lambda}(1+\lambda)\sin\sqrt{\lambda} + \lambda\left(1 - \frac{\lambda}{3}\right)\cos\sqrt{\lambda}\right),\\
D(\lambda;3) &= \frac{126}{\lambda^4}\left(\sqrt{\lambda}\left(2 - 2\lambda + \frac{\lambda^2}{2}\right)\sin\sqrt{\lambda} - \lambda\left(2 - \frac{4\lambda}{3} + \frac{\lambda^2}{10}\right)\cos\sqrt{\lambda}\right),\\
D(\lambda;4) &= \frac{864}{\lambda^5}\left(\lambda(-12 + 2\lambda) + \sqrt{\lambda}\left(6 + 6\lambda - 2\lambda^2 + \frac{\lambda^3}{6}\right)\sin\sqrt{\lambda}\right.\\
&\qquad\left. + \lambda\left(6 - 4\lambda + \frac{7\lambda^2}{10} - \frac{\lambda^3}{42}\right)\cos\sqrt{\lambda}\right).
\end{aligned}$$

We have

$$\begin{aligned}
\mathrm{E}(S(1;c)) &= \frac{1}{10} + \frac{1}{350}c, \quad & \mathrm{E}(S(2;c)) &= \frac{1}{7} + \frac{277}{31{,}752}c,\\
\mathrm{E}(S(3;c)) &= \frac{17}{90} + \frac{829}{44{,}550}c, \quad & \mathrm{E}(S(4;c)) &= \frac{5}{22} + \frac{557}{18{,}876}c.
\end{aligned}$$

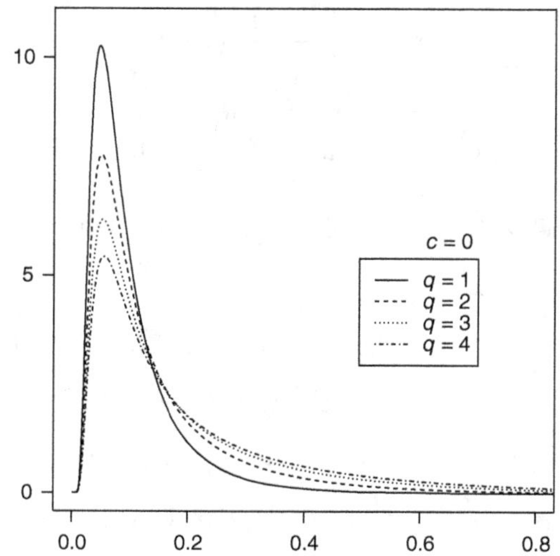

Figure 5.8 Probability densities of $S(q;c)$ for $q = 1, 2, 3, 4$ and $c = 0$.

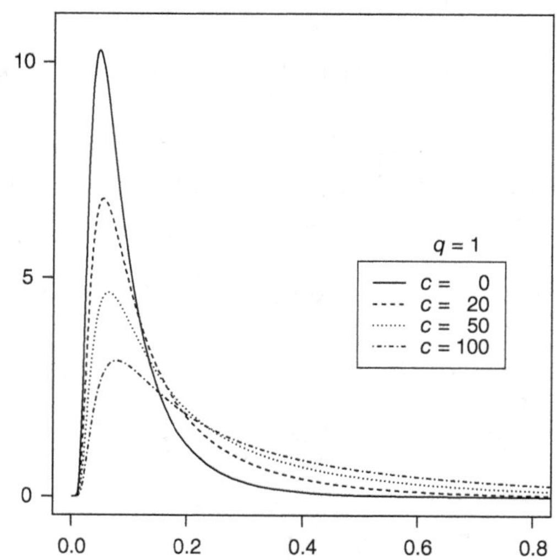

Figure 5.9 Probability densities of $S(q;c)$ for $q = 1$.

Figure 5.8 plots the probability densities of $S(q;c)$ for $q = 1, 2, 3, 4$ under $H_0 : c = 0$, whereas Figure 5.9 shows those for $q = 1$ and $c = 0, 20, 50, 100$. These two figures are parallel to Figures 5.3 and 5.4, where the densities of $R(q;c)$ are drawn using the same q and c. It is seen that the present distributions are more spread with larger means.

Case (c): $x_j = 1,\ z_j = (j, j^2)'$.

We have

$$A_n = C'MC, \quad M = I_n - Z(Z'Z)^{-1}Z',$$

where $Z = (\boldsymbol{d}, \boldsymbol{f})$, $\boldsymbol{d} = (1, 2, \ldots, n)'$, $\boldsymbol{f} = (1^2, 2^2, \ldots, n^2)'$. Then it can be shown (Exercise 5.4.4) that $A_n(j,k)/n$ converges uniformly to

$$\begin{aligned} K(s,t) &= 1 - \max(s,t) + \frac{2}{9}(1-s)(1-t) \\ &\quad \times \left(5\left(s^2 + t^2 - 8s^2t^2 + st(s+t)\right) - 4(1+s)(1+t)\right). \end{aligned} \tag{5.39}$$

The FD of $K(s,t)$ was presented earlier in (2.184), which was efficiently obtained from Theorem 2.21 by putting

$$K(s,t) = 1 - \max(s,t) - \sum_{j=1}^{3} \psi_j(s)\psi_j(t),$$

where

$$\psi_1(t) = \sqrt{2}\left(1 - t^2\right), \quad \psi_2(t) = -\frac{\sqrt{10}}{3}\, i\left(1 - t^3\right), \quad \psi_3(t) = \sqrt{10}\left(t^2 - t^3\right).$$

We now have the following theorem parallel to Theorem 5.5.

Theorem 5.8 *Consider the state space model*

$$y_j = \beta_j + \gamma_1\, j + \gamma_2\, j^2 + \varepsilon_j, \quad \beta_j = \beta_{j-1} + \eta_j, \quad \beta_0 = 0,$$

where $\{\varepsilon_j\}$ and $\{\eta_j\}$ satisfy the conditions described in Theorem 5.5. Then it holds that, as $n \to \infty$ under $\rho = c/n^2$,

$$\frac{1}{n}\, S_n = \frac{\boldsymbol{y}'MCC'M\boldsymbol{y}/\left(n^2\sigma_\varepsilon^2\right)}{\boldsymbol{y}'M\boldsymbol{y}/\left(n\sigma_\varepsilon^2\right)} \Rightarrow S(c),$$

$$S(c) = \int_0^1 \int_0^1 \left(K(s,t) + cK^{(2)}(s,t)\right) dW(s)\, dW(t), \tag{5.40}$$

$$E(S(c)) = \frac{11}{210} + \frac{179}{396{,}900}\, c,$$

where $K(s,t)$ is defined in (5.39) and $K^{(2)}(s,t)$ is the iterated kernel of $K(s,t)$. Moreover, the c.f. of $S(c)$ is given by (5.8) with γ replaced by c, and the FD $D(\lambda)$ of $K(s,t)$ is given by

$$D(\lambda) = \frac{80}{\lambda^4}\left(12(2+\lambda) - 4\sqrt{\lambda}\,(6+\lambda)\sin\sqrt{\lambda} - \left(24 - \lambda^2\right)\cos\sqrt{\lambda}\right). \tag{5.41}$$

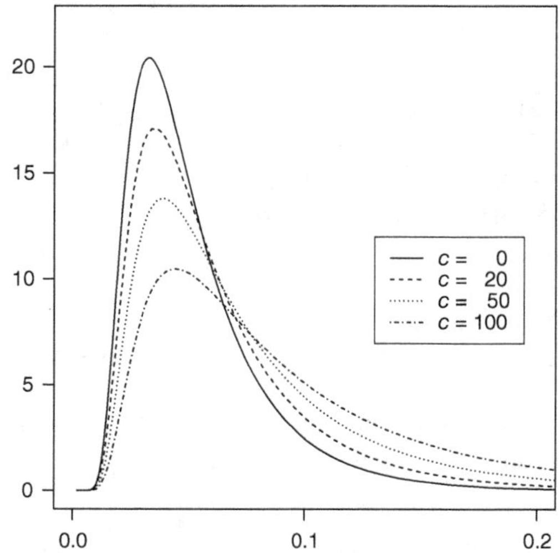

Figure 5.10 Probability densities of $S(c)$ for $c = 0$, 20, 50, 100.

Figure 5.10 plots the probability densities of $S(c)$ for $c = 0, 20, 50, 100$, which is parallel to Figure 5.5, where the densities of $R(c)$ are presented using the same values of c. A similar distributional property to that described before is observed, although the present distributions are more widely spread with larger means.

Exercises for Section 5.4

Exercise 5.4.1 Consider the c.f. of $S(m;c)$ given by

$$\mathrm{E}\left(e^{i\theta S(m;c)}\right) = \left[D\left(i\theta + \sqrt{-\theta^2 + 2ic\theta}\right) D\left(i\theta - \sqrt{-\theta^2 + 2ic\theta}\right)\right]^{-1/2},$$

where

$$D(\lambda) = \Gamma(-\nu+1)\, J_{-\nu}\left(\frac{\sqrt{\lambda}}{m+1}\right) \Bigg/ \left(\frac{\sqrt{\lambda}}{2(m+1)}\right)^{-\nu}, \quad \nu = \frac{2m+1}{2(m+1)}$$

and $J_\nu(z)$ is the Bessel function of the first kind. Then prove that

$$\lim_{m\to\infty} \mathrm{E}\left(e^{i\theta(m+1)S(m;c)}\right) = (1 - i\theta)^{-1/2}.$$

Exercise 5.4.2 Prove that

$$\begin{aligned}
\mathrm{E}(R(m;0)) &= \frac{1}{2(m+1)(4m+3)},\\
\mathrm{Var}(R(m;0)) &= \frac{1}{(m+1)(6m+5)(4m+3)^2},\\
\mathrm{E}(S(m;0)) &= \frac{1}{2(m+1)}, \quad \mathrm{Var}(S(m;0)) = \frac{1}{(m+1)(2m+3)},
\end{aligned}$$

where $R(m;c)$ and $S(m;c)$ are given in Theorems 5.3 and 5.6, respectively.

Exercise 5.4.3 Consider $A_n = C'MC$, where C is an $n \times n$ matrix whose (j,k)th element is 0 for $j < k$ and is 1 for $j \geq k$, $M = I_n - zz'/z'z$, and $z = (1^q, 2^q, \ldots, n^q)'$ with q a positive integer. Denoting by $A_n(j,k)$ the (j,k)th element of $C'MC$, show that $A_n(j,k)/n$ converges uniformly to

$$K(s,t;q) = 1 - \max(s,t) - \frac{2q+1}{(q+1)^2}\left(1 - s^{q+1}\right)\left(1 - t^{q+1}\right).$$

Exercise 5.4.4 Consider $A_n = C'MC$, where C is the same as in Exercise 5.4.3, $M = I_n - Z(Z'Z)^{-1}Z'$, $Z = (\boldsymbol{d}, \boldsymbol{f})$, $\boldsymbol{d} = (1,2,\ldots,n)'$, and $\boldsymbol{f} = (1^2, 2^2, \ldots, n^2)'$. Letting $A_n(j,k)$ be the (j,k)th element of $C'MC$, show that $A_n(j,k)/n$ converges uniformly to

$$\begin{aligned}
K(s,t) = {} & 1 - \max(s,t) + \frac{2}{9}(1-s)(1-t)\\
& \times \left(5\left(s^2 + t^2 - 8s^2t^2 + st(s+t)\right) - 4(1+s)(1+t)\right).
\end{aligned}$$

5.5 Comparison of Powers under the Two Initial Conditions

We have considered two cases for the initial value β_0 of the random walk $\{\beta_j\}$. One is the case where β_0 is an unknown constant to be estimated and the other the case where β_0 is a known constant assumed to be 0. In this section we compare, under the two different initial conditions, the limiting local powers of the corresponding LBI tests for the three cases of regressor variables given in (5.14). It is expected that the powers will be higher when β_0 is known because more information is contained in the state space model, which we show by computing powers.

The LBI test considered in this chapter rejects H_0 when the test statistic takes large values. Let T be a limiting test statistic under a sequence of local alternatives and $\phi(\theta)$ be the c.f. of T. Then the local power of the test is computed by Lévy's inversion formula described in (4.16) as

$$P(T > x(\alpha)) = 1 - \frac{1}{\pi}\int_0^\infty \mathrm{Re}\left[\frac{1 - e^{-i\theta x(\alpha)}}{i\theta}\phi(\theta)\right] d\theta, \tag{5.42}$$

where $x(\alpha)$ is the upper $100\alpha\%$ significance point of the distribution of T under H_0.

Here we present tables and figures of limiting local powers at the significance level $\alpha = 0.05$ for three cases Cases (a), (b), and (c) of the regressors. The state space model for Case (a) is given by

$$y_j = \beta_j\, j^m + \varepsilon_j, \quad \beta_j = \beta_{j-1} + \eta_j \quad (j = 1, \ldots, n),$$

where $m > -1/2$. For this model, the c.f. of the LBI statistic $R(m; c)$ for β_0 unknown is given in Theorem 5.3, whereas that of $S(m; c)$ for β_0 known is in Theorem 5.6.

Table 5.1 reports, for $m = 0, 1, 2, 3$, limiting local percentage powers under $\rho = c/n^{2m+2}$ for the two tests – with (1) β_0 unknown and (2) β_0 known – together with upper 5% points of $(m+1)^2 R(m; 0)$ and $(m+1)S(m; 0)$. Figure 5.11 plots powers of the two tests for $m = 0, 1, 2$. It is seen from Table 5.1 and Figure 5.11 that the powers are higher when β_0 is known. It can also be seen that the powers decrease as m becomes large. This is because, as m becomes large, ρ becomes smaller and closer to $\rho = 0$, the value under H_0.

Table 5.1 *Limiting percentage powers of the LBI tests at the 5% level: Case (a).*

	$x(0.05)$	$c = 10$	50	100	200	500
$m = 0$						
(1)	0.46136	17.14	45.56	61.06	75.37	89.46
(2)	1.65574	38.68	68.39	78.24	86.13	93.55
$m = 1$						
(1)	0.38306	7.51	17.42	27.57	41.60	62.69
(2)	1.76929	27.61	57.64	68.86	77.75	86.38
$m = 2$						
(1)	0.36207	6.02	10.26	15.47	24.56	42.75
(2)	1.81462	21.91	50.74	62.99	73.04	82.81
$m = 3$						
(1)	0.35233	5.55	7.83	10.75	16.39	30.08
(2)	1.83905	18.42	45.56	58.42	69.35	80.19

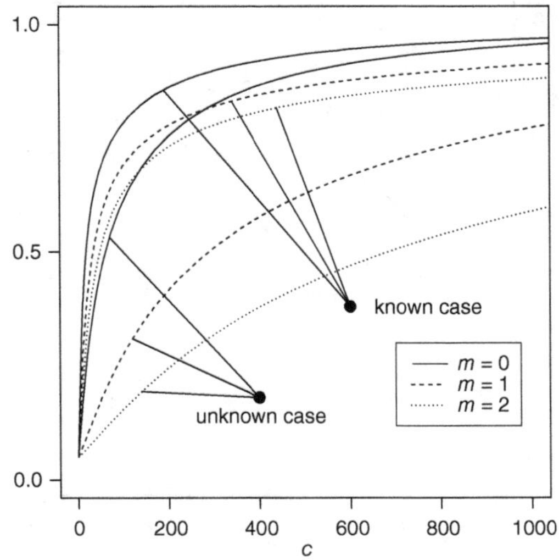

Figure 5.11 Limiting local powers of the LBI tests: Case (a).

Table 5.2 *Limiting percentage powers of the LBI tests: Cases (b) and (c).*

	$x(0.05)$	$c = 10$	50	100	200	500
				Case (b)		
$q = 1$						
(1)	0.14789	8.33	22.34	36.71	55.69	80.06
(2)	0.24810	11.24	32.30	48.35	65.74	85.02
$q = 2$						
(1)	0.16416	8.85	24.12	38.62	56.93	80.14
(2)	0.40511	15.84	42.11	56.71	70.53	85.57
$q = 3$						
(1)	0.19394	9.76	27.19	42.00	59.45	80.83
(2)	0.57380	20.25	49.22	62.65	74.47	86.93
				Case (c)		
(1)	0.08595	6.70	14.38	24.16	40.72	69.10
(2)	0.11050	7.34	17.63	29.53	47.48	74.33

The state space model for Case (b) is given by

$$y_j = \beta_j + \gamma j^q + \varepsilon_j, \quad \beta_j = \beta_{j-1} + \eta_j \quad (j = 1, \ldots, n).$$

For this model, the c.f. of the LBI statistic $R(q;c)$ for β_0 unknown is given in Theorem 5.4, whereas that of $S(q;c)$ for β_0 known is in Theorem 5.7.

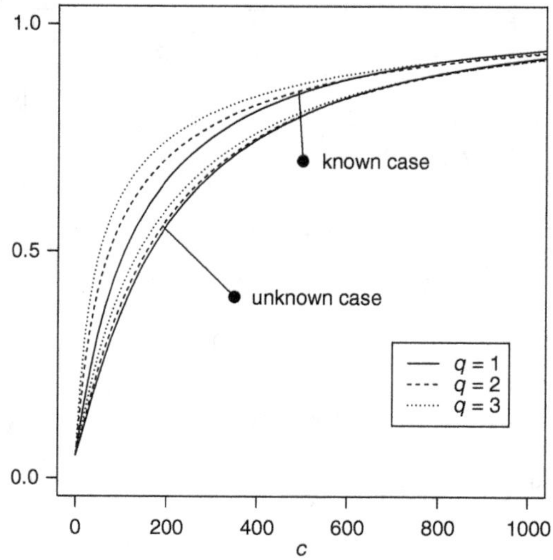

Figure 5.12 Limiting local powers of the LBI tests: Case (b).

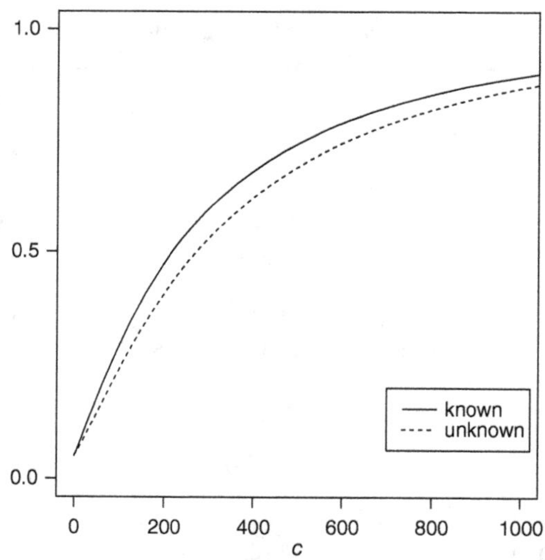

Figure 5.13 Limiting local powers of the LBI tests: Case (c).

Table 5.2 reports, for $q = 1, 2, 3$, limiting local percentage powers under $\rho = c/n^2$ for the two tests associated with $R(q;c)$ and $S(q;c)$ together with upper 5% points of $R(q;0)$ and $S(q;0)$. Figure 5.12 plots powers of the two tests for $q = 1, 2, 3$. It is seen from Table 5.2 and Figure 5.12 that the powers are higher when β_0 is known, as was expected. It is also seen that the powers are higher for higher q under each initial condition, although the difference eventually fades away.

The state space model for Case (c) is given by

$$y_j = \beta_j + \gamma_1 j + \gamma_2 j^2 + \varepsilon_j, \quad \beta_j = \beta_{j-1} + \eta_j \quad (j = 1, \ldots, n).$$

For this model, the c.f. of the LBI statistic $R(c)$ for β_0 unknown is given in Theorem 5.5, whereas that of $S(c)$ for β_0 known is in Theorem 5.8. Limiting local percentage powers under $\rho = c/n^2$ are reported for the two tests at the bottom of Table 5.2 together with upper 5% points of $R(0)$ and $S(0)$.

Figure 5.13 plots powers of the two tests. It is observed from Table 5.2 and Figure 5.13 that the powers are higher when β_0 is known, as in Cases (a) and (b).

6

Fredholm Determinants in the Moving Average Model

An MA unit root test in the MA(1) model was introduced in Section 1.5.3, where the model was unconditional in the sense that the initial value of the error process has the same distribution as the future values. The unconditional MA(1) model becomes noninvertible under the null distribution, and the LBIU statistic exhibits a nonstationary behavior. This chapter extends the model into two directions. One is to add a trend term to the model, and the other to consider the LBI test in the conditional model, which restricts the initial value of the error process. We derive the FDs under a sequence of local alternatives. It is found that the associated FDs for the LBIU tests in unconditional models have a close relationship with those for the LBI tests in state space models with the initial value of the state equation unknown. It is also shown that the associated FDs for the LBI tests in conditional models have no similarity to those for the LBI tests in state space models with the initial value known. Moreover, we compute and compare the limiting local powers of the two tests in the unconditional and conditional MA(1) models.

6.1 LBIU Unit Root Tests in Unconditional Models

Let us consider

$$y_j = \boldsymbol{\beta}'\boldsymbol{x}_j + \varepsilon_j - \alpha\varepsilon_{j-1} \quad (j = 1, \ldots, n), \tag{6.1}$$

where $\boldsymbol{\beta}$ and $\boldsymbol{x}_j$ are $p \times 1$ coefficient and regression vectors, respectively, and $\varepsilon_j \ (j = 0, 1, \ldots, n) \sim \text{NID}(0, \sigma_\varepsilon^2)$. The error term $\{\varepsilon_j - \alpha\varepsilon_{j-1}\}$ follows the unconditional MA(1) process in the sense that the initial value ε_0 has the same distribution as $\{\varepsilon_j\}$ $(j = 1, 2, \ldots)$. The model (6.1) can be rewritten as

$$\boldsymbol{y} = X\boldsymbol{\beta} + \boldsymbol{\eta} \ \sim \ \text{N}\big(X\boldsymbol{\beta}, \, \sigma_\varepsilon^2\Omega(\alpha)\big), \tag{6.2}$$

where $\boldsymbol{y} = (y_1, \ldots, y_n)'$, $X = (\boldsymbol{x}_1, \ldots, \boldsymbol{x}_n)'$, $\boldsymbol{\eta} = (\varepsilon_1 - \alpha\varepsilon_0, \ldots, \varepsilon_n - \alpha\varepsilon_{n-1})'$, and

$$\Omega(\alpha) = \begin{pmatrix} 1+\alpha^2 & -\alpha & & & 0 \\ -\alpha & 1+\alpha^2 & \cdot & & \\ & \cdot & \cdot & \cdot & \\ & & \cdot & \cdot & -\alpha \\ 0 & & & -\alpha & 1+\alpha^2 \end{pmatrix}. \tag{6.3}$$

The testing problem considered here is

$$H_0 : \alpha = 1 \quad \text{versus} \quad H_1 : \alpha < 1,$$

which was dealt with in Section 1.5.3 for $X = \boldsymbol{e} = (1, \ldots, 1)'$, and the LBIU test was suggested using the LM principle. The situation remains unchanged for general X.

The log-likelihood for $\boldsymbol{y}$ is given by

$$L(\boldsymbol{\delta}) = -\frac{n}{2}\log 2\pi\sigma_\varepsilon^2 - \frac{1}{2}\log|\Omega(\alpha)| - \frac{1}{2\sigma^2}(\boldsymbol{y} - X\boldsymbol{\beta})'\Omega^{-1}(\alpha)(\boldsymbol{y} - X\boldsymbol{\beta}), \tag{6.4}$$

where $\boldsymbol{\delta} = (\alpha, \boldsymbol{\beta}', \sigma_\varepsilon^2)'$. Noting that

$$\left.\frac{d\Omega(\alpha)}{d\alpha}\right|_{\alpha=1} = \Omega(1) = \Omega, \qquad \left.\frac{d^2\Omega(\alpha)}{d\alpha^2}\right|_{\alpha=1} = 2I_n,$$

it can be easily checked that $\partial L(\boldsymbol{\delta})/\partial\alpha|_{H_0} = 0$, which yields (Exercise 6.1.1)

$$\left.\frac{\partial^2 L(\boldsymbol{\delta})}{\partial\alpha^2}\right|_{H_0} = n\,\frac{\tilde{\boldsymbol{\eta}}'\Omega^{-2}\tilde{\boldsymbol{\eta}}}{\tilde{\boldsymbol{\eta}}'\Omega^{-1}\tilde{\boldsymbol{\eta}}} - \frac{n(n+5)}{6}, \tag{6.5}$$

where

$$\tilde{\boldsymbol{\eta}} = \tilde{N}\boldsymbol{y} = \tilde{N}\boldsymbol{\eta}, \quad \tilde{N} = I_n - X\left(X'\Omega^{-1}X\right)^{-1}X'\Omega^{-1}. \tag{6.6}$$

We can now derive the LBIU test, which rejects H_0 when

$$R_n = \frac{1}{n}\frac{\tilde{\boldsymbol{\eta}}'\Omega^{-2}\tilde{\boldsymbol{\eta}}}{\tilde{\boldsymbol{\eta}}'\Omega^{-1}\tilde{\boldsymbol{\eta}}} = \frac{1}{n}\frac{\boldsymbol{y}'\tilde{N}'\Omega^{-2}\tilde{N}\boldsymbol{y}}{\boldsymbol{y}'\tilde{N}'\Omega^{-1}\tilde{N}\boldsymbol{y}} = \frac{1}{n}\frac{\boldsymbol{\eta}'\tilde{N}'\Omega^{-2}\tilde{N}\boldsymbol{\eta}}{\boldsymbol{\eta}'\tilde{N}'\Omega^{-1}\tilde{N}\boldsymbol{\eta}} \tag{6.7}$$

takes large values.

Let us consider the asymptotic distribution of R_n under a sequence of local alternatives

$$H_1(c, \kappa) : \alpha = 1 - \frac{c}{n^\kappa}, \tag{6.8}$$

where c and κ are positive constants. Under $H_1(c, \kappa)$, the following relation is useful and is effectively used in subsequent discussions:

$$\Omega(\alpha) = \Omega(1) + (\alpha - 1) \left. \frac{d\Omega(\alpha)}{d\alpha} \right|_{\alpha=1} + \frac{(\alpha-1)^2}{2} \left. \frac{d^2\Omega(\alpha)}{d\alpha^2} \right|_{\alpha=1} = \alpha\Omega + (1-\alpha)^2 I_n. \tag{6.9}$$

To derive the distribution we specify the regression vector $\boldsymbol{x}_j$ in (6.1). Here we consider four cases:

$$\text{(A) } \boldsymbol{x}_j = 0, \quad \text{(B) } \boldsymbol{x}_j = 1, \quad \text{(C) } \boldsymbol{x}_j = j, \quad \text{(D) } \boldsymbol{x}_j = (1, j)'. \tag{6.10}$$

Note that the asymptotic distributions for Cases (A) and (B) were derived under H_0 in Section 1.5.3.

Let us consider first the case where $\alpha = 1 - c/n^\kappa$ with $\kappa = 1$ in (6.8). Since it follows from (6.9) that

$$\Omega(\alpha) = \left(1 - \frac{c}{n}\right)\Omega + \frac{c^2}{n^2} I_n \tag{6.11}$$

and

$$\begin{aligned}
\tilde{\boldsymbol{\eta}}'\Omega^{-1}\tilde{\boldsymbol{\eta}} &= \boldsymbol{\eta}'\tilde{N}'\Omega^{-1}\tilde{N}\boldsymbol{\eta} \\
&\stackrel{\mathcal{D}}{=} \boldsymbol{\zeta}'\Omega^{1/2}(\alpha)\tilde{N}'\Omega^{-1}\tilde{N}\Omega^{1/2}(\alpha)\boldsymbol{\zeta} \\
&\stackrel{\mathcal{D}}{=} \boldsymbol{\zeta}'M\Omega^{-1/2}\Omega(\alpha)\Omega^{-1/2}M\boldsymbol{\zeta},
\end{aligned}$$

where $\boldsymbol{\zeta} \sim \mathrm{N}(\mathbf{0}, \sigma_\varepsilon^2 I_n)$ and $M = \Omega^{-1/2}\tilde{N}\Omega^{1/2} = I_n - \Omega^{-1/2}X(X'\Omega^{-1}X)^{-1}$ $\Omega^{-1/2}$, it can be shown (Exercise 6.1.2) that

$$\frac{1}{n}\tilde{\boldsymbol{\eta}}'\Omega^{-1}\tilde{\boldsymbol{\eta}} \to \sigma_\varepsilon^2 \quad \text{in probability} \tag{6.12}$$

for Cases (A) through (D). Thus we concentrate on $\tilde{\boldsymbol{\eta}}'\Omega^{-2}\tilde{\boldsymbol{\eta}}$ to derive the distribution of R_n in (6.7). It holds that

$$\begin{aligned}
\tilde{\boldsymbol{\eta}}'\Omega^{-2}\tilde{\boldsymbol{\eta}} &= \boldsymbol{\eta}'\tilde{N}'\Omega^{-2}\tilde{N}\boldsymbol{\eta} = \boldsymbol{\eta}'A_n^2\boldsymbol{\eta} \\
&\stackrel{\mathcal{D}}{=} \boldsymbol{\zeta}'A_n\Omega(\alpha)A_n\boldsymbol{\zeta} \\
&\stackrel{\mathcal{D}}{=} \alpha\,\boldsymbol{\zeta}'A_n\,\boldsymbol{\zeta} + \frac{c^2}{n^2}\boldsymbol{\zeta}'A_n^2\,\boldsymbol{\zeta},
\end{aligned} \tag{6.13}$$

where

$$\begin{aligned}
A_n &= \Omega^{-1}\tilde{N} = \tilde{N}'\Omega^{-1} = A_n\Omega A_n = A_n\tilde{N} \qquad (6.14)\\
&= \Omega^{-1} - \Omega^{-1}X(X'\Omega^{-1}X)^{-1}X'\Omega^{-1},
\end{aligned}$$

$$\Omega^{-1} = CC' - \frac{1}{n+1}C\boldsymbol{e}\boldsymbol{e}'C'. \tag{6.15}$$

Here $\boldsymbol{e} = (1, \ldots, 1)' : n \times 1$, and C is the random walk-generating matrix defined in (1.93) whose (j,k)th element is 0 for $j < k$ and 1 for $j \geq k$. Note that the matrix A_n is nonnegative definite and takes the following form for Cases (A) through (D):

$$A_n = \begin{cases} \Omega^{-1} & \text{(Case (A))}, \\ \Omega^{-1} - \Omega^{-1}\boldsymbol{e}\boldsymbol{e}'\Omega^{-1}/\boldsymbol{e}'\Omega^{-1}\boldsymbol{e} & \text{(Case (B))}, \\ \Omega^{-1} - \Omega^{-1}\boldsymbol{d}\boldsymbol{d}'\Omega^{-1}/\boldsymbol{d}'\Omega^{-1}\boldsymbol{d} & \text{(Case (C))}, \\ \Omega^{-1} - \Omega^{-1}(\boldsymbol{e},\boldsymbol{d})\left((\boldsymbol{e},\boldsymbol{d})'\Omega^{-1}(\boldsymbol{e},\boldsymbol{d})\right)^{-1}(\boldsymbol{e},\boldsymbol{d})'\Omega^{-1} & \text{(Case (D))}, \end{cases}$$

where $\boldsymbol{d} = (1, 2, \ldots, n)'$. It holds that

$$\begin{aligned} \boldsymbol{e}'\Omega^{-1}\boldsymbol{e} &= \frac{n(n+1)(n+2)}{12} = \frac{n^3}{12} + O\left(n^2\right), \\ \boldsymbol{e}'\Omega^{-1}\boldsymbol{d} &= \frac{n^4}{4} + O\left(n^3\right), \\ \boldsymbol{d}'\Omega^{-1}\boldsymbol{d} &= \frac{n^5}{45} + O\left(n^4\right), \\ \left[\Omega^{-1}\right](j,k) &= \min(j,k) - \frac{1}{n+1}\,jk, \\ \left[\Omega^{-1}\boldsymbol{e}\boldsymbol{e}'\Omega^{-1}\right](j,k) &= \frac{1}{4}\,jk(n-j+1)(n-k+1), \\ \left[\Omega^{-1}\boldsymbol{e}\boldsymbol{d}'\Omega^{-1}\right](j,k) &= \frac{1}{12}\,jk(n-j)\left(n^2-k^2\right) + O\left(n^4\right), \\ \left[\Omega^{-1}\boldsymbol{d}\boldsymbol{d}'\Omega^{-1}\right](j,k) &= \frac{1}{36}\,jk\left(n^2-j^2\right)\left(n^2-k^2\right) + O\left(n^5\right), \end{aligned}$$

where $[H](j,k)$ is the (j,k)th element of the matrix H.

Let us denote by $K(s,t;M)$ the uniform limit of $A_n(j,k)/n$ for Case (M) (M=A,B,C,D) such that

$$\lim_{n\to\infty}\max_{1\leq j,k\leq n}\left|\frac{1}{n}A_n(j,k) - K\left(\frac{j}{n},\frac{k}{n};M\right)\right| = 0. \tag{6.16}$$

Then it follows from the above discussions that

$$\begin{aligned} K(s,t;A) &= \min(s,t) - st, \\ K(s,t;B) &= \min(s,t) - st - 3st(1-s)(1-t), \\ K(s,t;C) &= \min(s,t) - st - \frac{5}{4}st\left(1-s^2\right)\left(1-t^2\right), \\ K(s,t;D) &= \min(s,t) - st - 2st(1-s)(1-t)(4-5s-5t+10st). \end{aligned}$$

The FDs $D(\lambda; M)$ of $K(s,t;M)$ are already available from Chapter 5. We have

$$\begin{aligned} D(\lambda; A) &= \frac{\sin\sqrt{\lambda}}{\sqrt{\lambda}}, \\ D(\lambda; B) &= \frac{12}{\lambda^2}\left(2 - \sqrt{\lambda}\sin\sqrt{\lambda} - 2\cos\sqrt{\lambda}\right), \\ D(\lambda; C) &= \frac{15}{\lambda^3}\left(\sqrt{\lambda}\,(3-\lambda)\sin\sqrt{\lambda} - 3\lambda\cos\sqrt{\lambda}\right), \\ D(\lambda; D) &= \frac{720}{\lambda^4}\left(4(6+\lambda) - \sqrt{\lambda}(24-\lambda)\sin\sqrt{\lambda} - 8(3-\lambda)\cos\sqrt{\lambda}\right). \end{aligned}$$

Note that $D(\lambda; B)$ and $D(\lambda; C)$ can be computed easily by Darling's formula in (2.83), whereas $D(\lambda; D)$ can be computed from Theorem 2.20 by putting

$$K(s,t;D) = \min(s,t) - st - \sum_{j=1}^{3} \psi_j(s)\psi_j(t),$$

where

$$\psi_1(t) = \sqrt{2}it(1-t), \quad \psi_2(t) = \sqrt{10}t^2(1-t), \quad \psi_3(t) = \sqrt{10}t(1-t)^2.$$

Using Theorem 5.2, we now obtain the following theorem (Tanaka, 1996, chapter 10).

Theorem 6.1 *Let $R_n(M)$ be the LBIU statistic defined in (6.7) for Case (M) (M = A,B,C,D). Then it holds that, as $n \to \infty$ under $\alpha = 1 - c/n$,*

$$R_n(M) \Rightarrow R(M;c) = \int_0^1 \int_0^1 \left(K(s,t;M) + c^2 K^{(2)}(s,t;M)\right) dW(s)\,dW(t), \tag{6.17}$$

where $K^{(2)}(s,t;M)$ is the iterated kernel of $K(s,t;M)$, and the c.f. of $R(M;c)$ is given by

$$E\left(e^{i\theta R(M;c)}\right) = \left[D\left(i\theta + \sqrt{-\theta^2 + 2ic^2\theta}; M\right) D\left(i\theta - \sqrt{-\theta^2 + 2ic^2\theta}; M\right)\right]^{-1/2}. \tag{6.18}$$

We also have

$$\begin{aligned} E(R(A;c)) &= \frac{1}{6} + \frac{1}{90}c^2, \quad E(R(B;c)) = \frac{1}{15} + \frac{11}{12{,}600}c^2, \\ E(R(C;c)) &= \frac{1}{14} + \frac{1}{882}c^2, \quad E(R(D;c)) = \frac{3}{70} + \frac{11}{44{,}100}c^2. \end{aligned}$$

The present LBIU tests have a close relationship with the LBI tests in the state space model discussed in Chapter 5. More specifically, the test statistics under a local alternative yield essentially the same FD for the following combinations of the two models, where y_j follows the MA(1) model, whereas x_j follows the state space model with $\beta_j = \beta_{j-1} + \eta_j$ and β_0 unknown:

$$D(\lambda; A): \quad y_j = \varepsilon_j - \alpha\varepsilon_{j-1} \quad \Leftrightarrow \quad x_j = \beta_j + \varepsilon_j,$$

$$D(\lambda; B): \quad y_j = \mu + \varepsilon_j - \alpha\varepsilon_{j-1} \quad \Leftrightarrow \quad x_j = \beta_j + \gamma\, j + \varepsilon_j,$$

$$D(\lambda; C): \quad y_j = \delta\, j + \varepsilon_j - \alpha\varepsilon_{j-1} \quad \Leftrightarrow \quad x_j = \beta_j + \gamma\, j^2 + \varepsilon_j,$$

$$D(\lambda; D): \quad y_j = \mu + \delta\, j + \varepsilon_j - \alpha\varepsilon_{j-1} \quad \Leftrightarrow \quad x_j = \beta_j + \gamma_1\, j + \gamma_2\, j^2 + \varepsilon_j.$$

The equivalence for $D(\lambda; A)$ follows from (5.18) with $m = 0$, that for $D(\lambda; B)$ from (5.38) with $q = 1$, that for $D(\lambda; C)$ from (5.38) with $q = 2$, and that for $D(\lambda; D)$ from (5.40), where $\rho = \mathrm{Var}(\eta_j)/\mathrm{Var}(\varepsilon_j)$ is defined as c^2/n^2 rather than c/n^2. A question naturally arises whether the LBI test in the state space model with the initial value β_0 known has an equivalent FD in the conditional MA(1) model where the initial value ε_0 of the error process is set at 0. The answer is in the negative, which we shall show in the next section.

Figure 6.1 plots probability densities of $R(M; c)$ for $M = A, B, C, D$ under $H_0 : c = 0$. The distributions are shifted to the left and more concentrated as regressors are added, where it is noted that Case (B) and Case (C) are quite close to each other. Power functions will be drawn in Section 6.3, comparing them with those for the conditional model. So far we have assumed the local alternative to be of the form $H_1(c, \kappa) : \alpha = 1 - c/n^{\kappa}$ with $\kappa = 1$. We can deal

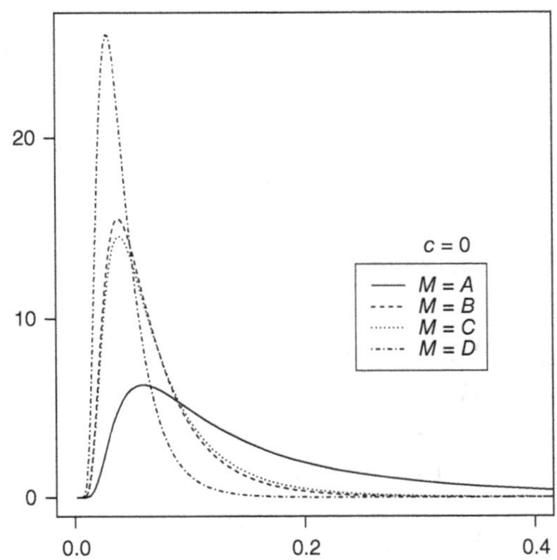

Figure 6.1 Probability densities of $R(M; c)$ for $c = 0$.

with the other values of κ. Let us consider, for τ (≥ 0) to be determined later,

$$P\left(\frac{1}{n^\tau} R_n \leq x\right) = P\left(\frac{1}{n^{\tau+1}} \frac{\boldsymbol{y}'\tilde{N}'\Omega^{-2}\tilde{N}\boldsymbol{y}}{\boldsymbol{y}'\tilde{N}'\Omega^{-1}\tilde{N}\boldsymbol{y}} \leq x\right)$$
$$= P\left(\frac{\boldsymbol{\zeta}'\left[\dfrac{\alpha}{n^{\tau+2}} A_n^2\Omega + \dfrac{c^2}{n^{\tau+2\kappa+2}} A_n^2\right]\boldsymbol{\zeta}}{\boldsymbol{\zeta}'\left[\dfrac{\alpha}{n} A_n\Omega + \dfrac{c^2}{n^{2\kappa+1}} A_n\right]\boldsymbol{\zeta}} \leq x\right),$$

where $\tilde{N} = I_n - X(X'\Omega^{-1}X)^{-1}X'\Omega^{-1}$, $A_n = \Omega^{-1}\tilde{N}$ and $\boldsymbol{\zeta} \sim \mathrm{N}(\mathbf{0}, \sigma_\varepsilon^2 I_n)$. Then we can derive the limiting distribution of R_n/n^τ as $n \to \infty$ under $\alpha = 1 - c/n^\kappa$ as follows (Exercise 6.1.3).

$$\frac{1}{n^\tau} R_n \Rightarrow \begin{cases} \displaystyle\int_0^1\int_0^1 K(s,t)\,dW(s)\,dW(t) & (\kappa > 1, \tau = 0), \\[2ex] \displaystyle\int_0^1\int_0^1 [K(s,t) + c^2 K^{(2)}(s,t)]\,dW(s)\,dW(t) & (\kappa = 1, \tau = 0), \\[2ex] \displaystyle c^2\int_0^1\int_0^1 K^{(2)}(s,t)\,dW(s)\,dW(t) & (1/2 < \kappa < 1, \tau = 2 - 2\kappa), \\[2ex] \dfrac{\displaystyle c^2\int_0^1\int_0^1 K^{(2)}(s,t)\,dW(s)\,dW(t)}{\displaystyle 1 + c^2\int_0^1\int_0^1 K(s,t)\,dW(s)\,dW(t)} & (\kappa = 1/2, \tau = 1), \\[2ex] \dfrac{\displaystyle\int_0^1\int_0^1 K^{(2)}(s,t)\,dW(s)\,dW(t)}{\displaystyle\int_0^1\int_0^1 K(s,t)\,dW(s)\,dW(t)} & (0 \leq \kappa < 1/2, \tau = 1), \end{cases}$$

where $K(s,t) = K(s,t;M)$ is the kernel given in (6.17), and $K^{(2)}(s,t)$ is the iterated kernel of $K(s,t)$.

The results are quite similar to those of the LBI statistic discussed in the last chapter. We can conclude from the above results that

(a) When $\kappa > 1$, the distribution of the LBIU statistic R_n is asymptotically the same as the null distribution, which means that the LBIU test has no asymptotic power when α is within an n^{-1} neighborhood of the unit root.

(b) When $\kappa = 1$, that is, when α is away from 1 on the order of n^{-1}, the LBIU statistic has a nondegenerate limiting distribution and the test is consistent. The c.f. of the limiting distribution is given in Theorem 6.1 for the present value of κ.

(c) When $\kappa < 1$, that is, when α is away from 1 by more than the order of n^{-1}, the LBIU statistic diverges. In particular, the statistic is of stochastic order n when $\kappa \le 1/2$ including the fixed alternative ($\kappa = 0$).

Exercises for Section 6.1

Exercise 6.1.1 Show that

$$\left.\frac{\partial^2 L(\boldsymbol{\delta})}{\partial \alpha^2}\right|_{H_0} = n\,\frac{\tilde{\boldsymbol{\eta}}'\Omega^{-2}\tilde{\boldsymbol{\eta}}}{\tilde{\boldsymbol{\eta}}'\Omega^{-1}\tilde{\boldsymbol{\eta}}} - \frac{n(n+5)}{6},$$

where $L(\boldsymbol{\delta})$ is defined in (6.4).

Exercise 6.1.2 Prove that, as $n \to \infty$ under $\alpha = 1 - c/n$,

$$\frac{1}{n}\,\tilde{\boldsymbol{\eta}}'\Omega^{-1}\tilde{\boldsymbol{\eta}} \;\to\; \sigma_\varepsilon^2 \quad \text{in probability}$$

for Cases (A) through (D) in (6.10).

Exercise 6.1.3 Prove that, as $n \to \infty$ under $\alpha = 1 - c/n^\kappa$,

$$\frac{1}{n^\tau} R_n \Rightarrow \begin{cases} c^2 \displaystyle\int_0^1\!\!\int_0^1 K^{(2)}(s,t)\,dW(s)\,dW(t) & (1/2 < \kappa < 1, \tau = 2 - 2\kappa), \\[2ex] \dfrac{c^2 \displaystyle\int_0^1\!\!\int_0^1 K^{(2)}(s,t)\,dW(s)\,dW(t)}{1 + c^2 \displaystyle\int_0^1\!\!\int_0^1 K(s,t)\,dW(s)\,dW(t)} & (\kappa = 1/2, \tau = 1), \\[2ex] \dfrac{\displaystyle\int_0^1\!\!\int_0^1 K^{(2)}(s,t)\,dW(s)\,dW(t)}{\displaystyle\int_0^1\!\!\int_0^1 K(s,t)\,dW(s)\,dW(t)} & (0 \le \kappa < 1/2, \tau = 1), \end{cases}$$

where R_n is the LBIU statistic defined in (6.7).

6.2 LBI Unit Root Tests in Conditional Models

In this section we consider the conditional MA(1) model

$$y_j = \boldsymbol{\beta}'\boldsymbol{x}_j + \varepsilon_j - \alpha\varepsilon_{j-1} \quad (j = 1, \ldots, n), \tag{6.19}$$

where $\{\boldsymbol{x}_j\}$ is a $p \times 1$ regression vector, and $\varepsilon_0 = 0$ and $\varepsilon_1, \ldots, \varepsilon_n \sim \mathrm{NID}(0, \sigma_\varepsilon^2)$. Note that the initial value ε_0 of the error term is set at 0, unlike in the unconditional model, and the present model can be rewritten as

$$\boldsymbol{y} = X\boldsymbol{\beta} + C^{-1}(\alpha)\boldsymbol{\varepsilon} \sim \mathrm{N}\left(X\boldsymbol{\beta},\, \sigma_\varepsilon^2 \Psi(\alpha)\right), \tag{6.20}$$

where $\boldsymbol{\varepsilon} = (\varepsilon_1, \ldots, \varepsilon_n)' \sim \mathrm{N}(\mathbf{0}, \sigma_\varepsilon^2 I_n)$ and

$$\Psi(\alpha) = \begin{pmatrix} 1 & -\alpha & & & 0 \\ -\alpha & 1+\alpha^2 & \cdot & & \\ & \cdot & \cdot & \cdot & \\ & & \cdot & \cdot & -\alpha \\ 0 & & & -\alpha & 1+\alpha^2 \end{pmatrix}, \tag{6.21}$$

$$C^{-1}(\alpha) = \begin{pmatrix} 1 & & & \\ -\alpha & \cdot & 0 & \\ & \cdot & \cdot & \\ & & \cdot & \cdot \\ 0 & & -\alpha & 1 \end{pmatrix}. \tag{6.22}$$

Note the differences between the model discussed in the last section and the present model, which may be described as follows.

(i) The error process $\{\varepsilon_j - \alpha\varepsilon_{j-1}\}$ in the previous model is stationary, whereas it is nonstationary in the present model because of the initial condition $\varepsilon_0 = 0$.

(ii) The difference of $\mathrm{Var}(\boldsymbol{y})$ between the two models is only in the $(1, 1)$st element. In fact, $\sigma_\varepsilon^2\Omega(\alpha) - \sigma_\varepsilon^2\Psi(\alpha) = \sigma_\varepsilon^2\alpha^2\boldsymbol{e}_1\boldsymbol{e}_1'$, where $\sigma_\varepsilon^2\Omega(\alpha)$ is $\mathrm{Var}(\boldsymbol{y})$ in the previous model and $\boldsymbol{e}_1 = (1, 0, \ldots, 0)'$; $n \times 1$. It holds that $\Omega(\alpha) = \alpha\Omega(1) + (1-\alpha)^2 I_n$, whereas

$$\Psi(\alpha) = \Omega(\alpha) - \alpha^2\boldsymbol{e}_1\boldsymbol{e}_1' = \alpha\Psi(1) + (1-\alpha)^2 I_n + \alpha(1-\alpha)\boldsymbol{e}_1\boldsymbol{e}_1'. \tag{6.23}$$

(iii) The previous model becomes noninvertible when $\alpha = 1$, but the present model is still invertible even if $\alpha = 1$ because there is a one-to-one correspondence between $\{\boldsymbol{y}_j\}$ and $\{\boldsymbol{\varepsilon}_j\}$.

Our concern here is to examine if the above differences yield different properties of the unit root test

$$H_0 : \alpha = 1 \quad \text{versus} \quad H_1 : \alpha < 1.$$

We can use the LM principle to derive the LBI test. The log-likelihood for $\boldsymbol{y}$ is given by

$$\ell(\boldsymbol{\delta}) = -\frac{n}{2}\log 2\pi\sigma_\varepsilon^2 - \frac{1}{2\sigma_\varepsilon^2}(\boldsymbol{y} - X\boldsymbol{\beta})'\Psi^{-1}(\alpha)(\boldsymbol{y} - X\boldsymbol{\beta}), \tag{6.24}$$

where $\boldsymbol{\delta} = (\alpha, \sigma_\varepsilon^2, \boldsymbol{\beta}')'$ and $|\Psi(\alpha)| = 1$. Let $\tilde{\boldsymbol{\delta}}$ be the MLE of $\boldsymbol{\delta}$ under H_0. Then we have, putting $C = C(1)$,

$$\tilde{\sigma}_\varepsilon^2 = \frac{1}{n}\boldsymbol{y}'\tilde{N}'C'C\tilde{N}\boldsymbol{y}, \quad \boldsymbol{y} - X\tilde{\boldsymbol{\beta}} = \tilde{N}\boldsymbol{y}, \quad \tilde{N} = I_n - X(X'C'CX)^{-1}X'C'C.$$

We also have (Exercise 6.2.1)

$$\left.\frac{d\Psi^{-1}(\alpha)}{d\alpha}\right|_{\alpha=1} = \left.\frac{dC'(\alpha)C(\alpha)}{d\alpha}\right|_{\alpha=1} = C'\boldsymbol{e}\boldsymbol{e}'C - C'C,$$

where $\boldsymbol{e} = (1, \ldots, 1)' : n \times 1$. Thus we obtain

$$\begin{aligned}\left.\frac{\partial\ell(\boldsymbol{\delta})}{\partial\alpha}\right|_{H_0} &= -\frac{n}{2}\frac{\boldsymbol{y}'\tilde{N}'(C'\boldsymbol{e}\boldsymbol{e}'C - C'C)\tilde{N}\boldsymbol{y}}{\boldsymbol{y}'\tilde{N}'C'C\tilde{N}\boldsymbol{y}} \\ &= -\frac{n}{2}\frac{\boldsymbol{y}'\tilde{N}'C'\boldsymbol{e}\boldsymbol{e}'C\tilde{N}\boldsymbol{y}}{\boldsymbol{y}'\tilde{N}'C'C\tilde{N}\boldsymbol{y}} + \frac{n}{2}.\end{aligned} \tag{6.25}$$

Defining $M = I_n - CX(X'C'CX)^{-1}X'C' = M' = M^2$ so that $C\tilde{N} = MC$, we can now derive the LBI test, which rejects $H_0 : \alpha = 1$ when

$$S_n = \frac{\boldsymbol{y}'C'M\boldsymbol{e}\boldsymbol{e}'MC\boldsymbol{y}}{\boldsymbol{y}'C'MC\boldsymbol{y}} \tag{6.26}$$

takes large values.

Let us consider four cases of the regression vector $\boldsymbol{x}_j$ in (6.19). We take up the same cases as in the unconditional model:

$$\text{(A) } \boldsymbol{x}_j = 0, \quad \text{(B) } \boldsymbol{x}_j = 1, \quad \text{(C) } \boldsymbol{x}_j = j, \quad \text{(D) } \boldsymbol{x}_j = (1, j)'. \tag{6.27}$$

We first derive the null distribution of S_n in (6.26). Noting that $MC\boldsymbol{y} = M\boldsymbol{\varepsilon}$ under H_0, we consider

$$S_n\Big|H_0 = \frac{\boldsymbol{\varepsilon}'M\boldsymbol{e}\boldsymbol{e}'M\boldsymbol{\varepsilon}}{\boldsymbol{\varepsilon}'M\boldsymbol{\varepsilon}} = \boldsymbol{e}'M\boldsymbol{e} \times \frac{\boldsymbol{\varepsilon}'A\boldsymbol{\varepsilon}}{\boldsymbol{\varepsilon}'A\boldsymbol{\varepsilon} + \boldsymbol{\varepsilon}'B\boldsymbol{\varepsilon}},$$

where

$$A = M\boldsymbol{e}\boldsymbol{e}'M/\boldsymbol{e}'M\boldsymbol{e} = A^2, \quad \mathrm{tr}(A) = 1, \quad B = M - A = B^2, \quad \mathrm{tr}(B) = n-p-1.$$

Since it holds that $\boldsymbol{\varepsilon}'A\boldsymbol{\varepsilon}/\sigma_\varepsilon^2 \sim \chi^2(1)$, $\boldsymbol{\varepsilon}'B\boldsymbol{\varepsilon}/\sigma_\varepsilon^2 \sim \chi^2(n-p-1)$ and $AB = 0$, we can deduce that

$$\frac{S_n \big| H_0}{\boldsymbol{e}'M\boldsymbol{e}} \sim \text{Be}\left(\frac{1}{2}, \frac{n-p-1}{2}\right),$$

where Be(a, b) stands for the beta distribution. As $n \to \infty$, it holds that

$$S_n \big| H_0 = \frac{1}{n}\boldsymbol{e}'M\boldsymbol{e} \times \frac{\boldsymbol{\varepsilon}'A\boldsymbol{\varepsilon}}{\frac{1}{n}(\boldsymbol{\varepsilon}'A\boldsymbol{\varepsilon} + \boldsymbol{\varepsilon}'B\boldsymbol{\varepsilon})} \Rightarrow \gamma Z^2, \tag{6.28}$$

where $Z \sim \text{N}(0, 1)$ and $\gamma = \lim_{n\to\infty} \boldsymbol{e}'M\boldsymbol{e}/n$ is given (Exercise 6.2.2) for each case by

$$\text{(A)}\ \gamma = 1, \quad \text{(B)}\ \gamma = \frac{1}{4}, \quad \text{(C)}\ \gamma = \frac{4}{9}, \quad \text{(D)}\ \gamma = \frac{1}{9}.$$

Thus the c.f. of the limiting null distribution of S_n is given by $(1 - 2i\gamma\theta)^{-1/2}$. The associated kernel is degenerate and is given by $K(s,t) = \gamma$.

We next derive the limiting distribution of S_n in (6.26) as $n \to \infty$ under

$$H_1(c) : \alpha = 1 - \frac{c}{n},$$

where c is a positive constant. We can show (Exercise 6.2.3) that, under $H_1(c)$,

$$\frac{1}{n}\boldsymbol{y}'C'MC\boldsymbol{y} \to \sigma_\varepsilon^2 \quad \text{in probability}, \tag{6.29}$$

for Cases (A) through (D). Let us consider

$$\begin{aligned}\frac{1}{n\sigma_\varepsilon^2}\boldsymbol{y}'C'M\boldsymbol{e}\boldsymbol{e}'MC\boldsymbol{y} &= \frac{1}{n\sigma_\varepsilon^2}\boldsymbol{\varepsilon}'(C'(\alpha))^{-1}C'M\boldsymbol{e}\boldsymbol{e}'MCC^{-1}(\alpha)\boldsymbol{\varepsilon}\\ &\stackrel{\mathcal{D}}{=} \frac{1}{n\sigma_\varepsilon^2}\left(\boldsymbol{e}'MCC^{-1}(\alpha)\boldsymbol{\varepsilon}\right)^2 \sim \gamma_n\,\chi^2(1),\end{aligned}$$

where

$$\gamma_n = \frac{1}{n}\boldsymbol{e}'MC\Psi(\alpha)C'M\boldsymbol{e}. \tag{6.30}$$

Using the expansion of $\Psi(\alpha)$ given in (6.23) and computing $\lim_{n\to\infty}\gamma_n$ for each case, we obtain the following theorem (Exercise 6.2.4).

Theorem 6.2 *Let $S_n(M)$ be the LBI statistic defined in (6.26) for Case (M) (M = A,B,C,D). Then it holds that, as $n \to \infty$ under $\alpha = 1 - c/n$,*

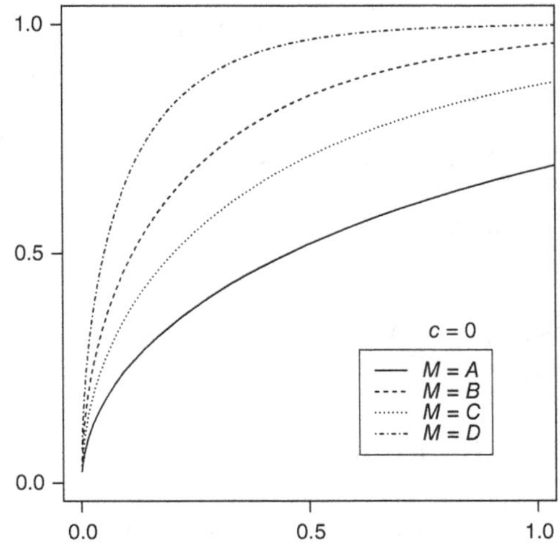

Figure 6.2 Distribution functions of $S(M;c)$ for $c=0$.

$$S_n(M) \;\Rightarrow\; S(M;c) = \gamma(M;c)\int_0^1\!\!\int_0^1 dW(s)\,dW(t) \;\sim\; \gamma(M;c)\,Z^2, \tag{6.31}$$

where $Z \sim N(0,1)$ *and*

$$\gamma(A;c) = 1 + c + \frac{1}{3}c^2, \quad \gamma(B;c) = \frac{1}{4} + \frac{1}{16}c + \frac{1}{120}c^2,$$
$$\gamma(C;c) = \frac{4}{9} + \frac{16}{81}c + \frac{2}{63}c^2, \quad \gamma(D;c) = \frac{1}{9} + \frac{1}{81}c + \frac{1}{945}c^2.$$

The FD $D(\lambda;M,c)$ *associated with* $S(M;c)$ *is* $D(\lambda;M,c) = 1 - \gamma(M;c)\lambda$ *and the c.f. of* $S(M;c)$ *is given by*

$$E\left(e^{i\theta S(M;c)}\right) = (D(2i\theta;M,c))^{-1/2} = (1 - 2i\theta\gamma(M;c))^{-1/2}.$$

It follows that $\mathrm{E}(S(M;c)) = \gamma(M;c)$ and $\mathrm{Var}(S(M;c)) = 2\gamma^2(M;c)$ for $M = A,B,C,D$. Figure 6.2 plots distribution functions of $S(M;c)$ for $M = A,B,C,D$ under $H_0 : c = 0$. These are all distributed as constant multiples of Z^2 with $Z \sim \mathrm{N}(0,1)$, where it can be seen that $S(D;0)$ is most concentrated toward the origin, followed by $S(B;0)$, $S(C;0)$, and $S(A;0)$.

It is notable that the present LBI test has no similarity to the LBI test in the conditional state space model discussed in Section 5.4, unlike the LBIU test in the unconditional MA(1) model. In the next section we compare the powers of the LBI and LBIU tests.

Exercises for Section 6.2

Exercise 6.2.1 Show that

$$\left.\frac{d\Psi^{-1}(\alpha)}{d\alpha}\right|_{\alpha=1} = \left.\frac{dC'(\alpha)C(\alpha)}{d\alpha}\right|_{\alpha=1} = C'\boldsymbol{e}\boldsymbol{e}'C - C'C,$$

where $\Psi(\alpha)$ is defined in (6.21).

Exercise 6.2.2 Compute $\gamma = \lim_{n\to\infty} \boldsymbol{e}'M\boldsymbol{e}/n$ defined in (6.28) for Cases (A) through (D) in (6.27).

Exercise 6.2.3 Using (6.23), prove that, under $\alpha = 1 - c/n$,

$$\frac{1}{n}\,\boldsymbol{y}'C'MC\boldsymbol{y} \;\to\; \sigma_\varepsilon^2 \quad \text{in probability}$$

for Cases (A) through (D) in (6.27), where $\boldsymbol{y}$ is given in (6.20).

Exercise 6.2.4 Using (6.23), compute

$$\gamma(M;c) = \lim_{n\to\infty} \gamma_n(c) = \lim_{n\to\infty} \frac{1}{n}\,\boldsymbol{e}'MC\Psi(\alpha)C'M\boldsymbol{e}$$

for Cases (A) through (D) in (6.27).

6.3 Comparison of Powers of the Two Tests

So far we have dealt with the LBIU test in the unconditional MA(1) model and the LBI test in the conditional MA(1) model. The corresponding test statistics were derived from the log-likelihood, where the LBIU statistic is based on its second derivative, whereas the LBI statistic is based on its first derivative. Both derivatives are evaluated at the MLE under $H_0 : \alpha = 1$.

In this section we compute and compare the limiting local powers of the two tests. Before doing so we consider the local MLE $\hat{\alpha}$ of the coefficient α in the unconditional and conditional MA(1) models when the true value of α is given by $\alpha = 1 - c/n$. The behavior of $\hat{\alpha}$ is found to be quite peculiar. In fact $P(\hat{\alpha} = 1)$ is positive as $n \to \infty$ under $\alpha = 1 - c/n$. It is also found that the limiting probability has an interesting relationship with the limiting power of the LBIU and LBI tests, as is seen in the following.

Let us denote by $L(\delta; M)$ the log-likelihood for Case (M) (M = A,B,C,D) in the unconditional model. Then it follows from (6.5), (6.7), and Theorem 6.1 that the limiting probability of $\hat{\alpha} = 1$ for Case (M) is given by

Table 6.1 *Limiting probabilities of the local MLE at unity.*

	$c = 0$	0.5	1	2	5	10	20
$P_1(A;c)$	0.657	0.651	0.633	0.570	0.347	0.143	0.029
$P_2(A;c)$	0.683	0.573	0.487	0.369	0.208	0.119	0.064
$P_1(B;c)$	0.968	0.967	0.966	0.958	0.898	0.686	0.295
$P_2(B;c)$	0.954	0.940	0.923	0.882	0.745	0.556	0.351
$P_1(C;c)$	0.952	0.952	0.949	0.938	0.856	0.620	0.254
$P_2(C;c)$	0.866	0.822	0.777	0.691	0.497	0.328	0.191
$P_1(D;c)$	0.999	0.999	0.999	0.998	0.995	0.961	0.713
$P_2(D;c)$	0.997	0.996	0.995	0.992	0.975	0.913	0.742

$$
\begin{aligned}
P_1(M;c) &= \lim_{n\to\infty} P\left(\left.\frac{\partial L^2(\delta;M)}{\partial \alpha^2}\right|_{H_0} < 0 \right) = P\left(R(M;c) < \frac{1}{6} \right) \\
&= P\left(\int_0^1 \int_0^1 \left(K(s,t;M) + c^2 K^{(2)}(s,t;M) \right) dW(s)\, dW(t) < \frac{1}{6} \right).
\end{aligned}
$$

Similarly, denote by $\ell(\delta; M)$ the log-likelihood for Case (M) in the conditional model. Then it follows from (6.25), (6.26), and Theorem 6.2 that the corresponding probability for Case (M) is given by

$$
\begin{aligned}
P_2(M;c) &= \lim_{n\to\infty} P\left(\left.\frac{\partial \ell(\delta;M)}{\partial \alpha}\right|_{H_0} > 0 \right) = P(S(M;c) < 1) \\
&= P\left(Z^2 < \frac{1}{\gamma(M;c)} \right).
\end{aligned}
$$

Table 6.1 presents $P_1(M;c)$ and $P_2(M;c)$ for M = A,B,C,D and various values of c. Figures 6.3 and 6.4 draw $P_1(M;c)$ and $P_2(M;c)$, respectively. It may be concluded from these results that

(a) Probabilities for each case decrease as c gets large, which is anticipated because the parameter diverges from unity.
(b) Probabilities are the highest for Case (D), followed by (B), (C), and (A) in both unconditional and conditional models.
(c) Probabilities for Case (A) in both models fade away quickly, but those for the other cases, Case (D) in particular, remain high even if the true value of α diverges from unity. This causes the estimation accuracy to decrease.

We now move on to the computation of powers. The limiting powers of the LBIU test can be computed by Lévy's inversion formula given in (5.42). For

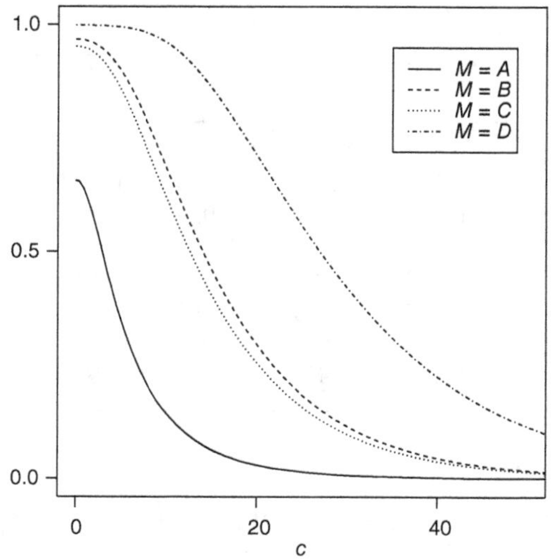

Figure 6.3 Limiting probabilities: Unconditional model.

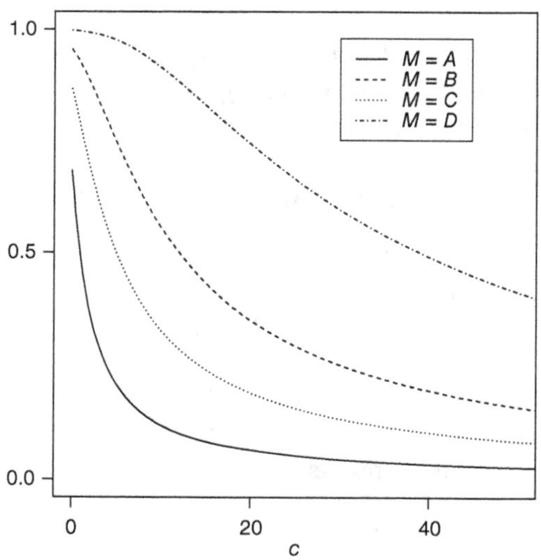

Figure 6.4 Limiting probabilities: Conditional model.

the LBI test the computation of powers is much easier because it is essentially that for the $\chi^2(1)$ distribution.

Figure 6.5 presents graphs of local powers at the 5% significance level for Cases (A), (B), (C), and (D). It is seen that, among the LBI tests, the power for Case (A) is the highest, followed by (C), (B), and (D). The same is true

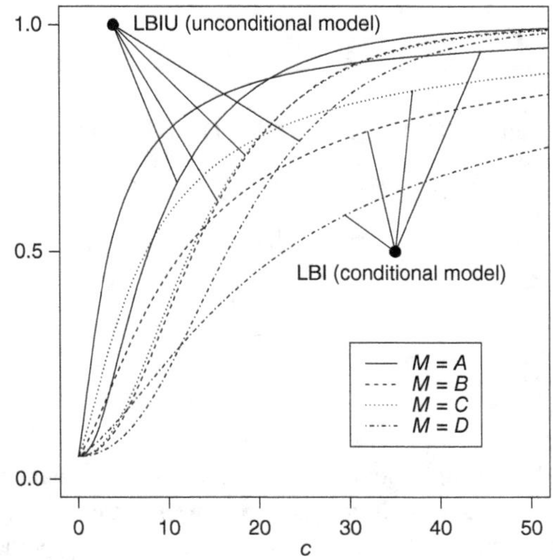

Figure 6.5 Limiting local powers of the LBIU and LBI tests.

for the LBIU tests, although those for (B) and (C) are very close to each other. It is interesting to note that this ordered relation of powers is the reverse of the ordered relation of probabilities discussed above. It is also seen for each case that, when the alternative is close to the null distribution, the LBI tests are superior to the LBIU tests, but they are eventually inferior. This is somewhat unexpected, and the information on the initial value is of little value. This also contrasts with the LBI tests in the unconditional and conditional state space models discussed in Chapter 5.

7

Fredholm Determinants in the Autoregressive Model

The LBI and LBIU unit root tests for the AR model were introduced in Section 1.5.4, together with the usual unit root test based on the LSE. The FDs associated with test statistics under the null hypothesis were also presented. In this chapter we derive the associated FDs under a sequence of local alternatives to compute and graphically present the limiting local powers of these AR unit root tests. It is found that the distributional behavior of test statistics is quite different from those discussed in previous chapters. In fact, the Ornstein–Uhlenbeck (O–U) process emerges under the local alternative, which has never occurred in previous problems. For this situation we introduce Girsanov's theorem concerning the transformation of measures induced by the O–U process to compute the associated c.f., where some examples are presented that include cointegration distributions.

7.1 LSE-Based Unit Root Tests

In this section we consider the unit root tests suggested by Dickey and Fuller (1979) and obtain the limiting powers of those tests under a sequence of local alternatives. For this purpose we deal with the following three models:

$$\text{Model } A: \quad y_j = \rho\, y_{j-1} + \varepsilon_j, \tag{7.1}$$

$$\text{Model } B: \quad y_j = \rho\, y_{j-1} + \alpha + \varepsilon_j, \tag{7.2}$$

$$\text{Model } C: \quad y_j = \rho\, y_{j-1} + \beta + \gamma\, j + \varepsilon_j, \tag{7.3}$$

where $y_0 = 0$ and $\{\varepsilon_j\} \sim \text{i.i.d.}(0, \sigma_\varepsilon^2)$. Model B contains a constant parameter, whereas Model C contains constant and trend parameters. These parameters have some restrictions. For example, Model C arises from

$$y_j = \beta_1 + \beta_2\, j + u_j, \quad u_j = \rho\, u_{j-1} + \varepsilon_j, \quad u_0 = 0, \tag{7.4}$$

which is equivalent to

$$y_j = \rho\, y_{j-1} + \beta_1(1-\rho) + \beta_2\rho + \beta_2(1-\rho)\, j + \varepsilon_j.$$

This is Model C in (7.3) with $\beta = \beta_1(1-\rho) + \beta_2\rho$ and $\gamma = \beta_2(1-\rho)$. Since we are concerned with a situation where ρ is close to 1, γ must be close to 0, although β is not restrictive. Similarly, the model (7.4) with $\beta_2 = 0$ is equivalent to Model B with $\alpha = \beta_1(1-\rho)$. Thus α must be close to 0. These seemingly curious restrictions are evident when $\rho = 1$ and were discussed in Section 1.5.4. We shall use the model representation (7.4) for Model C and (7.4) with $\beta_2 = 0$ for Model B in subsequent discussions.

Our concern here is to test

$$H_0 : \ \rho = 1 \quad \text{versus} \quad H_1(c,\kappa) : \ \rho = 1 - \frac{c}{n^{\kappa}} < 1, \tag{7.5}$$

where c and κ are positive constants. The LSE-based unit root test rejects H_0 when $n(\hat\rho - 1)$ takes small values, where $\hat\rho$ is the LSE of ρ applied to each model.

We use the FCLT and CMT introduced in Chapter 1 to examine the asymptotic behavior of $n(\hat\rho - 1)$. Assume first that $\kappa = 1$ in (7.5). For $u_j = \rho u_{j-1} + \varepsilon_j$ with $u_0 = 0$ and $\rho = 1 - c/n$ in (7.4), we construct the partial sum process

$$Y_n(t) = \frac{1}{\sqrt{n}\sigma_\varepsilon}\, u_{[nt]} + (nt - [nt])\, \frac{u_{[nt]+1} - u_{[nt]}}{\sqrt{n}\sigma_\varepsilon}. \tag{7.6}$$

Then it follows from the FCLT for $\{Y_n(t)\}$ (Bobkoski, 1983; Phillips, 1987b) that, as $n \to \infty$ under $\rho = 1 - c/n$,

$$\{Y_n(t)\} \Rightarrow \{Y(t)\}, \quad dY(t) = -cY(t)\,dt + dW(t), \quad Y(0) = 0, \tag{7.7}$$

where $\{Y(t)\}$ is the O–U process. The FCLT coupled with the CMT gives

$$\frac{1}{n\sigma_\varepsilon^2}\, u_n^2 = Y_n^2(1) \Rightarrow Y^2(1) = \left(\int_0^1 e^{-c(1-t)}\, dW(t)\right)^2, \tag{7.8}$$

$$\begin{aligned}\frac{1}{n^2\sigma_\varepsilon^2} \sum_{j=1}^{n} \left(u_j - \frac{j}{n} u_n\right)^2 &= \frac{1}{n} \sum_{j=1}^{n} \left(Y_n\left(\frac{j}{n}\right) - \frac{j}{n} Y_n(1)\right)^2 \\ &\Rightarrow \int_0^1 (Y(t) - tY(1))^2\, dt. \end{aligned} \tag{7.9}$$

We now consider the limiting distribution of $n(\hat\rho - 1)$ for each model as $n \to \infty$ under $\rho = 1 - c/n$. Here we present the integral expression in terms of the O–U process. We have (Tanaka, 1996, chapter 7), for each model M ($M = A, B, C$),

$$n(\hat{\rho} - 1) \Rightarrow Q(M; c) = \frac{\int_0^1 H(t)\, dY(t)}{\int_0^1 H^2(t)\, dt}, \tag{7.10}$$

where

$$H(t) = \begin{cases} Y(t) & \text{(Model } A\text{)}, \\ Y(t) - \int_0^1 Y(s)\, ds & \text{(Model } B\text{)}, \\ Y(t) + (6t - 4) \int_0^1 Y(s)\, ds - (12t - 6) \int_0^1 sY(s)\, ds & \text{(Model } C\text{)}. \end{cases}$$

The distribution of $Q(M; c)$ in (7.10) is generically called the near-unit root distribution, where $Q(A; c)$ was earlier presented in (1.66). It is seen that the limiting expressions appearing above take the ratio form $Q(M; c) = U(M)/V(M)$ with $V(M) > 0$ for Model M ($M = A, B, C$), unlike the test statistics dealt with in previous chapters. To compute the distribution function of $U(M)/V(M)$, we can use Imhof's (1961) formula given by

$$\begin{aligned} P\left(\frac{U(M)}{V(M)} < x\right) &= P\Big(xV(M) - U(M) > 0\Big) \\ &= \frac{1}{2} + \frac{1}{\pi} \int_0^\infty \frac{\mathrm{Im}[\phi(\theta; M, c, x)]}{\theta}\, d\theta, \end{aligned} \tag{7.11}$$

where $\phi(\theta; M, c, x)$ is the c.f. of $xV(M) - U(M)$. It can be shown (Nabeya and Tanaka, 1990a, 1990b) that

$$xV(M) - U(M) = \int_0^1 \int_0^1 K(s, t; M, c, x)\, dW(s)\, dW(t) + \frac{1}{2},$$

where $K(s, t; M, c, x) = xK_D(s, t; M, c) - K_N(s, t; M, c)$ with $K_D(s, t; M, c)$ associated with $V(M)$ and $K_N(s, t; M, c)$ associated with $U(M) + 1/2$. It is found in Nabeya and Tanaka (1990a) that

$$K(s, t; A, c, x) = \frac{x}{2c}\left(e^{-c|s-t|} - e^{-c(2-s-t)}\right) - \frac{1}{2} e^{-c(2-s-t)},$$

$$\begin{aligned} K(s, t; B, c, x) = x &\left[\frac{1}{2c}\left(e^{-c|s-t|} - e^{-c(2-s-t)}\right) - \frac{1}{c^2} g(s) g(t)\right] \\ &- \frac{1}{2}\left[e^{-c(2-s-t)} - \frac{1}{c}\left(e^{-c(1-s)} g(t) + e^{-c(1-t)} g(s)\right)\right], \end{aligned}$$

$$K(s,t;C,c,x) = x\left[\frac{1}{2c}\left(e^{-c|s-t|} - e^{-c(2-s-t)}\right) - \frac{4(c^2-3c+3)}{c^4}\,g(s)g(t)\right.$$
$$\left. + \frac{6(c-2)}{c^3}\,(g(s)h(t)+g(t)h(s)) - \frac{12}{c^2}\,h(s)h(t)\right]$$
$$-\frac{1}{2}\left[p(s)p(t) - \frac{3}{c^4}\,q(s)q(t)\right],$$

where

$$g(t) = 1 - e^{-c(1-t)}, \quad h(t) = t - e^{-c(1-t)},$$
$$p(t) = e^{-c(1-t)} + \frac{2}{c}\,g(t) - \frac{6}{c^2}\,(g(t) + ch(t)),$$
$$q(t) = 3c(1-t) - (c+3)g(t).$$

Thus the c.f. of $xV(M) - U(M)$ is given by

$$\phi(\theta; M, c, x) = e^{i\theta/2}\Big(D(2i\theta; M, c, x)\Big)^{-1/2},$$

where $D(\lambda; M, c, x)$ given in Theorem 7.1 is the FD of $K(s,t;M,c,x)$. Note that $D(\lambda; A, c, x)$ was earlier presented in (2.117) with $c = -\gamma$.

We have the following theorem established in Nabeya and Tanaka (1990a, 1990b) concerning the limiting distribution of $n(\hat{\rho} - 1)$ and the associated FD for Model M.

Theorem 7.1 *For Model M (M = A, B, C), it holds that, as $n \to \infty$ under $\rho = 1 - c/n$,*

$$\lim_{n\to\infty} P\Big(n(\hat{\rho}-1) < x\Big) = \frac{1}{2} + \frac{1}{\pi}\int_0^\infty \frac{\mathrm{Im}\left[e^{i\theta/2}\Big(D(2i\theta; M, c, x)\Big)^{-1/2}\right]}{\theta}\,d\theta,$$

where $D(\lambda; M, c, x)$ is the FD of $K(s,t;M,c,x)$ given by

$$D(\lambda; A, c, x) = e^{-c}\left[\cos\mu + \left(\frac{\lambda}{2} + c\right)\frac{\sin\mu}{\mu}\right],$$
$$D(\lambda; B, c, x) = e^{-c}\left[\frac{\lambda^2/4 + \lambda x - \lambda c^2/2 - c^3}{\mu^2}\,\frac{\sin\mu}{\mu} - \frac{c^2}{\mu^2}\cos\mu\right.$$
$$\left. + \frac{\lambda^2/2 - 2\lambda cx - \lambda c^2}{\mu^4}\,(\cos\mu - 1)\right],$$

$$D(\lambda; C, c, x)$$
$$= e^{-c}\left[\frac{c^5 + \lambda\left(c^4 - 2\lambda(c^2+3c+27) - 8x(c^2-3c-3)\right)/2}{\mu^4}\frac{\sin\mu}{\mu}\right.$$
$$+\left(\frac{c^4}{\mu^4} + \frac{4\lambda\left(c^3(c+2x) - 2\lambda(c^2+3c+6)\right)}{\mu^6}\right)\cos\mu$$
$$-\frac{12\lambda\left(c^4 - 2\lambda^2 x + 2\lambda(c+1)(x^2-3)\right)}{\mu^6}\left(\frac{\cos\mu}{\mu^2} + \frac{\sin\mu}{\mu} - \frac{1}{\mu^2}\right)$$
$$\left.+\frac{2\lambda\left(c^4 - 2\lambda(c^2+3c-3) + 2c^2x(c+3)\right)}{\mu^6}\right],$$

with $\mu = \sqrt{\lambda x - c^2}$.

We briefly discuss how to derive $D(\lambda; M, c, x)$. We already mentioned that $D(\lambda; A, c, x)$ was obtained in (2.117). To derive $D(\lambda; B, c, x)$, we consider the integral equation

$$f(t) = \lambda \int_0^1 K(1-s, 1-t; B, c, x) f(s)\, ds, \tag{7.12}$$

where

$$K(1-s, 1-t; B, c, x) = b_0 e^{-c|s-t|} + b_1(s) + b_2(s)e^{-ct},$$

with

$$b_0 = \frac{x}{2c}, \quad b_1(s) = -\frac{x}{c^2} + \left(\frac{x}{c^2} + \frac{1}{2c}\right)e^{-cs},$$
$$b_2(s) = \frac{x}{c^2} - \left(\frac{x}{c^2} + \frac{x}{2c} + \frac{1}{c} + \frac{1}{2}\right)e^{-cs}.$$

It can be shown (Exercise 7.1.1) that the integral equation (7.12) is equivalent to

$$f''(t) + (\lambda x - c^2) f(t) = \lambda a \tag{7.13}$$

with the two boundary conditions

$$f'(1) + cf(1) = -\frac{\lambda}{c}a,$$

$$f(0) = \frac{\lambda}{2}\left[-\int_0^1 e^{-cs} f(s)\, ds + \frac{1}{c}\int_0^1 \left(1 - e^{-cs}\right) f(s)\, ds\right],$$

where

$$a = x\int_0^1 \left(1 - e^{-cs}\right) f(s)\, ds - \frac{c}{2}\int_0^1 e^{-cs} f(s)\, ds. \tag{7.14}$$

The general solution to (7.13) is given by

$$f(t) = c_1 \cos \mu t + c_2 \sin \mu t + \frac{\lambda}{\mu^2} a,$$

where $\mu = \sqrt{\lambda x - c^2}$. Then, substituting $f(t)$ into the above two boundary conditions and the right-hand side of a in (7.14), we obtain a system of linear homogeneous equations $M(\lambda)\,\boldsymbol{c} = \boldsymbol{0}$ in $\boldsymbol{c} = (a, c_1, c_2)'$, where $M(\lambda)$ represents a highly complicated expression, which is omitted here. In any case, computing $|M(\lambda)|$ and finding a candidate for the FD, we arrive at $D(\lambda; B, c, x)$ given in Theorem 7.1 after some manipulations. The derivation of $D(\lambda; C, c, x)$ proceeds in much the same way, but it is quite laborious to arrive at $D(\lambda; C, c, x)$. In Section 7.5 we shall take an alternative approach for the derivation of the FD by directly using the integral expressions given in (7.10), where we employ Girsanov's theorem concerning the transformation of measures induced by the O–U process.

Moments of the limiting distribution of $n(\hat{\rho} - 1)$, that is, moments of $Q(M; c) = U(M)/V(M)$ in (7.10), can be obtained following Evans and Savin (1981). In fact, using the relation

$$\Gamma(\alpha) = \int_0^\infty x^{\alpha-1} e^{-x}\, dx \quad (\alpha > 0),$$

we have, for $\alpha > 0$ and $k = 0, 1, 2, \ldots,$

$$\begin{aligned}
\mathrm{E}\left(\frac{U^k(M)}{V^\alpha(M)}\right) &= \frac{1}{\Gamma(\alpha)}\mathrm{E}\left[\int_0^\infty V^{-\alpha}(M) x^{\alpha-1} U^k(M) e^{-x}\, dx\right] \\
&= \frac{1}{\Gamma(\alpha)}\mathrm{E}\left[\int_0^\infty \theta_2^{\alpha-1} U^k(M) e^{-\theta_2 V(M)}\, d\theta_2\right] \\
&= \frac{1}{\Gamma(\alpha)}\int_0^\infty \theta_2^{\alpha-1} \left.\frac{\partial^k \psi(\theta_1, -\theta_2; M)}{\partial \theta_1^k}\right|_{\theta_1=0} d\theta_2, \qquad (7.15)
\end{aligned}$$

where $\psi(\theta_1, -\theta_2; M)$ is the joint *moment generating function* (m.g.f.) of $U(M)$ and $-V(M)$, which can be obtained from the c.f. $\phi(\theta; M, c, x)$ of $xV(M) - U(M)$ as

$$\begin{aligned}
\psi\Big(\theta_1, -\theta_2; M\Big) &= \mathrm{E}\left(e^{\theta_1 U(M) - \theta_2 V(M)}\right) = \phi\Big(i\theta_1; M, c, \theta_2/\theta_1\Big) \\
&= e^{-\theta_1/2}\Big(D(-2\theta_1; M, c, \theta_2/\theta_1)\Big)^{-1/2}.
\end{aligned}$$

For example, we have

$$\psi\Big(\theta_1, -\theta_2; A\Big) = e^{(c-\theta_1)/2}\left[\cosh\nu + (c - \theta_1)\frac{\sinh\nu}{\nu}\right]^{-1/2},$$

where $\nu = \sqrt{2\theta_2 + c^2}$. Then, using the formula in (7.15), we can compute moments of $Q(A; c)$. The same is true for the other models.

Table 7.1 reports the means and variances of $Q(M; c)$ for $M = A, B, C$ and various values of c. According to the *central limit theorem* (CLT) discussed below, $Q(M; c)$ tends to $\mathrm{N}(-c, 2c)$ as $c \to \infty$, which is reflected to some extent in the results of Table 7.1.

Figure 7.1 shows probability densities of $Q(M; c)$ for $M = A, B, C$ under $H_0 : c = 0$, where $Q(A; 0)$ corresponds to the usual unit root distribution. The computation of the distribution $F(x; M, c) = P(Q(M; c) < x)$ can be done by Imhof's (1960) formula given in (7.11), and the density can be computed

Table 7.1 *Means and variances of near-unit root distributions.*

c	0	5	10	20	50
Model A					
Mean	−1.781	−6.976	−11.991	−21.997	−51.999
Var	10.112	19.272	29.145	49.075	109.030
Model B					
Mean	−5.379	−9.934	−14.573	−24.317	−54.135
Var	20.347	30.125	39.215	58.324	117.589
Model C					
Mean	−10.246	−13.100	−17.383	−26.798	−56.345
Var	36.407	41.391	49.788	68.120	126.473

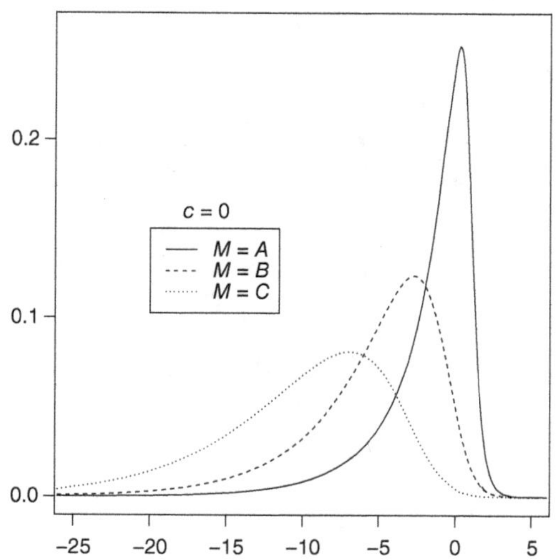

Figure 7.1 Probability densities of $Q(M; c)$ for $M = A, B, C$ and $c = 0$.

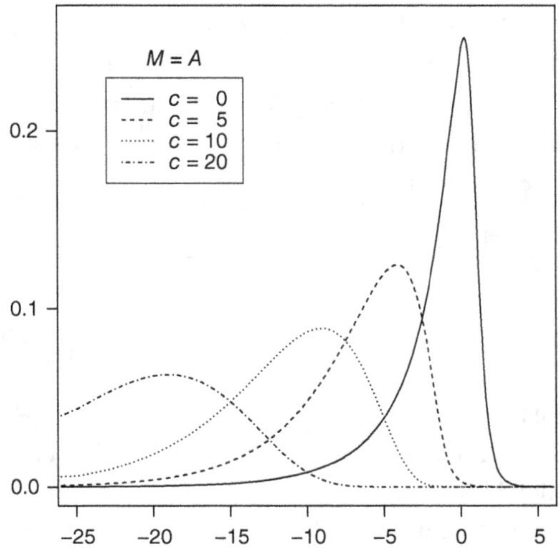

Figure 7.2 Probability densities of $Q(A;c)$ for $c = 0, 5, 10, 20$.

by numerical derivatives. The distribution is shifted to the left as the model complexity increases. Figure 7.2 shows probability densities of $Q(M;c)$ for $M = A$ and $c = 0, 5, 10, 20$ (see Figures 45 and 46 in Appendix B for probability densities of $Q(B;c)$ and $Q(C;c)$).

To examine the effect of c as $c \to \infty$ on the asymptotic distribution of $Q(M;c) = U(M)/V(M)$, we take up the case of $M = A$ and consider first $U(A)/cV(A)$. Since the joint c.f. of $U(A)$ and $V(A)$ is given from Theorem 7.1 by

$$\phi(\theta_1, \theta_2) = \mathrm{E}\left(e^{i\theta_1 U(A) + i\theta_2 V(A)}\right) = e^{(c-i\theta_1)/2}\left(\cos\mu + (c - i\theta_1)\frac{\sin\mu}{\mu}\right)^{-1/2},$$

where $\mu = \sqrt{2i\theta_2 - c^2}$, putting $\nu = \sqrt{c^2 - 2ic\theta_2} = c - i\theta_2 + O(c^{-1})$, we obtain

$$\begin{aligned}\phi(\theta_1, c\theta_2) &= \mathrm{E}\left(e^{i\theta_1 U + ic\theta_2 V}\right) \\ &= e^{(c-i\theta_1)/2}\left(\cosh\nu + (c - i\theta_1)\frac{\sinh\nu}{\nu}\right)^{-1/2} \\ &= e^{(c-i\theta_1)/2}\left(e^{\nu} + (c - i\theta_1 - \nu)\frac{\sinh\nu}{\nu}\right)^{-1/2} \\ &\to \exp\left\{\frac{-i\theta_1 + i\theta_2}{2}\right\} \quad (c \to \infty).\end{aligned}$$

Thus $U(A) \to -1/2$ and $cV(A) \to 1/2$ in probability as $c \to \infty$, which implies that $Q(A;c)/c = U(A)/(cV(A)) \to -1$, so that $Q(A;c)$ diverges to $-c$ as $c \to \infty$, and the test based on $Q(A;c)$ is consistent.

To deduce another distributional property, treating c as a sample size, it can be shown (Exercise 7.1.2) that, as $c \to \infty$,

$$\begin{aligned}\sqrt{c}\left(\frac{U(A)}{cV(A)}+1\right) &= \frac{1}{\sqrt{c}}\,(Q(A;c)+c)\\ &= \frac{\sqrt{c}\,(U(A)+cV(A))}{cV(A)} \Rightarrow \mathrm{N}(0,2). \qquad (7.16)\end{aligned}$$

This result implies that $Q(A;c)$ tends to $\mathrm{N}(-c, 2c)$ as $c \to \infty$, which is observed in Figure 7.2. It can be shown that the above distributional properties also hold for Model $M = B, C$.

We have so far assumed that $\rho = 1 - c/n^\kappa$ with $\kappa = 1$. It is possible to deal with the other values of κ. For simplicity, consider Model A. Then we have

$$\frac{1}{n^\gamma}\, n(\hat{\rho}-1) \Rightarrow \begin{cases} \dfrac{\int_0^1 W(t)\,dW(t)}{\int_0^1 W^2(t)\,dt} & (\kappa > 1,\ \gamma = 0),\\[2ex] \dfrac{\int_0^1 Y(t)\,dY(t)}{\int_0^1 Y^2(t)\,dt} & (\kappa = 1,\ \gamma = 0),\\[2ex] -c & (0 < \kappa < 1,\ \gamma = 1-\kappa),\\[2ex] -c & (\kappa = 0,\ \gamma = 1,\ 0 < c < 2),\end{cases}$$

where $\{Y(t)\}$ is the O–U process defined in (7.7). When $\kappa > 1$, ρ is much closer to 1, so that the limiting behavior of the statistic is the same as the case of $\rho = 1$. When $\kappa = 1$, the limiting statistic corresponds to the near-unit root distribution, which diverges to $-c$ as $c \to \infty$.

Let us consider the case of $0 < \kappa < 1$. This corresponds to moderate deviations from a unit root discussed in Phillips and Magdalinos (2007), who proved that, for $0 < \kappa < 1$,

$$\operatorname{plim}_{n\to\infty}\left(\frac{1}{n^{\kappa+1}}\sum_{j=1}^{n} u_j^2\right) = \frac{\sigma_\varepsilon^2}{2c}, \quad \frac{1}{n^{(\kappa+1)/2}}\sum_{j=1}^{n} u_{j-1}\varepsilon_j \Rightarrow \mathrm{N}\left(0, \frac{\sigma_\varepsilon^4}{2c}\right). \tag{7.17}$$

Since we have

$$\frac{1}{n^\gamma} n(\hat{\rho}-1) = \frac{\frac{1}{n^{\gamma+\kappa}} \sum_{j=2}^n u_{j-1}(u_j - \rho u_{j-1} + (\rho-1)u_{j-1})}{\frac{1}{n^{\kappa+1}} \sum_{j=2}^n u_{j-1}^2}$$

$$= \frac{\frac{1}{n^{\gamma+\kappa}} \sum_{j=2}^n u_{j-1}\varepsilon_j - \frac{c}{n^{\gamma+2\kappa}} \sum_{j=2}^n u_{j-1}^2}{\frac{1}{n^{\kappa+1}} \sum_{j=2}^n u_{j-1}^2},$$

it holds that $n(\hat{\rho}-1)/n^\gamma \to -c$ in probability for $0 < \kappa < 1$ by putting $\gamma = 1-\kappa$. When $\kappa = 0$, which corresponds to the fixed alternative $H_1 : \rho = 1-c$, we assume that $0 < c < 2$ to ensure the stationarity of $\{y_j\}$. Then it holds that $\hat{\rho} - 1 \to \rho - 1 = -c$ in probability.

It may be concluded from the above results that

(a) When $\kappa > 1$, the distribution of the LSE-based statistic $n(\hat{\rho}-1)$ under $H_1 : \rho = 1 - c/n^\kappa$ is asymptotically the same as that under $H_0 : \rho = 1$, which means that the LSE-based test has no asymptotic power when ρ is within an n^{-1} neighborhood of the unit root.
(b) When $\kappa = 1$, that is, when ρ is away from 1 on the order of n^{-1}, the statistic has a nondegenerate limiting distribution and the test is consistent.
(c) When $\kappa < 1$, that is, when ρ is away from 1 by more than the order of n^{-1}, the statistic diverges to $-\infty$ with the order of $n^{1-\kappa}$. In particular, the statistic is of stochastic order n under the fixed alternative.

Exercises for Section 7.1

Exercise 7.1.1 Show that the integral equation (7.12) is equivalent to

$$f''(t) + \left(\lambda x - c^2\right) f(t) = \lambda a$$

with the two boundary conditions

$$f'(1) + cf(1) = -\frac{\lambda}{c} a,$$

$$f(0) = \frac{\lambda}{2}\left[-\int_0^1 e^{-cs} f(s)\, ds + \frac{1}{c}\int_0^1 (1-e^{-cs}) f(s)\, ds \right],$$

where

$$a = x \int_0^1 (1-e^{-cs}) f(s)\, ds - \frac{c}{2}\int_0^1 e^{-cs} f(s)\, ds.$$

Exercise 7.1.2 Prove that, as $c \to \infty$,

$$\frac{1}{\sqrt{2c}}\left(\frac{\int_0^1 Y(t)\,dY(t)}{\int_0^1 Y^2(t)\,dt} + \sqrt{c}\right) \Rightarrow \mathrm{N}(0,1),$$

where $dY(t) = -cY(t)\,dt + dW(t)$ with $Y(0) = 0$.

7.2 LBI Unit Root Tests in Autoregressive Models

Let us consider the following model:

$$y_j = \boldsymbol{\beta}'\boldsymbol{x}_j + u_j, \quad u_j = \rho\, u_{j-1} + \varepsilon_j \qquad (j = 1, \ldots, n), \tag{7.18}$$

where $\boldsymbol{\beta}$ and $\boldsymbol{x}_j$ are $p \times 1$ coefficient and regression vectors, respectively, and $\{u_j\}$ is an error term which follows the AR(1) process with $\{\varepsilon_j\} \sim \mathrm{NID}(0, \sigma_\varepsilon^2)$. The AR(1) coefficient parameter ρ is assumed to take any value, and the initial value u_0 of $\{u_j\}$ is assumed to be $u_0 = 0$. This model may be rewritten as

$$\boldsymbol{y} = X\boldsymbol{\beta} + \boldsymbol{u}, \qquad \boldsymbol{u} = C(\rho)\,\boldsymbol{\varepsilon}, \tag{7.19}$$

where

$$\boldsymbol{y} = \begin{pmatrix} y_1 \\ y_2 \\ \vdots \\ y_n \end{pmatrix}, \quad X = \begin{pmatrix} \boldsymbol{x}_1' \\ \boldsymbol{x}_2' \\ \vdots \\ \boldsymbol{x}_n' \end{pmatrix}, \quad \boldsymbol{u} = \begin{pmatrix} u_1 \\ u_2 \\ \vdots \\ u_n \end{pmatrix}, \quad \boldsymbol{\varepsilon} = \begin{pmatrix} \varepsilon_1 \\ \varepsilon_2 \\ \vdots \\ \varepsilon_n \end{pmatrix},$$

$$C(\rho) = \begin{pmatrix} 1 & & & & 0 \\ \rho & 1 & & & \\ \cdot & & \cdot & & \\ \cdot & & & \cdot & \\ \cdot & & & & \cdot \\ \rho^{n-1} & \cdot & \cdot & \rho & 1 \end{pmatrix}. \tag{7.20}$$

Here $C(1) = C$ is the random walk-generating matrix defined in (1.93).

Our purpose here is to consider the testing problem

$$H_0 : \rho = 1 \quad \text{versus} \quad H_1(c) : \rho = 1 - \frac{c}{n} < 1, \tag{7.21}$$

where c is any positive constant.

We now derive the LBI test by using the LM principle. The log-likelihood for $\boldsymbol{y}$ is given by

$$L(\boldsymbol{\delta}) = -\frac{n}{2}\log 2\pi\sigma_\varepsilon^2 - \frac{1}{2\sigma_\varepsilon^2}(\boldsymbol{y} - X\boldsymbol{\beta})'\Phi^{-1}(\rho)(\boldsymbol{y} - X\boldsymbol{\beta}), \tag{7.22}$$

where $\boldsymbol{\delta} = (\rho, \boldsymbol{\beta}', \sigma_\varepsilon^2)'$ and $\Phi(\rho) = C(\rho)C'(\rho)$ with $|\Phi(\rho)| = 1$. Then, noting that

$$\left.\frac{d\Phi(\rho)}{d\rho}\right|_{\rho=1} = \boldsymbol{d}\boldsymbol{d}' - CC', \quad \boldsymbol{d} = \begin{pmatrix} 1 \\ 2 \\ \vdots \\ n \end{pmatrix},$$

where this relation is given in (1.132), we have

$$\begin{aligned}
\left.\frac{\partial L(\boldsymbol{\delta})}{\partial \rho}\right|_{H_0} &= \left.\frac{1}{2\sigma_\varepsilon^2}(\boldsymbol{y} - X\boldsymbol{\beta})'\Phi^{-1}(\rho)\frac{d\Phi(\rho)}{d\rho}\Phi^{-1}(\rho)(\boldsymbol{y} - X\boldsymbol{\beta})\right|_{H_0} \\
&= \frac{1}{2\tilde{\sigma}_\varepsilon^2}\tilde{\boldsymbol{u}}'(CC')^{-1}(\boldsymbol{d}\boldsymbol{d}' - CC')(CC')^{-1}\tilde{\boldsymbol{u}},
\end{aligned}$$

where $\tilde{\boldsymbol{u}} = N\boldsymbol{y} = N\boldsymbol{u}$ and

$$\tilde{\sigma}_\varepsilon^2 = \frac{1}{n}\boldsymbol{y}'N'(CC')^{-1}N\boldsymbol{y}, \quad N = I_n - X(X'(CC')^{-1}X)^{-1}X'(CC')^{-1}.$$

Thus the LBI test rejects H_0 when

$$R_n = \frac{\boldsymbol{y}'N'(CC')^{-1}\boldsymbol{d}\boldsymbol{d}'(CC')^{-1}N\boldsymbol{y}}{\boldsymbol{y}'N'(CC')^{-1}N\boldsymbol{y}} = \frac{\boldsymbol{y}'\tilde{M}\boldsymbol{d}\boldsymbol{d}'\tilde{M}\boldsymbol{y}}{\boldsymbol{y}'\tilde{M}\boldsymbol{y}} \tag{7.23}$$

takes small values, where

$$\tilde{M} = (CC')^{-1} - (CC')^{-1}X(X'(CC')^{-1}X)^{-1}X'(CC')^{-1}.$$

Since it holds that $\tilde{M}X = 0$, R_n reduces to 0 if $\boldsymbol{d}$ belongs to the column space of X. Avoiding this case, let us consider two cases of $\boldsymbol{x}_j$ in (7.18):

$$\text{Case 1: } \boldsymbol{x}_j = 0, \quad \text{Case 2: } \boldsymbol{x}_j = 1, \tag{7.24}$$

and derive the asymptotic distribution of R_n under $\rho = 1 - c/n$.

For Case 1, it holds that $\tilde{M} = (CC')^{-1}$ and

$$\begin{aligned}
\tilde{M}\boldsymbol{y} &= (C')^{-1}(u_1, u_2 - u_1, u_3 - u_2, \ldots, u_n - u_{n-1})', \\
\boldsymbol{y}'\tilde{M}\boldsymbol{d}\boldsymbol{d}'\tilde{M}\boldsymbol{y} &= u_n^2, \\
\boldsymbol{y}'\tilde{M}\boldsymbol{y} &= \sum_{j=1}^{n}(u_j - u_{j-1})^2.
\end{aligned}$$

Since $\sum_{j=1}^{n}(u_j - u_{j-1})^2/n\sigma_\varepsilon^2 \to 1$ in probability under $\rho = 1 - c/n$, it follows from (7.8) that

$$R_n = \frac{u_n^2/n\sigma_\varepsilon^2}{\sum_{j=1}^{n}(u_j - u_{j-1})^2/n\sigma_\varepsilon^2} \Rightarrow R(1;c) = Y^2(1), \tag{7.25}$$

where $\{Y(t)\}$ is the O–U process defined in (7.7). Note that $R(1;c) \sim \chi^2(1)$ under $H_0 : \rho = 1$ since $Y(1) = W(1)$ under H_0. For Case 2, we have $R_n \Rightarrow R(2;c) = Y^2(1)$, which has the same distribution as $R(1;c)$ in (7.25) (Exercise 7.2.1). Computing the c.f. of $Y^2(1)$, we have the following theorem (Exercise 7.2.2).

Theorem 7.2 *Let R_n be the LBI statistic defined in (7.23). Then it holds that, as $n \to \infty$ under $\rho = 1 - c/n$, $R_n \Rightarrow R(1;c) = R(2;c)$, where*

$$\begin{aligned} R(1;c) &= \int_0^1 \int_0^1 e^{-c(2-s-t)}\, dW(s)\, dW(t) \qquad (7.26)\\ &\sim \frac{1-e^{-2c}}{2c} Z^2, \quad Z \sim N(0,1), \\ E(R(1;c)) &= \frac{1-e^{-2c}}{2c}, \quad \mathrm{Var}(R(1;c)) = 2\left(\frac{1-e^{-2c}}{2c}\right)^2. \end{aligned}$$

Moreover, the c.f. of $R(1;c)$ is given by $E\left(e^{i\theta R(1;c)}\right) = (D(2i\theta))^{-1/2}$, where $D(\lambda)$ is the FD of $e^{-c(2-s-t)}$ given by

$$D(\lambda) = 1 - \frac{1-e^{-2c}}{2c}\lambda.$$

It is seen that the kernel $e^{-c(2-s-t)}$ associated with the statistics $R(1;c)$ and $R(2;c)$ is degenerate since it has only one eigenvalue $2c/(1-e^{-2c})$, which reduces to 1 under H_0. Asymptotic normality is not expected for these statistics as $c \to \infty$.

Figure 7.3 shows probability densities of $R(1;c)$ for $c = 0, 10, 20, 50$, where $\mathrm{E}(R(1;c)) = 1, 0.1, 0.05, 0.025$, respectively, for these values of c. The distributions are all monotone and shrink toward the origin as c gets large. The limiting local power of the LBI test under $H_1 : \rho = 1 - c/n$ will be graphically presented in Section 7.4. It is found that the power performance is quite poor in comparison with the LSE-based test discussed in Section 7.1.

We have so far assumed that $\rho = 1 - c/n^\kappa$ with $\kappa = 1$. We can also deal with the other values of κ. Let us consider the behavior of the LBI test under $\rho = 1 - c/n^\kappa$ with $\kappa \geq 0$. It holds that

$$u_n = \rho\, u_{n-1} + \varepsilon_n = \sum_{j=1}^{n} \rho^{n-j}\varepsilon_j \sim \mathrm{N}\left(0, \frac{1-\rho^{2n}}{1-\rho^2}\sigma_\varepsilon^2\right),$$

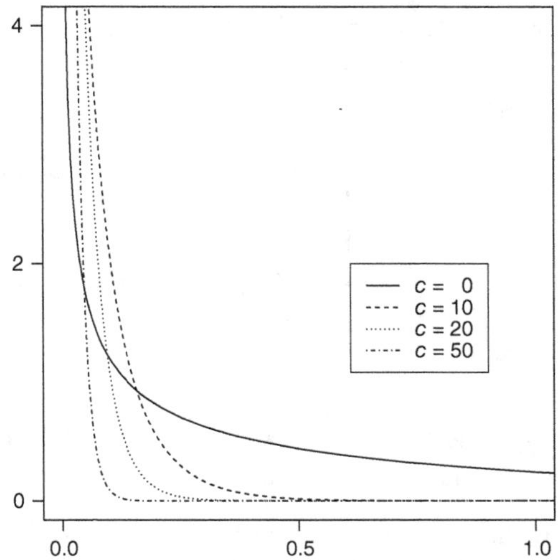

Figure 7.3 Probability densities of $R(1;c)$ for $c = 0,\ 10,\ 20,\ 50$.

where $\dfrac{u_n^2}{n\sigma_\varepsilon^2} \sim \dfrac{1-\rho^{2n}}{n(1-\rho^2)}\,\chi^2(1)$, and

$$\frac{1-\rho^{2n}}{n(1-\rho^2)} = \begin{cases} 1+o(1) & (\kappa > 1), \\[2ex] \dfrac{1-e^{-2c}}{2c} + o(1) & (\kappa = 1), \\[2ex] \dfrac{1}{2c}\, n^{\kappa-1} + O(n^{-1}) & (0<\kappa<1), \\[2ex] \dfrac{1}{c(2-c)}\,\dfrac{1}{n} + o(n^{-1}) & (\kappa = 0,\ 0<c<2). \end{cases}$$

We also have

$$\begin{aligned} \frac{1}{n}\sum_{j=2}^{n}(u_j - u_{j-1})^2 &= \frac{1}{n}\sum_{j=2}^{n}\big((\rho-1)u_{j-1}+\varepsilon_j\big)^2 \\ &= \frac{c^2}{n^{2\kappa+1}}\sum_{j=2}^{n}u_{j-1}^2 - \frac{2c}{n^{\kappa+1}}\sum_{j=2}^{n}u_{j-1}\varepsilon_j + \frac{1}{n}\sum_{j=2}^{n}\varepsilon_j^2 \\ &\to \begin{cases} \sigma_\varepsilon^2 & (\kappa>0), \\[2ex] \dfrac{2}{2-c}\,\sigma_\varepsilon^2 & (\kappa=0,\ 0<c<2). \end{cases} \end{aligned}$$

Thus the LBI statistic R_n for Case 1 follows

$$n^\gamma R_n = \frac{n^\gamma u_n^2}{\sum_{j=1}^n (u_j - u_{j-1})^2} \Rightarrow \begin{cases} Z^2 & (\kappa > 1,\ \gamma = 0), \\ \dfrac{1-e^{-2c}}{2c} Z^2 & (\kappa = 1,\ \gamma = 0), \\ \dfrac{1}{2c} Z^2 & (0 < \kappa < 1,\ \gamma = 1-\kappa), \\ \dfrac{1}{2c} Z^2 & (\kappa = 0,\ \gamma = 1,\ 0 < c < 2), \end{cases}$$

where $Z \sim \mathrm{N}(0,1)$. The LBI statistic for Case 2 has the same asymptotic behavior. It may be concluded from the above results that

(a) When $\kappa > 1$, the distribution of the LBI statistic R_n is asymptotically the same as the null distribution, which means that the LBI test has no asymptotic power when ρ is within an n^{-1} neighborhood of the unit root.
(b) When $\kappa = 1$, that is, when ρ is away from 1 on the order of n^{-1}, the statistic has a nondegenerate limiting distribution with $R(1;c) = R(2;c) = O_p(c^{-1})$ as $c \to \infty$ and the test is consistent.
(c) When $\kappa < 1$, that is, when ρ is away from 1 by more than the order of n^{-1}, the statistic converges to 0 with the order of $n^{\kappa-1}$. In particular, the statistic is of stochastic order n^{-1} under the fixed alternative.

Exercises for Section 7.2

Exercise 7.2.1 Consider R_n defined in (7.23) as

$$R_n = \frac{\boldsymbol{y}'\tilde{M}\boldsymbol{d}\boldsymbol{d}'\tilde{M}\boldsymbol{y}}{\boldsymbol{y}'\tilde{M}\boldsymbol{y}}.$$

Derive the limiting distribution of R_n as $n \to \infty$ under $\rho = 1 - c/n$ for the case of $\boldsymbol{x}_j = 1$.

Exercise 7.2.2 Compute the c.f. of

$$Y^2(1) = \int_0^1 \int_0^1 e^{-c(2-s-t)}\, dW(s)\, dW(t).$$

7.3 LBIU Unit Root Tests in Autoregressive Models

We next consider the LBIU test for the case where $\boldsymbol{d} = (1, 2, \ldots, n)'$ belongs to the column space of X. For the log-likelihood $L(\boldsymbol{\delta})$ defined in (7.22), it follows from (1.136) that

$$\left.\frac{\partial^2 L(\boldsymbol{\delta})}{\partial \rho^2}\right|_{H_0} = -n\frac{\boldsymbol{y}'N'(I_n - \boldsymbol{e}_n\boldsymbol{e}_n')N\boldsymbol{y}}{\boldsymbol{y}'N'(CC')^{-1}N\boldsymbol{y}},$$

where $N = I_n - X(X'(CC')^{-1}X)^{-1}X'(CC')^{-1}$ and $\boldsymbol{e}_n = (0, \ldots, 0, 1)'$. Since it holds that $N'(CC')^{-1}X = 0$, $\boldsymbol{y}'N'\boldsymbol{e}_n$ reduces to 0 if $\boldsymbol{e}_n$ belongs to the column space of $(CC')^{-1}X$ or if $CC'\boldsymbol{e}_n = \boldsymbol{d}$ belongs to the column space of X. In that case we reject $H_0 : \rho = 1$ when

$$S_n = \frac{1}{n}\frac{\boldsymbol{y}'N'N\boldsymbol{y}}{\boldsymbol{y}'N'(CC')^{-1}N\boldsymbol{y}} = \frac{1}{n}\frac{\boldsymbol{y}'N'N\boldsymbol{y}}{\boldsymbol{y}'\tilde{M}\boldsymbol{y}} \tag{7.27}$$

takes small values, where

$$\tilde{M} = (CC')^{-1} - (CC')^{-1}X(X'(CC')^{-1}X)^{-1}X'(CC')^{-1}.$$

Let us consider Case 3: $\boldsymbol{x}_j = (1, j)'$, so that $X = (\boldsymbol{e}, \boldsymbol{d})$, which yields, from the discussions in Section 1.5.4,

$$\boldsymbol{y}'N'N\boldsymbol{y} = \sum_{j=2}^{n}\left(u_j - u_1 - \frac{j-1}{n-1}(u_n - u_1)\right)^2,$$
$$\boldsymbol{y}'\tilde{M}\boldsymbol{y} = \sum_{j=2}^{n}\left(u_j - u_{j-1} - \frac{u_n - u_1}{n-1}\right)^2.$$

Here it holds that $\boldsymbol{y}'\tilde{M}\boldsymbol{y}/n\sigma_\varepsilon^2 \to 1$ in probability as $n \to \infty$ under $\rho = 1 - c/n$. However, it follows from the FCLT and CMT described in (7.9) that

$$\frac{1}{n^2\sigma_\varepsilon^2}\boldsymbol{y}'N'N\boldsymbol{y} = \frac{1}{n^2\sigma_\varepsilon^2}\sum_{j=1}^{n}\left(u_j - \frac{j}{n}u_n\right)^2 + o_p(1) \Rightarrow S(c),$$
$$S(c) = \int_0^1 (Y(t) - tY(1))^2\, dt. \tag{7.28}$$

The expression for $S(c)$ in (7.28) can be rewritten as

$$S(c) \stackrel{\mathcal{D}}{=} \int_0^1\int_0^1 \mathrm{Cov}\,(Y(s) - sY(1), Y(t) - tY(1))\, dW(s)\, dW(t)$$
$$= \int_0^1\int_0^1 K(s,t)\, dW(s)\, dW(t),$$

where

$$K(s,t) = L(s,t) - tL(s,1) - sL(1,t) + stL(1,1), \tag{7.29}$$
$$L(s,t) = \frac{1}{2c}\left(e^{-c|s-t|} - e^{-c(s+t)}\right).$$

Note that $K(s,t) = \min(s,t) - st$ with $L(s,t) = \min(s,t)$ under H_0.

To compute the FD of $K(s,t)$ in (7.29), we consider the integral equation with the kernel $K(s,t)$, which is shown to be equivalent to

$$f''(t) + (\lambda - c^2)f(t) = \lambda a t, \quad a = c^2 \int_0^1 (L(s,1) - sL(1,1))f(s)\,ds$$

with the two boundary conditions $f(0) = f(1) = 0$. The general solution is given by

$$f(t) = c_1 \cos \mu t + c_2 \sin \mu t + \frac{\lambda a}{\mu^2} t, \quad \mu = \sqrt{\lambda - c^2}.$$

Using our approach (see also Section 7.5 for another approach), we obtain (Nabeya and Tanaka, 1990b)

Theorem 7.3 *Let S_n be the LBIU statistic defined in (7.27) with $\boldsymbol{x}_j = (1, j)'$. Then it holds that, as $n \to \infty$ under $\rho = 1 - c/n$,*

$$\begin{aligned} S_n \Rightarrow S(c) &= \int_0^1 (Y(t) - tY(1))^2\,dt \\ &\stackrel{\mathcal{D}}{=} \int_0^1\int_0^1 K(s,t)\,dW(s)\,dW(t), \\ E(S(c)) &= \frac{2}{3c} - \frac{5}{4c^2} + \frac{1}{c^3} - \left(\frac{1}{6c} + \frac{3}{4c^2} + \frac{1}{c^3}\right)e^{-2c}, \end{aligned} \tag{7.30}$$

where $dY(t) = -cY(t)\,dt + dW(t)$ with $Y(0) = 0$, and $K(s,t)$ is defined in (7.29). Moreover, the c.f. of $S(c)$ is given by $E\left(e^{i\theta S(c)}\right) = (D(2i\theta))^{-1/2}$, where $D(\lambda)$ is the FD of $K(s,t)$ defined by

$$D(\lambda) = e^{-c}\left[\frac{c^4}{\mu^4}\cos\mu - \left(\frac{3c^3 - \lambda(c^2+3c+3)}{3\mu^2} + \frac{\lambda c^2}{\mu^4}\right)\frac{\sin\mu}{\mu}\right] \tag{7.31}$$

with $\mu = \sqrt{\lambda - c^2}$.

The statistic $S(c)$ in (7.30) is a quadratic functional of the O–U process. As was mentioned in Section 7.1, there is a simpler way of computing the associated FD, which will be demonstrated in Section 7.5. It is seen that $D(\lambda)$ reduces to $\sin\sqrt{\lambda}/\sqrt{\lambda}$ when $c = 0$.

Figure 7.4 shows probability densities of $S(c)$ for $c = 0, 5, 10, 20$. It is seen that the distribution is shifted to the left and shrinks to the origin as c gets large, where $\mathrm{E}(S(c)) = 0.167, 0.091, 0.055, 0.030$, respectively, for these values of c. Asymptotic normality is not expected as $c \to \infty$.

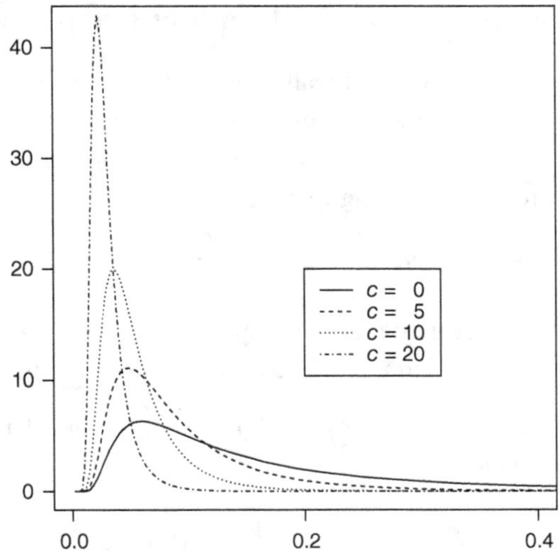

Figure 7.4 Probability densities of $S(c)$ for $c = 0, 5, 10, 20$.

Let us consider the asymptotic behavior of the LBIU statistic S_n with $\boldsymbol{x}_j = (1, j)'$ under $\rho = 1 - c/n^\kappa$ with $\kappa \geq 0$ and $c > 0$. We have

$$S_n = \frac{\sum_{j=2}^{n} \left(u_j - u_1 - \frac{j-1}{n-1}(u_n - u_1)\right)^2 \Big/ n^2}{\sum_{j=2}^{n} (u_j - u_{j-1})^2 \Big/ n - (u_n - u_1)^2 \Big/ n(n-1)}.$$

Noting that $\sum_{j=2}^{n} (u_j - u_{j-1})^2 = O_p(n)$ for $\kappa \geq 0$, and

$$u_n = \begin{cases} O_p\left(\sqrt{n}\right) & (\kappa \geq 1), \\ O_p\left(n^{\kappa/2}\right) & (0 \leq \kappa < 1), \end{cases} \qquad \sum_{j=1}^{n} u_j^2 = \begin{cases} O_p\left(n^2\right) & (\kappa \geq 1), \\ O_p\left(n^{\kappa+1}\right) & (0 \leq \kappa < 1), \end{cases}$$

it holds that

$$S_n = \begin{cases} O_p(1) & (\kappa \geq 1), \\ O_p\left(n^{\kappa-1}\right) & (0 \leq \kappa < 1). \end{cases}$$

The stochastic order of the LBIU statistic S_n is the same as the LBI statistic R_n discussed in the last section. Thus the test is inconsistent for $\kappa > 1$. When $\kappa = 1$, $S_n \Rightarrow S(c)$ with $S(c) = O_p(c^{-1})$ as $c \to \infty$, so that the test is consistent. For $\kappa < 1$, $S_n = O_p(n^{\kappa-1}) \to 0$ as $n \to \infty$ for any c.

7.4 Comparison of the Limiting Local Powers

In this section we compare graphically the performance of the LSE-based, LBI, and LBIU tests discussed in the previous sections of the chapter. For this purpose we compute the limiting local powers of these tests at the 5% significance level for the following four cases:

Case 1: LSE-based and LBI tests for $\boldsymbol{x}_j = 0$.
Case 2: LSE-based and LBI tests for $\boldsymbol{x}_j = 1$.
Case 3: LSE-based and LBIU tests for $\boldsymbol{x}_j = (1, j)'$.
Case 4: LSE-based test and the test based on S_n in (7.27) for $\boldsymbol{x}_j = 1$.

Note that the S_n-test applied to $\boldsymbol{x}_j = 1$ in Case 4 is not LBIU but is still applicable, and it holds that

$$S_n \Rightarrow T(c) = \int_0^1 Y^2(t)\, dt, \tag{7.32}$$

where $\{Y(t)\}$ is the O–U process defined in (7.7) by $dY(t) = -cY(t)\, dt + dW(t)$ with $Y(0) = 0$. The c.f. of $T(c)$ is available from (2.115) as

$$\mathrm{E}\left(e^{i\theta T(c)}\right) = e^{c/2}\left(\cos\mu + c\frac{\sin\mu}{\mu}\right)^{-1/2},$$

where $\mu = \sqrt{2i\theta - c^2}$.

Figure 7.5 presents four figures as Cases 1, 2, 3, and 4, where the solid lines are for LSE-based tests and the broken lines for the other tests. Powers of the LSE-based tests were computed by Imhof's formula given in (7.11), whereas those of the LBIU tests were by Lévy's inversion formula in (5.42). Computation of powers of the LBI tests was easily done by noting that the test statistic is distributed as a constant multiple of $\chi^2(1)$. As Cases 1 and 2 show, the LBI tests behave very badly and are not recommended. The optimality of the LBI tests is only local. On the other hand, as Case 3 shows, the LBIU test is superior to the LSE-based test. Case 4 shows that the test based on $T(c)$ in (7.32) is quite good, unlike the LBI test in Case 2, and is highly recommended.

7.5 Computation via Girsanov's Theorem

The derivation of the FD discussed so far starts with expressing the quadratic Brownian functional as the Riemann–Stieltjes double integral of a kernel with respect to Bm. We then construct the integral equation with the kernel and find the equivalent differential equation with appropriate boundary conditions. Solving the differential equation, the FD can be derived from the boundary conditions.

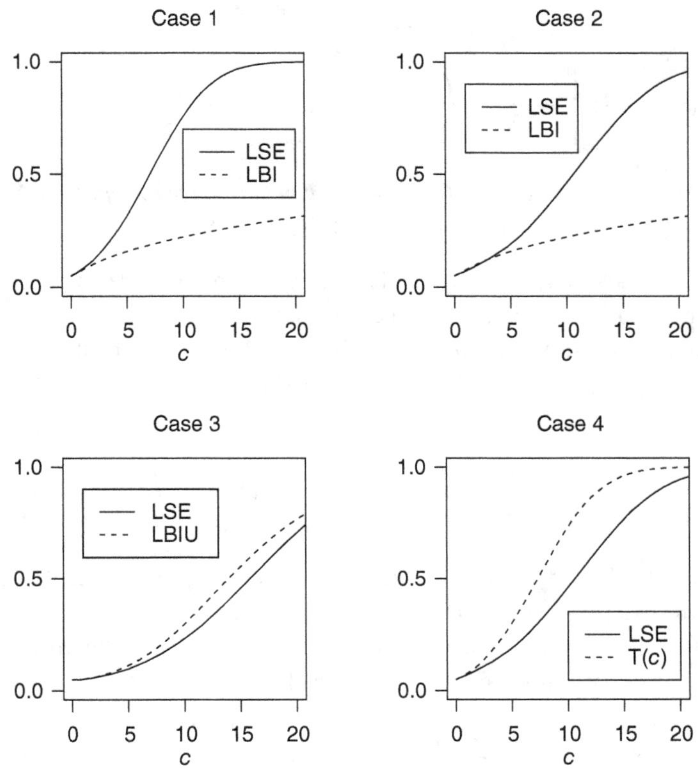

Figure 7.5 Power comparison of the unit root tests.

In this section we suggest another approach specialized in the O–U process. For this purpose, the following theorem due to Girsanov (1960) (see also Liptser and Shiryaev (2001a, p.290)) is useful.

Theorem 7.4 (Girsanov's theorem) *Let* $Y^\alpha = \{Y^\alpha(t) : 0 \le t \le 1\}$ *and* $Y^\beta = \{Y^\beta(t) : 0 \le t \le 1\}$ *be the O–U processes in* $C = C[0, 1]$ *defined by*

$$dY^\alpha(t) = \alpha Y^\alpha(t)\,dt + dW(t), \tag{7.33}$$

$$dY^\beta(t) = \beta Y^\beta(t)\,dt + dW(t), \qquad Y^\alpha(0) = Y^\beta(0). \tag{7.34}$$

Let μ_Y^α *and* μ_Y^β *be probability measures on* $(C, \mathcal{B}(C))$ *induced by* Y^α *and* Y^β*, respectively, by the relation*

$$\mu_Y^\alpha(A) = P(\omega : Y^\alpha \in A), \quad \mu_Y^\beta(A) = P(\omega : Y^\beta \in A), \quad A \in \mathcal{B}(C).$$

Then measures μ_Y^α *and* μ_Y^β *are equivalent, and*

$$\frac{d\mu_Y^\alpha}{d\mu_Y^\beta}(Y^\gamma) = \exp\left[\frac{\beta^2-\alpha^2}{2}\int_0^1 \left(Y^\gamma(t)\right)^2 dt - (\beta-\alpha)\int_0^1 Y^\gamma(t)\, dY^\gamma(t)\right], \tag{7.35}$$

where the left-hand side is the Radon–Nikodym derivative evaluated at $Y^\gamma \in C$ *with* $dY^\gamma(t) = \gamma Y^\gamma(t)\, dt + dW(t)$ *and* $Y^\gamma(0) = Y^\alpha(0)$.

The Radon–Nikodym derivative in (7.35) may be interpreted as the likelihood ratio under contiguity. To see this, suppose that $y_j = \rho_n y_{j-1} + \varepsilon_j$ $(j = 1, \ldots, n)$ with $y_0 = 0$ and $\{\varepsilon_j\} \sim \mathrm{NID}(0, 1)$. Let $\ell_n(\alpha)$ be the likelihood for $y_1, \ldots, y_n$ under $\rho_n = 1 - \alpha/n$. Then it holds (Exercise 7.5.1) that, as $n \to \infty$,

$$\left.\frac{\ell_n(\alpha)}{\ell_n(\beta)}\right|_\gamma \Rightarrow \exp\left[\frac{\beta^2-\alpha^2}{2}\int_0^1 (X^\gamma(t))^2\, dt - (\beta-\alpha)\int_0^1 X^\gamma(t)\, dX^\gamma(t)\right], \tag{7.36}$$

where the likelihood ratio is evaluated at $\rho_n = 1 - \gamma/n$, and $dX^\gamma(t) = \gamma X^\gamma(t)\, dt + dW(t)$ with $X^\gamma(0) = 0$.

The Radon–Nikodym derivative can also be used to obtain the MLE of α in (7.33) from the likelihood for $\{Y_H^\alpha(t)\}$ defined by

$$\frac{d\mu_Y^\alpha}{d\mu_Y^0}(Y^\alpha) = \exp\left[-\frac{\alpha^2}{2}\int_0^1 \left(Y^\alpha(t)\right)^2 dt + \alpha\int_0^1 Y^\alpha(t)\, dY^\alpha(t)\right].$$

The MLE $\tilde{\alpha}$ of α is obtained as

$$\tilde{\alpha} = \int_0^1 Y^\alpha(t)\, dY^\alpha(t) \Big/ \int_0^1 \left(Y^\alpha(t)\right)^2 dt,$$

which is seen to coincide with the limit in distribution of $n(\hat{\rho} - 1)$ for Model A described in (7.10).

On the basis of Theorem 7.4 we can compute the c.f.s associated with various quadratic functionals of the O–U process. Perron (1991) was the first to attempt to apply Girsanov's theorem to the present problem. Here we consider four examples, Examples 7.5 through 7.8. The last example is concerned with the *cointegration distribution* , which is the distribution arising from the regression of an I(1) process on another I(1) process (Engle and Granger, 1987).

Example 7.5 Let us consider $dY(t) = -cY(t)\, dt + dW(t)$ with $Y(0) = a$, so that

$$Y(t) = ae^{-ct} + e^{-ct}\int_0^t e^{cs}\, dW(s)$$

is the O–U process starting from $Y(0) = a$. Then we must deal with

$$\begin{aligned} S &= \int_0^1 Y^2(t)\,dt \\ &= \int_0^1 \int_0^1 K(s,t)\,dW(s)\,dW(t) + 2a \int_0^1 K(0,t)\,dW(t) \\ &\quad + a^2 K(0,0), \end{aligned} \tag{7.37}$$

where

$$K(s,t) = \frac{1}{2c}\left(e^{-c|s-t|} - e^{-c(2-s-t)}\right).$$

The FD of $K(s,t)$ was derived by our approach discussed in Chapter 2 and is given in (2.115) as

$$D(\lambda) = e^{-c}\left(\cos\mu + c\frac{\sin\mu}{\mu}\right), \tag{7.38}$$

where $\mu = \sqrt{\lambda - c^2}$. Then, using the property of the resolvent discussed in Chapter 3, the c.f. of S was obtained in (3.24) and (3.34) as

$$\mathrm{E}\left(e^{i\theta S}\right) = (D(2i\theta))^{-1/2} \exp\left[\frac{ia^2\theta \sin\nu/\nu}{\cos\nu + c\sin\nu/\nu}\right], \tag{7.39}$$

where $\nu = \sqrt{2i\theta - c^2}$.

Here we derive (7.39) using Girsanov's theorem. Let us put

$$dY^\beta(t) = \beta Y^\beta(t)\,dt + dW(t), \quad Y^\beta(0) = Y(0) = a.$$

Theorem 7.4 (see also Liptser and Shiryaev (2001b, p.230)) leads us to

$$\begin{aligned} &\mathrm{E}\left(e^{\theta S}\right) \\ &= \mathrm{E}\left[\exp\left\{\theta \int_0^1 (Y(t))^2\,dt\right\}\right] = \mathrm{E}\left[\exp\left\{\theta \int_0^1 \left(Y^\beta(t)\right)^2 dt\right\} \frac{d\mu_Y}{d\mu_Y^\beta}(Y^\beta)\right] \\ &= \mathrm{E}\left[\exp\left\{\left(\theta + \frac{\beta^2 - c^2}{2}\right) \int_0^1 \left(Y^\beta(t)\right)^2 dt - (\beta + c) \int_0^1 Y^\beta(t)\,dY^\beta(t)\right\}\right] \\ &= \mathrm{E}\left[\exp\left\{-(\beta + c) \int_0^1 Y^\beta(t)\,dY^\beta(t)\right\}\right] \\ &= \mathrm{E}\left[\exp\left\{\frac{-\beta - c}{2}\left(\left(Y^\beta(1)\right)^2 - a^2 - 1\right)\right\}\right] \\ &= \exp\left[\frac{1}{2}(a^2 + 1)(\beta + c)\right] \mathrm{E}\left[\exp\left\{\frac{-\beta - c}{2}\left(Y^\beta(1)\right)^2\right\}\right], \end{aligned} \tag{7.40}$$

where we put $\beta = \sqrt{c^2 - 2\theta}$. The above computational process can be explained as follows.

(a) Girsanov's theorem is used in the second equality, which transforms the measure μ_Y^c induced by Y^c into the measure μ_Y^β induced by Y^β.
(b) The parameter β in the third equality is under our control. We choose $\beta = \sqrt{c^2 - 2\theta}$ so that the term involving $\int_0^1 (Y^\beta(t))^2\,dt$ vanishes.
(c) The fifth equality follows from the property of the Itô integral described in (1.61), that is,

$$\int_0^1 Y^\beta(t)\,dY^\beta(t) = \frac{1}{2}\left((Y^\beta(1))^2 - a^2 - 1\right).$$

We now concentrate on the last expression in (7.40). Since

$$Y^\beta(1) = ae^\beta + e^\beta \int_0^1 e^{-\beta s}\,dW(s) \sim \mathrm{N}\left(ae^\beta, \frac{e^{2\beta}-1}{2\beta}\right),$$

we obtain, putting $\kappa = (-\beta - c)/2$,

$$\mathrm{E}\left(e^{\kappa(Y^\beta(1))^2}\right) = \left(1 + \frac{\kappa(1-e^{2\beta})}{\beta}\right)^{-1/2} \exp\left[\frac{\kappa a^2 e^{2\beta}}{1+\kappa(1-e^{2\beta})/\beta}\right],$$

so that

$$\begin{aligned}
\mathrm{E}\left(e^{\theta S}\right) &= \exp\left[\frac{c}{2} + \frac{\beta}{2} - \kappa a^2\right]\left(1 + \frac{\kappa(1-e^{2\beta})}{\beta}\right)^{-1/2} \\
&\quad \times \exp\left[\frac{\kappa a^2 e^{2\beta}}{1+\kappa(1-e^{2\beta})/\beta}\right] \\
&= e^{c/2}\left\{e^{-\beta}\left(1 + \frac{\kappa(1-e^{2\beta})}{\beta}\right)\right\}^{-1/2} \\
&\quad \times \exp\left[-\kappa a^2 + \frac{\kappa a^2 e^{2\beta}}{1+\kappa(1-e^{2\beta})/\beta}\right] \\
&= e^{c/2}\left(\cosh\beta + c\frac{\sinh\beta}{\beta}\right)^{-1/2} \exp\left[\frac{a^2\theta \sinh\beta/\beta}{\cosh\beta + c\sinh\beta/\beta}\right] \\
&= (D(2\theta))^{-1/2} \exp\left[\frac{a^2\theta\sin\mu/\mu}{\cos\mu + c\sin\mu/\mu}\right], \qquad (7.41)
\end{aligned}$$

where $\mu = \sqrt{2\theta - c^2}$. Thus, using Girsanov's theorem, we arrive at (7.39) without any knowledge of the FD and the resolvent.

It can be seen that, when $a = c = 0$, we have $Y(t) = W(t)$, and the result in (7.41) yields

$$\mathrm{E}\left(\exp\left\{\theta\int_0^1 W^2(t)\,dt\right\}\right) = \left(\cos\sqrt{2\theta}\right)^{-1/2}.$$

Example 7.6 Let us consider

$$S = \int_0^1 (X(t) - tX(1))^2\,dt = \int_0^1\int_0^1 K(s,t)\,dW(s)\,dW(t), \tag{7.42}$$

where $dX(t) = -cX(t)\,dt + dW(t)$ with $X(0) = 0$, whereas the kernel $K(s,t)$ is defined in (7.29) as $K(s,t;c)$, and its FD was given in (7.31) by the usual approach.

Here we use Girsanov's theorem. Putting $dX^\beta(t) = -\beta X^\beta(t)\,dt + dW(t)$ with $X^\beta(0) = X(0) = 0$ and proceeding in the same way as in Example 7.5, we have

$$\begin{aligned}
\mathrm{E}\left(e^{\theta S}\right) &= \mathrm{E}\left[\exp\left\{\theta\int_0^1 (X(t) - tX(1))^2\,dt\right\}\right] \\
&= \mathrm{E}\left[\exp\left\{\theta\int_0^1 \left(X^\beta(t) - tX^\beta(1)\right)^2 dt\right\}\frac{d\mu_X}{d\mu_X^\beta}(X^\beta)\right] \\
&= \mathrm{E}\left[\exp\left\{\left(\theta + \frac{\beta^2 - c^2}{2}\right)\int_0^1 \left(X^\beta(t)\right)^2 dt + (\beta - c)\right.\right. \\
&\qquad \left.\left.\times\int_0^1 X^\beta(t)\,dX^\beta(t) - 2\theta X^\beta(1)\int_0^1 tX^\beta(t)\,dt + \frac{\theta}{3}\left(X^\beta(1)\right)^2\right\}\right] \\
&= e^{(c-\beta)/2}\,\mathrm{E}\left[\exp\left\{\left(\frac{\beta - c}{2} + \frac{\theta}{3}\right)\left(X^\beta(1)\right)^2 - 2\theta X^\beta(1)\right.\right. \\
&\qquad \left.\left.\times\int_0^1 tX^\beta(t)\,dt\right\}\right] \\
&= e^{(c-\beta)/2}\,\mathrm{E}\left[\exp\left\{\boldsymbol{W}'A\boldsymbol{W}\right\}\right],
\end{aligned}$$

where $\beta = \sqrt{c^2 - 2\theta}$ and

$$\boldsymbol{W} = \begin{pmatrix} X^\beta(1) \\ \displaystyle\int_0^1 tX^\beta(t)\,dt \end{pmatrix}, \quad A = \begin{pmatrix} \dfrac{\beta - c}{2} + \dfrac{\theta}{3} & -\theta \\ -\theta & 0 \end{pmatrix}.$$

It holds that $\boldsymbol{W} \sim \mathrm{N}(\boldsymbol{0},\ \Sigma)$, where the (j,k)th element σ_{jk} of Σ is given by

$$\sigma_{11} = \frac{1-e^{-2\beta}}{2\beta}, \quad \sigma_{12} = \frac{\beta-1+(\beta+1)e^{-2\beta}}{2\beta^3},$$
$$\sigma_{22} = \frac{2\beta^3-3\beta^2+3-3(\beta+1)^2e^{-2\beta}}{6\beta^5}.$$

Then we have

$$\begin{aligned}
&\mathrm{E}\left(e^{\theta S_2}\right) \\
&= e^{(c-\beta)/2}\,\mathrm{E}\Big[\exp\left\{\boldsymbol{W}'A\boldsymbol{W}\right\}\Big] = e^{(c-\beta)/2}\Big|I_2 - 2\Sigma A\Big|^{-1/2} \\
&= e^{c/2}\Big[\frac{1}{6\beta^5}\Big\{(e^{\beta}+e^{-\beta})(3\beta^5+12\beta^3\theta+12\beta\theta^2) \\
&\qquad +(e^{\beta}-e^{-\beta})(3\beta^4c-2\beta^4\theta-4\beta^2\theta^2-12\beta^2\theta-12\theta^2)\Big\}\Big]^{-1/2} \\
&= e^{c/2}\left[\cosh\beta\left(1+\frac{4\theta}{\beta^2}+\frac{4\theta^2}{\beta^4}\right)+\frac{\sinh\beta}{\beta}\left(c-\frac{2\theta}{3}-\frac{4\theta^2}{3\beta^2}-\frac{4\theta}{\beta^2}-\frac{4\theta^2}{\beta^4}\right)\right]^{-1/2},
\end{aligned}$$

which leads to (7.31).

It can be seen that when $c = 0$, we have $X(t) = W(t)$, and the above result yields

$$\mathrm{E}\left[\exp\left\{\theta\int_0^1 (W(t)-tW(1))^2\,dt\right\}\right] = \left(\frac{\sin\sqrt{2\theta}}{\sqrt{2\theta}}\right)^{-1/2}.$$

Example 7.7 Here we consider the distribution of

$$Q(C;c) = \frac{\displaystyle\int_0^1 H(t)\,dX(t)}{\displaystyle\int_0^1 H^2(t)\,dt} = \frac{U}{V}, \tag{7.43}$$

where $dX(t) = -cX(t)\,dt + dW(t)$ with $X(0) = 0$ and

$$H(t) = X(t) + (6t-4)\int_0^1 X(s)\,ds - (12t-6)\int_0^1 sX(s)\,ds.$$

The statistic $Q(C;c)$ is the limit in distribution of the unit root statistic $n(\hat{\rho}-1)$ for Model C discussed in Section 7.1. Putting

$$X_1 = X(1), \quad X_2 = \int_0^1 X(s)\,ds, \quad X_3 = \int_0^1 sX(s)\,ds,$$

we have $Q(C;c) = U/V$, where

$$U = -\frac{1}{2} + \frac{1}{2}X_1^2 - 6X_2^2 + 2X_1X_2 - 6X_1X_3 + 12X_2X_3 = -\frac{1}{2} + \boldsymbol{X}'B_1\boldsymbol{X},$$

$$V = \int_0^1 X^2(t)\,dt - 4X_2^2 - 12X_3^2 + 12X_2X_3 = \int_0^1 X^2(t)\,dt + \boldsymbol{X}'B_2\boldsymbol{X},$$

$$\boldsymbol{X} = \begin{pmatrix} X_1 \\ X_2 \\ X_3 \end{pmatrix}, \quad B_1 = \begin{pmatrix} 1/2 & 1 & -3 \\ 1 & -6 & 6 \\ -3 & 6 & 0 \end{pmatrix}, \quad B_2 = \begin{pmatrix} 0 & 0 & 0 \\ 0 & -4 & 6 \\ 0 & 6 & -12 \end{pmatrix}.$$

Putting $dX^\beta(t) = -\beta X^\beta(t)\,dt + dW(t)$ with $X^\beta(0) = 0$, the joint m.g.f. of U and V is given (Exercise 7.5.2) by

$$\begin{aligned} m(\theta_1, \theta_2) &= \mathrm{E}\left(e^{\theta_1 U + \theta_2 V}\right) \\ &= e^{-\theta_1/2}\mathrm{E}\left[\exp\left\{\theta_1 \boldsymbol{X}'B_1\boldsymbol{X} + \theta_2 \int_0^1 X^2(t)\,dt + \theta_2 \boldsymbol{X}'B_2\boldsymbol{X}\right\}\right] \\ &= e^{-\theta_1/2}\mathrm{E}\left[\exp\left\{\left(\theta_2 + \frac{\beta^2 - c^2}{2}\right)\int_0^1 \left(X^\beta(t)\right)^2 dt + \theta_1 \boldsymbol{Z}'B_1\boldsymbol{Z}\right.\right. \\ &\qquad \left.\left. + \theta_2 \boldsymbol{Z}'B_2\boldsymbol{Z} + \frac{\beta - c}{2}\left(\left(X^\beta(1)\right)^2 - 1\right)\right\}\right] \\ &= e^{(c-\theta_1-\beta)/2}\mathrm{E}\left[\exp\{\boldsymbol{Z}'B\boldsymbol{Z}\}\right], \end{aligned} \tag{7.44}$$

where $\beta = \sqrt{c^2 - 2\theta_2}$, $\boldsymbol{Z} = \left(X^\beta(1), \int_0^1 X^\beta(s)\,ds, \int_0^1 sX^\beta(s)\,ds\right)'$, and

$$B = \begin{pmatrix} \dfrac{\theta_1 + \beta - c}{2} & \theta_1 & -3\theta_1 \\ \theta_1 & -6\theta_1 - 4\theta_2 & 6(\theta_1 + \theta_2) \\ -3\theta_1 & 6(\theta_1 + \theta_2) & -12\theta_2 \end{pmatrix}. \tag{7.45}$$

It holds that $\boldsymbol{Z} \sim \mathrm{N}(\boldsymbol{0}, \Omega)$, where the (j,k)th element ω_{jk} of Ω is given (Exercise 7.5.3) by

$$\omega_{11} = \sigma_{11}, \quad \omega_{22} = \frac{2\beta - (1 - e^{-\beta})(3 - e^{-\beta})}{2\beta^3}, \quad \omega_{33} = \sigma_{22},$$

$$\omega_{12} = \frac{(1 - e^{-\beta})^2}{2\beta^2}, \quad \omega_{13} = \sigma_{12}, \quad \omega_{23} = \frac{\beta^2 - (\beta + 1)(1 - e^{-\beta})^2}{2\beta^4}.$$

Then, using computerized algebra, we obtain

$$m\left(\theta_1,\theta_2\right) = e^{(c-\theta_1-\beta)/2}\left|I_3 - 2\Omega B\right|^{-1/2} = e^{(c-\theta_1)/2}\left[H(\theta_1,\theta_2)\right]^{-1/2},$$

where

$$\begin{aligned} H(\theta_1,\theta_2) = & \ \frac{c^5 - c^4\theta_1 - 4\theta_1^2\left(c^2+3c+27\right) - 8\theta_2\left(c^2-3c-3\right)}{\mu^4}\frac{\sin\mu}{\mu} \\ & + \frac{24\left(c^4\theta_1 + 8\theta_1^2\theta_2 + 4(c+1)\left(3\theta_1^2-\theta_2^2\right)\right)}{\mu^6}\left(\frac{\sin\mu}{\mu}+\frac{\cos\mu}{\mu^2}-\frac{1}{\mu^2}\right) \\ & + \left(\frac{c^4}{\mu^4} - \frac{8\left(c^3(c\theta_1 - 2\theta_2) + 4\theta_1^2\left(c^2+3c+6\right)\right)}{\mu^6}\right)\cos\mu \\ & - \frac{4\left(c^4\theta_1 + 4\theta_1^2\left(c^2+3c-3\right) - 2c^2\theta_2(c+3)\right)}{\mu^6}, \end{aligned}$$

with $\mu = \sqrt{2\theta_2 - c^2}$.

The distribution of $Q(C;c)$ and the limiting powers of the LSE-based unit root test for Model C can be computed from

$$\begin{aligned} P\left(Q(C;c) < x\right) &= P\left(\frac{U}{V} < x\right) = P\left(xV - U > 0\right) \\ &= \frac{1}{2} + \frac{1}{\pi}\int_0^\infty \frac{\operatorname{Im}\left[e^{i\theta/2}\left\{e^{-c}H(-i\theta, i\theta x)\right\}^{-1/2}\right]}{\theta}\,d\theta. \end{aligned}$$

It can be shown that, as $c \to \infty$,

$$\begin{aligned} m\left(\sqrt{\frac{c}{2}}\theta_1, c\sqrt{\frac{c}{2}}\theta_1 + c\theta_2\right) &= \exp\left[\exp\left\{\theta_1\sqrt{\frac{c}{2}}(U+cV) + c\theta_2 V\right\}\right] \\ &\to \exp\left[\frac{1}{2}\left(\frac{\theta_1^2}{4} + \theta_2\right)\right], \\ m(\theta_1, c\theta_2) &= \mathrm{E}\left(e^{\theta_1 U + c\theta_2 V}\right) \to \exp\left\{\frac{-\theta_1+\theta_2}{2}\right\}. \end{aligned}$$

Thus it holds that $(Q(C;c)+c)/\sqrt{2c} \Rightarrow \mathrm{N}(0,1)$ and $Q(C;c) \to -c$, so that the LSE-based test for Model C is consistent, which was earlier mentioned in Section 7.1.

We have assumed here that $X(0) = 0$ in the definition of $Q(C;c)$ in (7.43). We can also deal with the case of $X(0) = a$, where a is a nonzero constant or a random variable. For the derivation of the corresponding c.f. and more details, see Tanaka (1996, 2017).

Example 7.8 Here we consider the two processes x_j and y_j defined by

$$x_j = x_{j-1} + u_j, \quad y_j = \alpha + \beta x_j + v_j \quad (j = 1, \ldots, n), \tag{7.46}$$

where $x_0 = y_0 = 0$ and $(u_j, v_j)' \sim \text{i.i.d.}(\mathbf{0}, \Sigma)$ with

$$\Sigma = \begin{pmatrix} \mathrm{Var}(u_j) & \mathrm{Cov}(u_j, v_j) \\ \mathrm{Cov}(u_j, v_j) & \mathrm{Var}(v_j) \end{pmatrix} = \begin{pmatrix} \sigma_{11} & \sigma_{12} \\ \sigma_{12} & \sigma_{22} \end{pmatrix}.$$

It is seen that both x_j and y_j are I(1) processes, but $y_j - \beta x_j = \alpha + v_j$ becomes an i.i.d. process. This is a simple case of cointegration (Engle and Granger, 1987), where $(-\beta, 1)'$ is the cointegrating vector for the two-dimensional I(1) process $(x_j, y_j)'$.

Let us consider the asymptotic distribution of the LSE of β in (7.46). Assume first that $\alpha = 0$, and denote the LSE of β by $\hat{\beta}_1$. Then we have (Tanaka, 1996, chapter 11)

$$n(\hat{\beta}_1 - \beta) = \frac{1}{n} \sum_{j=1}^{n} x_j v_j \bigg/ \frac{1}{n^2} \sum_{j=1}^{n} x_j^2 \Rightarrow S_1 = (U_1 + U_2 + \sigma_{12}) / V, \tag{7.47}$$

where

$$U_1 = \sigma_{12} \int_0^1 W_1(t)\, dW_1(t),$$
$$U_2 = \sqrt{\sigma_{11}\sigma_{22} - \sigma_{12}^2} \int_0^1 W_1(t)\, dW_2(t),$$
$$V = \sigma_{11} \int_0^1 W_1^2(t)\, dt.$$

Here $\boldsymbol{W}(t) = \left(W_1(t), W_2(t)\right)'$ is standard two-dimensional Bm.

We now consider $P\left(S_1 \le x\right) = P\left(X_1 \ge 0\right)$, where

$$\begin{aligned} X_1 &= xV - U_1 - U_2 - \sigma_{12} \\ &= ax \int_0^1 W_1^2(t)\, dt - b \int_0^1 W_1(t)\, dW_1(t) \\ &\quad - c \int_0^1 W_1(t)\, dW_2(t) - b \end{aligned} \tag{7.48}$$

with

$$a = \sigma_{11}, \quad b = \sigma_{12}, \quad c = \sqrt{\sigma_{11}\sigma_{22} - \sigma_{12}^2}.$$

It holds that, given $\{W_1(t)\}$, X_1 is conditionally normal with

$$\mathrm{E}\Big(X_1 \,\Big|\, \{W_1(t)\}\Big) = ax\int_0^1 W_1^2(t)\,dt - \frac{b}{2}W_1^2(1) - \frac{b}{2},$$
$$\mathrm{Var}\Big(X_1 \,\Big|\, \{W_1(t)\}\Big) = c^2\int_0^1 W_1^2(t)\,dt.$$

Thus the m.g.f. $m_1(\theta;x)$ of X_1 is given by

$$\begin{aligned} m_1(\theta;x) &= \mathrm{E}\Big[\exp\Big\{\theta\mathrm{E}\Big(X_1 \,\Big|\, \{W_1(t)\}\Big) + \theta^2\mathrm{Var}\Big(X_1 \,\Big|\, \{W_1(t)\}\Big)\Big/2\Big\}\Big] \\ &= e^{-b\theta/2}\mathrm{E}\left[\exp\left\{\theta\left(ax + \frac{c^2\theta}{2}\right)\int_0^1 W_1^2(t)\,dt - \frac{b\theta}{2}W_1^2(1)\right\}\right]. \end{aligned}$$

Then Girsanov's theorem coupled with Imhof's formula yields (Exercise 7.5.4)

$$\begin{aligned} P\Big(n(\hat{\beta}_1 - \beta) \le x\Big) &\to P\Big(S_1 \le x\Big) \\ &= \frac{1}{2} + \frac{1}{\pi}\int_0^\infty \frac{1}{\theta}\,\mathrm{Im}\Big(\phi_1(\theta;x)\Big)\,d\theta, \end{aligned} \tag{7.49}$$

where $\phi_1(\theta;x)$ is the c.f. of X_1 in (7.48) and is given by

$$\phi_1(\theta;x) = e^{-ib\theta/2}\left[\cos\nu + ib\theta\frac{\sin\nu}{\nu}\right]^{-1/2}, \qquad \nu = \sqrt{i\theta(2ax + ic^2\theta)}.$$

The LSE $\hat{\beta}_2$ of β in (7.46) with $\alpha \ne 0$ can be dealt with similarly. We obtain

$$\begin{aligned} n\Big(\hat{\beta}_2 - \beta\Big) &= \frac{1}{n}\sum_{j=1}^n \Big(x_j - \bar{x}\Big)v_j \Bigg/ \frac{1}{n^2}\sum_{j=1}^n \Big(x_j - \bar{x}\Big)^2 \\ &\Rightarrow S_2 = \left(\bar{U}_1 + \bar{U}_2 + \sigma_{12}\right)/\bar{V}, \end{aligned} \tag{7.50}$$

where $\bar{x} = \sum_{j=1}^n x_j/n$ and

$$\bar{U}_1 = \sigma_{12}\int_0^1 \bar{W}_1(t)\,dW_1(t), \quad \bar{U}_2 = \sqrt{\sigma_{11}\sigma_{22} - \sigma_{12}^2}\int_0^1 \bar{W}_1(t)\,dW_2(t),$$
$$\bar{V} = \sigma_{11}\int_0^1 \bar{W}_1^2(t)\,dt, \quad \bar{W}_1(t) = W_1(t) - \int_0^1 W_1(s)\,ds.$$

We now consider $P\big(S_2 \le x\big) = P\big(X_2 \ge 0\big)$, where

$$\begin{aligned} X_2 &= x\bar{V} - \bar{U}_1 - \bar{U}_2 - \sigma_{12} \\ &= ax\int_0^1 \bar{W}_1^2(t)\,dt - b\int_0^1 \bar{W}_1(t)\,dW_1(t) \\ &\quad - c\int_0^1 \bar{W}_1(t)\,dW_2(t) - b. \end{aligned} \tag{7.51}$$

It holds that, given $\{W_1(t)\}$, X_2 is conditionally normal with

$$
\begin{aligned}
\mathrm{E}\Big(X_2 \,\Big|\, \big\{W_1(t)\big\}\Big) &= ax\int_0^1 \bar{W}_1^2(t)\,dt + bW_1(1)\int_0^1 W_1(t)\,dt \\
&\quad - \frac{b}{2}\Big(W_1^2(1)+1\Big), \\
\mathrm{Var}\Big(X_2 \,\Big|\, \big\{W_1(t)\big\}\Big) &= c^2\int_0^1 \bar{W}_1^2(t)\,dt.
\end{aligned}
$$

Thus the m.g.f. $m_2(\theta;x)$ of X_2 is given by

$$
\begin{aligned}
m_2(\theta;x) &= \mathrm{E}\Big[\exp\Big\{\theta\mathrm{E}\Big(X_2\,\Big|\,\big\{W_1(t)\big\}\Big) + \theta^2\mathrm{Var}\Big(X_2\,\Big|\,\big\{W_1(t)\big\}\Big)\Big/2\Big\}\Big] \\
&= e^{-b\theta/2}\mathrm{E}\Bigg[\exp\Bigg\{\theta\left(ax+\frac{c^2\theta}{2}\right)\int_0^1 \bar{W}_1^2(t)\,dt \\
&\qquad - \frac{b\theta}{2}W_1^2(1) + b\theta W_1(1)\int_0^1 W_1(t)\,dt\Bigg\}\Bigg].
\end{aligned}
$$

Then we obtain (Exercise 7.5.5)

$$
\begin{aligned}
P\Big(n(\hat{\beta}_2-\beta)\le x\Big) &\to P\Big(S_2\le x\Big) \\
&= \frac{1}{2}+\frac{1}{\pi}\int_0^\infty \frac{1}{\theta}\,\mathrm{Im}\Big(\phi_2(\theta;x)\Big)\,d\theta, \qquad (7.52)
\end{aligned}
$$

where $\phi_2(\theta;x)$ is the c.f. of X_2 in (7.51) and is given by

$$
\begin{aligned}
\phi_2(\theta;x) &= e^{-ib\theta/2}\left[\frac{\sin\nu}{\nu}+\frac{b^2\theta^2}{\nu^2}\left(\frac{2(1-\cos\nu)}{\nu^2}-\frac{\sin\nu}{\nu}\right)\right]^{-1/2}, \\
\nu &= \sqrt{i\theta(2ax+ic^2\theta)}.
\end{aligned}
$$

Figure 7.6 shows the limiting probability densities of $n(\hat{\beta}_k-\beta)$, that is, the densities of S_k $(k=1,2)$ for two cases, where Case 1: $\sigma_{11}=1, \sigma_{12}=0, \sigma_{22}=0.8$ and Case 2: $\sigma_{11}=1, \sigma_{12}=0.5, \sigma_{22}=0.8$. S_1 has a smaller variance than S_2, which can be proved by computing $\mathrm{Var}(S_k)$ (Tanaka, 1996, chapter 11). Case 1 is referred to as strictly exogeneous since $\sigma_{12}=0$ and the regressor x_j is uncorrelated with the error term v_j. In Case 2 the LSEs are biased because $\sigma_{12}\ne 0$ and $\mathrm{Cov}(x_j,v_j)\ne 0$. When $\sigma_{12}\ne 0$, the MLEs are unbiased and give a smaller variance than the LSEs, which is shown in Tanaka (1996, chapter 11), where more general cointegration distributions are discussed.

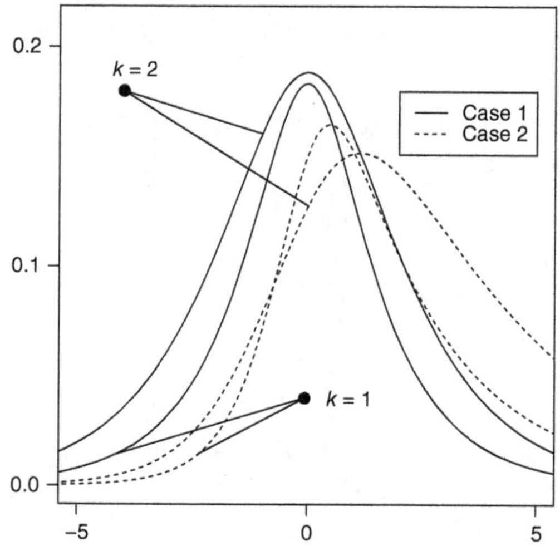

Figure 7.6 Cointegration distributions.

Exercises for Section 7.5

Exercise 7.5.1 Consider the near-random walk

$$y_j = \rho_n y_{j-1} + \varepsilon_j, \quad y_0 = 0 \quad (j = 1, \ldots, n),$$

where $\{\varepsilon_j\} \sim \text{NID}(0, 1)$. Defining by $\ell_n(\alpha)$ the likelihood for $y_1, \ldots, y_n$ under $\rho_n = 1 - \alpha/n$, show that

$$\left.\frac{\ell_n(\alpha)}{\ell_n(\beta)}\right|_{\gamma} \Rightarrow \exp\left[\frac{\beta^2 - \alpha^2}{2}\int_0^1 X^2(t)\,dt + (\beta - \alpha)\int_0^1 X(t)\,dX(t)\right],$$

where $dX(t) = -\gamma X(t)\,dt + dW(t)$ with $X(0) = 0$.

Exercise 7.5.2 Consider

$$Q(C; c) = \frac{\displaystyle\int_0^1 H(t)\,dX(t)}{\displaystyle\int_0^1 H^2(t)\,dt} = \frac{U}{V},$$

where $H(t)$ is defined in (7.43) and $dX(t) = -cX(t)\,dt + dW(t)$ with $X(0) = 0$. Show that $\text{E}\left(e^{\theta_1 U + \theta_2 V}\right)$ can be expressed as (7.44).

Exercise 7.5.3 Compute the covariance matrix of

$$\mathbf{Z} = \left(Z(1), \int_0^1 Z(t)\,dt, \int_0^1 tZ(t)\,dt\right),$$

where $dZ(t) = -\beta Z(t)\,dt + dW(t)$ with $Z(0) = 0$.

Exercise 7.5.4 Prove that the distribution function of S_1 in (7.47) can be computed following (7.49).

Exercise 7.5.5 Prove that the distribution function of S_2 in (7.50) can be computed following (7.52).

8

Fredholm Determinants for Fractional Brownian Motion

We have so far discussed the FD associated with quadratic plus linear or bilinear functionals of Bm or the ratio of those functionals. In this chapter we extend Bm to fractional Bm (fBm) and consider the fractional unit root distribution based on fBm. The derivation of the associated FD still remains to be done in the literature. We suggest an approximation to the distribution using a martingale suggested in Norros, Valkeila, and Virtamo (1999). Also dealt with is the multiple unit root distribution based on g-fold integrated Bm. We find some similarities between the two unit root distributions and present an interesting moment property.

8.1 Multiple Unit Root Distribution

In this section we deal with

$$y_j = \rho y_{j-1} + u_j, \quad (1-L)^g u_j = \varepsilon_j, \quad \{\varepsilon_j\} \sim \text{i.i.d.}\big(0, \sigma_\varepsilon^2\big) \quad (j = 1, \ldots, n), \tag{8.1}$$

where $g = 1, 2, \ldots$, $u_j = 0$ for $j \leq 0$, and the true value of ρ is unity. We exclude the case of $g = 0$ discussed in Chapter 7 because the subsequent discussion does not apply to this case. The model (8.1) is the AR(1) model with the error term following an I(g) process with $g \geq 1$. Since ρ is assumed to be unity, $\{y_j\}$ follows an I($g+1$) process given by $y_j = (1-L)^{-(g+1)}\varepsilon_j$.

Let us define the partial sum process of the I($g+1$) process $\{y_j\}$ by

$$Y_n^{(g+1)}(t) = \frac{1}{n^{g+1/2}\sigma_\varepsilon}\, y_{[nt]} + (nt - [nt])\frac{1}{n^{g+1/2}\sigma_\varepsilon}\, u_{[nt]+1}.$$

Then it follows from the FCLT for $\{Y_n^{(g+1)}(t)\}$ (Chan and Wei, 1988) that

$$\left\{Y_n^{(g+1)}(t)\right\} \Rightarrow \left\{F_g(t)\right\}.$$

The process $\{F_g(t)\}$ is g-fold integrated Bm defined by

$$F_g(t) = \int_0^t F_{g-1}(s)\,ds = \int_0^t \int_0^{t_1} \cdots \int_0^{t_{g-1}} W(t_g)\,dt_g\,dt_{g-1}\cdots dt_1$$
$$= \int_0^t \frac{(t-s)^g}{g!}\,dW(s), \tag{8.2}$$

where this last equality comes from (1.15), and it was pointed out that there is the following correspondence between $F_g(t)$ and the I$(g+1)$ process y_j:

$$F_g(t) = \int_0^t \frac{(t-s)^g}{g!}\,dW(s) \;\Leftrightarrow\; y_j = (1-L)^{-(g+1)}\varepsilon_j \approx \frac{1}{g!}\sum_{i=1}^{j}(j-i)^g\varepsilon_i.$$

For the model (8.1) we consider the asymptotic distribution of the LSE $\hat{\rho}$ of ρ given by

$$\hat{\rho} = \sum_{j=2}^{n} y_{j-1}y_j \bigg/ \sum_{j=2}^{n} y_{j-1}^2,$$
$$\hat{\rho} - 1 = \left(\frac{1}{2}\left(y_n^2 - y_1^2\right) - \frac{1}{2}\sum_{j=2}^{n} u_j^2\right) \bigg/ \sum_{j=2}^{n} y_{j-1}^2 . \tag{8.3}$$

Noting that

$$y_n^2 = O_p\big(n^{2g+1}\big), \quad \sum_{j=2}^{n} y_{j-1}^2 = O_p\big(n^{2g+2}\big), \quad \sum_{j=2}^{n} u_j^2 = O_p\big(n^{2g}\big),$$

and using the the FCLT and CMT, we obtain

$$n\big(\hat{\rho} - 1\big) \Rightarrow R(g) = \frac{1}{2}F_g^2(1) \bigg/ \int_0^1 F_g^2(t)\,dt. \tag{8.4}$$

Note that this result is not applicable to the case of $g = 0$. It is also noted that we obtain this result even if the i.i.d. error term $\{\varepsilon_j\}$ is replaced by a stationary sequence, which also does not hold when $g = 0$.

Let us consider the distribution of $R(g)$ in (8.4). We have

$$P\big(R(g) < x\big) = P\left(x\int_0^1 F_g^2(t)\,dt - \frac{1}{2}F_g^2(1) > 0\right) = P\big(S(g;x) > 0\big),$$

where

$$S(g;x) = x\int_0^1 F_g^2(t)\,dt - \frac{1}{2}F_g^2(1)$$
$$= \int_0^1 \int_0^1 K_g(s,t;x)\,dW(s)\,dW(t) \tag{8.5}$$
$$\stackrel{\mathcal{D}}{=} \int_0^1 \int_0^1 L_g(s,t;x)\,dW(s)\,dW(t), \tag{8.6}$$

$$K_g(s,t;x) = \frac{1}{(g!)^2}\left[x\int_{\max(s,t)}^1 ((u-s)(u-t))^g\,du - \frac{1}{2}\,((1-s)(1-t))^g\right], \tag{8.7}$$

$$\begin{aligned} L_g(s,t;x) &= K_g(1-s,1-t;x) \\ &= \frac{1}{(g!)^2}\left[x\int_0^{\min(s,t)} ((u-s)(u-t))^g\,du - \frac{1}{2}\,(st)^g\right]. \end{aligned} \tag{8.8}$$

Then we can compute the distribution of $R(g)$, using Imhof's formula, by

$$P\Big(R(g) < x\Big) = \frac{1}{2} + \frac{1}{\pi}\int_0^\infty \frac{\operatorname{Im}\left[\big(D_g(2i\theta;x)\big)^{-1/2}\right]}{\theta}\,d\theta, \tag{8.9}$$

where $D_g(\lambda;x)$ is the FD of $K_g(s,t;x)$ and $L_g(s,t;x)$.

To derive $D_g(\lambda;x)$, consider the integral equation with the kernel L_g given by

$$\begin{aligned} f(t) &= \lambda\int_0^1 L_g(s,t;x)f(s)\,ds \\ &= \lambda x\left[\int_0^t A_1(s,t)f(s)\,ds + \int_t^1 A_2(s,t)f(s)\,ds\right] \\ &\qquad - \frac{\lambda t^g}{2(g!)^2}\int_0^1 s^g f(s)\,ds, \end{aligned} \tag{8.10}$$

where

$$\begin{aligned} A_1(s,t) &= \frac{1}{(g!)^2}\int_0^s ((s-u)(t-u))^g\,du, \\ A_2(s,t) &= \frac{1}{(g!)^2}\int_0^t ((s-u)(t-u))^g\,du. \end{aligned}$$

Then we have, for $j = 0, 1, \ldots, g$,

$$\begin{aligned} f^{(j)}(t) &= \lambda x\left[\int_0^t \frac{\partial^j A_1(s,t)}{\partial t^j} f(s)\,ds + \int_t^1 \frac{\partial^j A_2(s,t)}{\partial t^j} f(s)\,ds\right] \\ &\qquad - \frac{\lambda g(g-1)\cdots(g-j+1)t^{g-j}}{2(g!)^2}\int_0^1 s^g f(s)\,ds, \end{aligned}$$

which yields

$$f^{(j)}(0) = 0 \quad (j = 0, 1, \ldots, g-1), \quad f^{(g)}(0) = -\frac{\lambda}{2g!}\int_0^1 s^g f(s)\,ds.$$

We also have, for $j = 1, 2, \ldots, g+1$,

$$
\begin{aligned}
f^{(g+j)}(t) &= \lambda x \int_t^1 \frac{\partial^{g+j} A_2(s,t)}{\partial t^{g+j}} f(s)\, ds \\
&= \lambda x \int_t^1 \frac{(-1)^{j-1}(s-t)^{g-j+1}}{(g-j+1)!} f(s)\, ds,
\end{aligned}
$$

which yields

$$
f^{(g+j)}(1) = 0 \quad (j = 1, \ldots, g+1),
$$

and

$$
f^{(2g+1)}(t) = \lambda x (-1)^g \int_t^1 f(s)\, ds, \quad f^{(2g+2)}(t) = \lambda x (-1)^{g+1} f(t).
$$

Thus the integral equation (8.10) leads to

$$
f^{(2g+2)}(t) + (-1)^g \lambda x f(t) = 0 \tag{8.11}
$$

with $2g+2$ boundary conditions

$$
\begin{aligned}
&f(0) = f'(0) = \cdots = f^{(g-1)}(0) = 0, \quad f^{(g)}(0) = -\frac{\lambda}{2g!} \int_0^1 s^g f(s)\, ds, \\
&f^{(g+1)}(1) = f^{(g+2)}(1) = \cdots = f^{(2g+1)}(1) = 0.
\end{aligned}
$$

Conversely, assume (8.11) with the above $2g+2$ boundary conditions and define

$$
\begin{aligned}
C_j(s,t) &= \frac{\partial^j A_1(s,t)}{\partial s^j} = \frac{g(g-1)\cdots(g-j+1)}{(g!)^2} \int_0^s (s-u)^{g-j}(t-u)^g\, du, \\
D_j(s,t) &= \frac{\partial^j A_2(s,t)}{\partial s^j} = \frac{g(g-1)\cdots(g-j+1)}{(g!)^2} \int_0^t (s-u)^{g-j}(t-u)^g\, du.
\end{aligned}
$$

It holds that $C_j(t,t) = D_j(t,t)$ and $C_j(0,t) = D_j(s,0) = D_{g+1+j}(s,t) = 0$ for $j = 0, 1, \ldots, g$, whereas

$$
C_{g+1}(s,t) = \frac{1}{g!}(t-s)^g, \quad C_{g+1+j}(t,t) = 0 \quad (j = 0, 1, \ldots, g-1),
$$

and $C_{2g+1}(s,t) = (-1)^g$. Then we have, by integration by parts,

$$\begin{aligned}
G &= \lambda \int_0^1 L_g(s,t;x) f(s)\, ds \\
&= (-1)^{g+1} \left[\int_0^t A_1(s,t) f^{(2g+2)}(s)\, ds + \int_t^1 A_2(s,t) f^{(2g+2)}(s)\, ds \right] \\
&\quad - \frac{\lambda t^g}{2(g!)^2} \int_0^1 s^g f(s)\, ds \\
&= (-1)^{g+2} \left[\int_0^t C_1(s,t) f^{(2g+1)}(s)\, ds + \int_t^1 D_1(s,t) f^{(2g+1)}(s)\, ds \right] \\
&\quad - \frac{\lambda t^g}{2(g!)^2} \int_0^1 s^g f(s)\, ds,
\end{aligned}$$

where we have used $A_1(t,t) = A_2(t,t)$, $A_1(0,t) = 0$, and $f^{(2g+1)}(1) = 0$. Proceeding further by integration by parts, we obtain

$$\begin{aligned}
G &= -\left[\int_0^t C_g(s,t) f^{(g+2)}(s)\, ds + \int_t^1 D_g(s,t) f^{(g+2)}(s)\, ds \right] \\
&\quad - \frac{\lambda t^g}{2(g!)^2} \int_0^1 s^g f(s)\, ds \\
&= \int_0^t C_{g+1}(s,t) f^{(g+1)}(s)\, ds - \frac{\lambda t^g}{2(g!)^2} \int_0^1 s^g f(s)\, ds \\
&= -\int_0^t C_{g+2}(s,t) f^{(g)}(s)\, ds - C_{g+1}(0,t) f^{(g)}(0) - \frac{\lambda t^g}{2(g!)^2} \int_0^1 s^g f(s)\, ds \\
&= -\int_0^t C_{g+2}(s,t) f^{(g)}(s)\, ds = (-1)^g \int_0^t C_{2g+1}(s,t) f'(s)\, ds \\
&= (-1)^{2g} \int_0^t f'(s)\, ds = f(t) - f(0) \\
&= f(t),
\end{aligned}$$

which gives the left-hand side of the integral equation (8.10) to be proved.

It can also be shown (Exercise 8.1.1) that the integral equation with the kernel $K_g(s,t;x)$ in (8.7) is equivalent to the differential equation (8.11) with $2g+2$ boundary conditions

$$\begin{aligned}
f(1) = f'(1) = \cdots = f^{(g-1)}(1) &= 0, \\
f^{(g)}(1) &= \frac{(-1)^{g-1}\lambda}{2g!} \int_0^1 (1-s)^g f(s)\, ds, \\
f^{(g+1)}(0) = f^{(g+2)}(0) = \cdots = f^{(2g+1)}(0) &= 0.
\end{aligned}$$

We now move on to derive the FD $D_g(\lambda; x)$ in (8.9) for $g = 1$ and 2 working with $L_g(s,t;x)$. For the case $g = 1$, we have, for $s \leq t$,

$$L_1(s,t;x) = \frac{x}{6}\left(-s^3 + 3s^2t\right) - \frac{1}{2}st,$$

and the integral equation with the kernel $L_1(s,t;x)$ is equivalent to

$$f^{(4)}(t) - \lambda x f(t) = 0$$

with the boundary conditions

$$f(0) = f''(1) = f'''(1) = 0, \quad f'(0) = -\frac{\lambda}{2}\int_0^1 sf(s)\,ds.$$

It follows from Section 2.4.5 that the general solution to the differential equation is given by

$$f(t) = c_1 \cos At + c_2 \sin At + c_3 \cos(iAt) + c_4 \sin(iAt), \quad A = (\lambda x)^{1/4}.$$

The above four boundary conditions yield a homogeneous equation $M(\lambda)\,\boldsymbol{c} = \mathbf{0}$, where $\boldsymbol{c} = (c_1, c_2, c_3, c_4)'$ and

$$M(\lambda) = \begin{pmatrix} 1 & 0 & 1 & 0 \\ M_{21} & M_{22} & M_{23} & M_{24} \\ -\cos A & -\sin A & \cosh A & i\sinh A \\ \sin A & -\cos A & \sinh A & i\cosh A \end{pmatrix},$$

$$\begin{aligned} M_{21} &= \frac{\lambda}{2A^2}(\cos A + A\sin A - 1), \quad M_{22} = A + \frac{\lambda}{2A^2}(-A\cos A + \sin A), \\ M_{23} &= \frac{\lambda}{2A^2}(-\cosh A + A\sinh A + 1), \\ M_{24} &= iA + \frac{\lambda}{2A^2}(iA\cosh A - i\sinh A). \end{aligned}$$

Computing $|M(\lambda)|$ we obtain, as the FD of $L_1(s,t;x)$,

$$D_1(\lambda; x) = \frac{1}{2}(1 + \cos A\cosh A) + \frac{A}{4x}(\cosh A\sin A - \cos A\sinh A). \quad (8.12)$$

This result was also obtained in Nabeya and Perron (1994).

When $g = 2$, we have, for $s \leq t$,

$$L_2(s,t;x) = \frac{x}{120}\left(s^5 - 5s^4t + 10s^3t^2\right) - \frac{1}{8}s^2t^2$$

and the differential equation

$$f^{(6)}(t) + \lambda x f(t) = 0$$

with the boundary conditions

$$f(0) = f'(0) = f'''(1) = f^{(4)}(1) = f^{(5)}(1) = 0,$$
$$f''(0) = -\frac{\lambda}{4}\int_0^1 s^2 f(s)\,ds.$$

It follows from Section 2.4.5 that the general solution to the differential equation is given by

$$f(t) = c_1 \cos Bt + c_2 \sin Bt + c_3 \cos B\omega t + c_4 \sin B\omega t + c_5 \cos B\omega^2 t + c_6 \sin B\omega^2 t,$$

where $B = (\lambda x)^{1/6}$ and $\omega = (1+\sqrt{3}i)/2$. The above six boundary conditions yield $M(\lambda)\,\boldsymbol{c} = \boldsymbol{0}$, where $\boldsymbol{c} = (c_1, \ldots, c_6)'$. Computing $|M(\lambda)|$ by computerized algebra, we obtain (Tanaka, 1996), as the FD of $L_2(s,t;x)$,

$$D_2(\lambda;x) = \frac{1}{18}\left[4\left(1 + \cos B + \cos B\omega + \cos B\omega^2\right) + 2\cos B \cos B\omega \cos B\omega^2\right.$$
$$+ \frac{1}{x}\left\{\frac{1}{2}\left(B\sin 2B + B\omega \sin 2B\omega + B\omega^2 \sin 2B\omega^2\right)\right.$$
$$\left.\left. + 2\left(B \sin B + B\omega \sin B\omega + B\omega^2 \sin B\omega^2\right)\right\}\right].$$

Figure 8.1 plots the probability densities of $R(g)$ for $g = 0, 1, 2$ computed by using (8.9) and numerical derivatives. Note that $R(0)$ is not the unit root distribution, but it is still a distribution and is plotted for comparison with $R(1)$ and $R(2)$, where the FD of $L_0(s,t;x) = x\min(s,t) - 1/2$ associated with $R(0)$ is given (Exercise 8.1.2) by

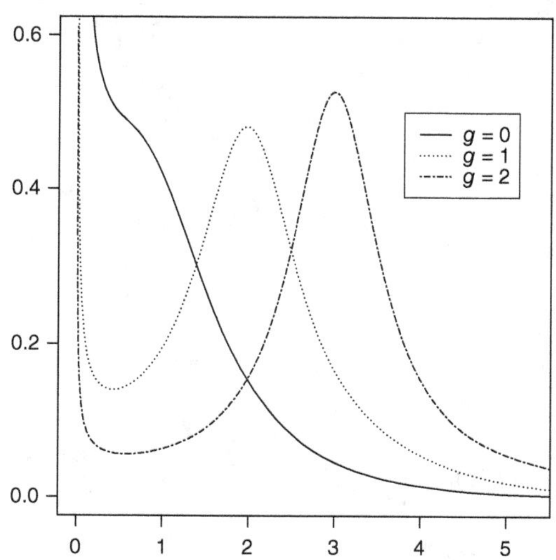

Figure 8.1 Probability densities of $R(g)$ for $g = 0, 1, 2$.

$$D_0(\lambda; x) = \cos\sqrt{\lambda x} + \frac{\lambda}{2}\frac{\sin\sqrt{\lambda x}}{\sqrt{\lambda x}}.$$

It can be seen that the distributions of $R(1)$ and $R(2)$ are bimodal. One peak is at the origin; the other is at about $x = 2$ for $R(1)$ and $x = 3$ for $R(2)$. The mean of $R(g)$ is found to be

$$\mathrm{E}(R(g)) = \mathrm{E}\left(\frac{1}{2}F_g^2(1)\Big/\int_0^1 F_g^2(t)\,dt\right) = g+1, \tag{8.13}$$

which was proved in Jin and Li (2015) for any nonnegative integer g. In particular, we have (Exercise 8.1.3)

$$\mathrm{E}\left(\frac{1}{2}W^2(1)\Big/\int_0^1 W^2(t)\,dt\right) = 1.$$

It holds in (8.13) that, although $R(g)$ and the denominator of $R(g)$ are not independent (Exercise 8.1.3),

$$\mathrm{E}(R(g)) = \mathrm{E}\left(\frac{1}{2}F_g^2(1)\right)\Big/\mathrm{E}\left(\int_0^1 F_g^2(t)\,dt\right) = g+1$$

because

$$\mathrm{E}\left(\frac{1}{2}F_g^2(1)\right) = \frac{1}{2(2g+1)(g!)^2},$$
$$\mathrm{E}\left(\int_0^1 F_g^2(t)\,dt\right) = \frac{1}{(2g+1)(2g+2)(g!)^2}.$$

Exercises for Section 8.1

Exercise 8.1.1 Prove that the integral equation

$$f(t) = \lambda\int_0^1 K_g(s,t;x)f(s)\,ds,$$

where $K_g(s,t;x)$ is defined in (8.7), is equivalent to

$$f^{(2g+2)}(t) + (-1)^g\lambda x f(t) = 0$$

with $2g+2$ boundary conditions

$$f(1) = f'(1) = \cdots = f^{(g-1)}(1) = 0,$$
$$f^{(g)}(1) = \frac{(-1)^{g-1}\lambda}{2g!}\int_0^1 (1-s)^g f(s)\,ds,$$
$$f^{(g+1)}(0) = f^{(g+2)}(0) = \cdots = f^{(2g+1)}(0) = 0.$$

Exercise 8.1.2 Show that the FD of $L_0(s,t;x) = x\min(s,t) - 1/2$ is given by

$$D_0(\lambda;x) = \cos\sqrt{\lambda x} + \frac{\lambda}{2}\frac{\sin\sqrt{\lambda x}}{\sqrt{\lambda x}}.$$

Exercise 8.1.3 Define

$$R = \frac{U}{V}, \quad U = \frac{1}{2}W^2(1), \quad V = \int_0^1 W^2(t)\,dt.$$

Then prove that

$$\mathrm{E}(R) = \mathrm{E}(U/V) = \mathrm{E}(U)/\mathrm{E}(V) = 1, \quad \mathrm{E}(RV) = \mathrm{E}(R)\mathrm{E}(V)$$

by using the formula

$$\mathrm{E}\left(\frac{U^k}{V^k}\right) = \frac{1}{(k-1)!}\int_0^\infty \theta_2^{k-1}\left.\frac{\partial^k\psi(\theta_1,-\theta_2)}{\partial\theta_1^k}\right|_{\theta_1=0} d\theta_2,$$

where $\psi(\theta_1,-\theta_2) = \mathrm{E}\left(e^{\theta_1 U-\theta_2 V}\right)$. Verify also that R and V are not independent, although $\mathrm{E}(RV) = \mathrm{E}(R)\mathrm{E}(V)$.

8.2 Fractional Brownian Motion and Some Properties

The last section was concerned with the g-fold integrated Bm $\{F_g(t)\}$ ($g = 1,2,\dots$) arising from the discrete-time model (8.1). This section extends (8.1) to the case where g is fractional. The extended model is

$$x_j = \rho x_{j-1} + u_j, \quad (1-L)^d u_j = \varepsilon_j, \quad \{\varepsilon_j\} \sim \text{i.i.d.}(0,\sigma_\varepsilon^2), \tag{8.14}$$

where the true value of ρ is unity, $x_0 = 0$, d is any positive real number, and $\{u_j\}$ is a process defined by

$$u_j = (1-L)^{-d}\varepsilon_j = \sum_{\ell=0}^{j-1}\psi_\ell\,\varepsilon_{j-\ell}, \quad \psi_\ell = \frac{\Gamma(\ell+d)}{\Gamma(d)\Gamma(\ell+1)}. \tag{8.15}$$

Then it holds (Marinucci and Robinson, 1999) that

$$X_n(t) = \frac{1}{\sigma_\varepsilon n^{d+1/2}}\left(x_{[nt]} + (nt-[nt])u_{[nt]+1}\right) \Rightarrow W_d(t), \tag{8.16}$$

where

$$W_d(t) = \int_0^t \frac{(t-u)^d}{\Gamma(d+1)}\,dW(u). \tag{8.17}$$

The result in (8.16) is the FCLT for $\{X_n(t)\}$ and can be intuitively explained as follows. We have

$$x_{[nt]} = \sum_{j=1}^{[nt]} u_j = \sum_{j=1}^{[nt]} \sum_{\ell=0}^{j-1} \psi_\ell \varepsilon_{j-\ell} = \sum_{j=1}^{[nt]} \left(\sum_{\ell=0}^{[nt]-j} \psi_\ell \right) \varepsilon_j,$$

where it follows from Sowell (1990) and Stirling's formula that

$$\sum_{\ell=0}^{k} \psi_\ell = \sum_{\ell=0}^{k} \frac{\Gamma(\ell+d)}{\Gamma(d)\Gamma(\ell+1)} = \frac{\Gamma(k+d+1)}{\Gamma(d+1)\Gamma(k+1)} \approx \frac{1}{\Gamma(d+1)} k^d, \tag{8.18}$$

which yields

$$x_{[nt]} \approx \frac{n^d}{\Gamma(d+1)} \sum_{j=1}^{[nt]} \left(t - \frac{j}{n} \right)^d \varepsilon_j.$$

Then, using the arguments in Helland (1982), we can establish that

$$\frac{1}{\sigma_\varepsilon n^{d+1/2}} x_{[nt]} \Rightarrow \frac{1}{\Gamma(d+1)} \int_0^t (t-u)^d \, dW(u) = W_d(t).$$

It can be seen that $W_d(t)$ is a natural extension of $F_g(t)$ dealt with in the last section. Note that the process $W_d(t)$ is well defined for any $d > -1/2$ and is called *type II fBm* (Marinucci and Robinson, 1999).

It might be thought that $W_d(t)$ can be handled easily as a direct extention of $F_g(t)$, but this is not the case. In fact, it is impossible to explicitly compute

$$\mathrm{Cov}(W_d(s), W_d(t)) = \frac{1}{\Gamma^2(d+1)} \int_0^{\min(s,t)} ((s-u)(t-u))^d \, du,$$

if d is a noninteger. It is difficult to proceed working with $W_d(t)$.

As an alternative, we introduce another type of fBm. For this purpose we restrict d in (8.14) to be $0 < d < 1/2$ and consider

$$y_j = \rho y_{j-1} + v_j, \quad (1-L)^d v_j = \varepsilon_j, \quad \{\varepsilon_j\} \sim \text{i.i.d.}\left(0, \sigma_\varepsilon^2\right), \tag{8.19}$$

where $y_0 = 0$ and the true value of ρ is unity; now

$$v_j = (1-L)^{-d} \varepsilon_j = \sum_{\ell=0}^{\infty} \psi_\ell \varepsilon_{j-\ell}, \quad \psi_\ell = \frac{\Gamma(\ell+d)}{\Gamma(d)\Gamma(\ell+1)}. \tag{8.20}$$

The process $\{v_j\}$ is stationary, and its spectrum is given by

$$f(\omega) = \frac{\sigma_\varepsilon^2}{2\pi} \left| 1 - e^{i\omega} \right|^{-2d} = \frac{\sigma_\varepsilon^2}{2\pi} \left(4 \sin^2 \frac{\omega}{2} \right)^{-d}.$$

It follows from Gradshteyn and Ryzhik (1965) and Hosking (1981) that

$$\gamma_v(h) = \text{Cov}(v_j, v_{j+h}) = \frac{\sigma_\varepsilon^2}{2\pi} \int_{-\pi}^{\pi} \left(4 \sin^2 \frac{\omega}{2}\right)^{-d} e^{ih\omega} \, d\omega$$
$$= \frac{\sigma_\varepsilon^2 \Gamma(1-2d)\Gamma(|h|+d)}{\Gamma(d)\Gamma(1-d)\Gamma(|h|+1-d)}.$$

Since it follows from Stirling's formula that, as $h \to \infty$,

$$\gamma_v(h) = \frac{\sigma_\varepsilon^2 \Gamma(1-2d)}{\Gamma(d)\Gamma(1-d)} h^{2d-1} + o(h^{2d-1}) \quad (0 < d < 1/2), \tag{8.21}$$

we have

$$\sum_{h=-\infty}^{\infty} |\gamma_v(h)| = \infty.$$

Because of this last property of the autocovariance, the stationary process $\{v_j\}$ is called a *long-memory process*. Using (8.21), it can also be shown (Exercise 8.2.1) that

$$\frac{1}{n^{2d+1}} \text{Var}(y_n) = \frac{1}{n^{2d+1}} \text{Var}\left(\sum_{j=1}^{n} v_j\right) \to \sigma_\varepsilon^2 G^2(d), \tag{8.22}$$

where

$$G^2(d) = \frac{\Gamma(1-2d)}{(2d+1)\Gamma(1+d)\Gamma(1-d)}.$$

We now describe the FCLT associated with this stationary long-memory process, which was earlier discussed in Davydov (1970) and developed in Taqqu (1975) (see also Giraitis, Koul, and Surgailis (2012) for more details). Here we present its simplified version. For $y_n = v_1 + \cdots + v_n$ with v_j defined in (8.20), it holds that, by putting $d = H - 1/2$,

$$Y_n(t) = \frac{1}{\sigma_\varepsilon n^H} \left(y_{[nt]} + (nt - [nt]) v_{[nt]+1}\right) \Rightarrow \frac{A(H)}{\Gamma(H+1/2)} B_H(t), \tag{8.23}$$

where $t \in [0, 1]$ and $H = d + 1/2$ $(1/2 < H < 1)$ is called the *Hurst index*, and $B_H(t)$ defined by

$$B_H(t) = \frac{1}{A(H)} \left[\int_{-\infty}^{0} \left\{(t-u)^d - (-u)^d\right\} dW(u) + \int_0^t (t-u)^d \, dW(u)\right] \tag{8.24}$$

is called *type I fBm*. Here $A(H)$ is a normalizing constant, so that $\text{Var}(B_H(1)) = 1$. In fact, it is defined by

$$A(H) = \left[\mathrm{Var} \left\{ \int_{-\infty}^{0} \left\{ (1-u)^d - (-u)^d \right\} dW(u) + \int_0^1 (1-u)^d \, dW(u) \right\} \right]^{1/2}$$
$$= \left[\frac{1}{2H} + \int_0^\infty \left((1+u)^{H-1/2} - u^{H-1/2} \right)^2 du \right]^{1/2}.$$

It follows from (8.22) and (8.23) that

$$A(H) = \Gamma(H+1/2) G(H-1/2) = \left[\frac{\Gamma(H+1/2)\Gamma(2-2H)}{2H\,\Gamma(3/2-H)} \right]^{1/2}. \quad (8.25)$$

The result in (8.23) can also be intuitively explained as follows. We have

$$\begin{aligned} y_{[nt]} &= \sum_{j=1}^{[nt]} v_j = \sum_{j=1}^{[nt]} \sum_{\ell=0}^{\infty} \psi_\ell \, \varepsilon_{j-\ell} \\ &= \sum_{j=-\infty}^{0} \left(\sum_{\ell=0}^{[nt]-j} \psi_\ell - \sum_{\ell=0}^{-j} \psi_\ell \right) \varepsilon_j + \sum_{j=1}^{[nt]} \left(\sum_{\ell=0}^{[nt]-j} \psi_\ell \right) \varepsilon_j \\ &\approx \frac{n^d}{\Gamma(d+1)} \left[\sum_{j=-n}^{0} \left\{ \left(t - \frac{j}{n} \right)^d - \left(-\frac{j}{n} \right)^d \right\} \varepsilon_j + \sum_{j=1}^{[nt]} \left(t - \frac{j}{n} \right)^d \varepsilon_j \right], \end{aligned}$$

where we used the relation in (8.18). Then we can establish

$$\begin{aligned} \frac{\Gamma(d+1)}{\sigma_\varepsilon n^{d+1/2}} y_{[nt]} &\Rightarrow \int_{-\infty}^{0} \left\{ (t-u)^d - (-u)^d \right\} dW(u) \\ &\quad + \int_0^t (t-u)^d \, dW(u) \qquad (8.26) \\ &= A(H) \, B_H(t) \qquad (H = d + 1/2), \end{aligned}$$

where $A(H)$ is defined in (8.25).

fBm was invented by Kolmogorov (1940) and largely developed by Mandelbrot and Van Ness (1968), who obtained the integral representation for $B_H(t)$ in (8.24). It can be seen that $B_H(t)$ is defined by introducing the increments of the Bm from the infinite past, which produces nice properties of $B_H(t)$ as described in the following. It is also recognized that $B_H(t)$ is well defined, with $H \in (0, 1)$ and $t \in [0, T]$ for any $T > 0$. In particular, $B_H(t)$ reduces to Bm when $H = 1/2$. The reason why $H > 1$ is not permissible will be made clear shortly.

In subsequent discussions we concentrate on type I fBm $B_H(t)$, which is assumed to be defined with $H \in (0, 1)$ and $t \in [0, T]$. We call $B_H(t)$ just fBm. Since $\{B_H(t)\}$ is a zero-mean Gaussian process, another definition may be given (Nourdin, 2012) by

Definition 8.1 (Fractional Brownian motion) The nonstationary stochastic process $\{B_H(t)\}$ defined with $H \in (0,1)$ and $t \in [0,T]$ is called fBm if it is a zero-mean Gaussian process with $B_H(0) = 0$ and

$$\Gamma_H(s,t) = \mathrm{Cov}(B_H(s), B_H(t)) = \frac{1}{2}\left[s^{2H} + t^{2H} - |s-t|^{2H}\right]. \tag{8.27}$$

Note that $\mathrm{E}\big(B_H^2(t)\big) = t^{2H}$. When $H > 1$, $\Gamma_H(s,t)$ is not positive definite. In fact, it holds (Nourdin, 2012) that, for $s = 1$, $t = 2$ and $\boldsymbol{a} = (a_1, a_2)' = (-2, 1)'$,

$$\boldsymbol{a}' \begin{pmatrix} \Gamma_H(s,s) & \Gamma_H(s,t) \\ \Gamma_H(s,t) & \Gamma_H(t,t) \end{pmatrix} \boldsymbol{a} = 4 - 2^{2H} < 0 \quad (H > 1).$$

However, $B_H(t)$ is well defined for $H = 1$. It holds that, for $H = 1$, $\mathrm{E}[(B_H(t) - tB_H(1))^2] = 0$, so that $B_H(t) = tB_H(1)$ with probability 1. This trivial case is excluded from our consideration.

To examine the memory property of $B_H(t)$ for $t \in [0,\infty)$, we consider

$$\Delta B_H(j) = B_H(j) - B_H(j-1) \quad (j = 1, 2, \ldots),$$

which is called *fractional Gaussian noise* (fGn) and is a zero-mean stationary Gaussian process because it holds (Exercise 8.2.2) that

$$\begin{aligned} \gamma_\Delta(h) &= \mathrm{Cov}(\Delta B_H(j), \Delta B_H(j+h)) \\ &= \frac{1}{2}\left[|h+1|^{2H} + |h-1|^{2H} - 2|h|^{2H}\right] \\ &= H(2H-1)|h|^{2H-2} + O(|h|^{2H-3}) \qquad (|h| \to \infty). \end{aligned}$$

It is recognized that fGn has negative correlation when $0 < H < 1/2$, positive correlation when $1/2 < H < 1$, and is independent when $H = 1/2$. It holds (Exercise 8.2.3) that

$$A = \sum_{h=-\infty}^{\infty} \gamma_\Delta(h) = \begin{cases} 0 & (0 < H < 1/2), \\ 1 & (H = 1/2), \\ \infty & (1/2 < H < 1). \end{cases}$$

Then we say that the fBm is a short-memory process when $A < \infty$ $(0 < H \leq 1/2)$ and is a long-memory process when $A = \infty$ $(1/2 < H < 1)$.

Some properties of $\{B_H(t)\}$ follow for $t \in [0,\infty)$ (Nourdin, 2012).

(a) **Self-similarity**

$$B_H(\gamma t) \overset{\mathcal{D}}{=} \gamma^H B_H(t) \text{ for any } \gamma > 0 \quad (0 < H < 1).$$

(b) **Stationarity of increments** (Exercise 8.2.4)

$$\mathrm{Var}\Big(B_H(s) - B_H(t)\Big) = |s-t|^{2H} \quad (0 < H < 1).$$

(c) **Memory property**

$$\sum_{h=0}^{\infty} \mathrm{Cov}\Big(\Delta B_H(j), \Delta B_H(j+h)\Big) = \infty \quad (1/2 < H < 1),$$

where $\Delta B_H(j) = B_H(j) - B_H(j-1)$.

(d) **Infinitesimal rule**

$$\mathrm{Cov}\Big(dB_H(s), dB_H(t)\Big) = H(2H-1)|s-t|^{2H-2}\,dsdt \quad (1/2 < H < 1).$$

(e) **Quadratic variation property** (Exercise 8.2.5)

$$\mathrm{plim}_{n\to\infty} \frac{1}{n^{1-2H}} \sum_{j=1}^{n} \Big(B_H(t_j) - B_H(t_{j-1})\Big)^2 = 1, \tag{8.28}$$

which implies that

$$\mathrm{plim}_{n\to\infty} \sum_{j=1}^{n} \left(B_H(t_j) - B_H(t_{j-1})\right)^2 = \begin{cases} \infty & (0 < H < 1/2), \\ 1 & (H = 1/2), \\ 0 & (1/2 < H < 1), \end{cases}$$

where $t_j = j/n$.

(f) **Nonsemimartingale property for $H \neq 1/2$**

The definition of a semimartingale will be given shortly. The Itô calculus was developed for semimartingales, which means that the Itô calculus is not applicable to the analysis of fBm. The fact that $\{B_H(t)\}$ is not a semimartingale for $H \neq 1/2$ is shown in Nourdin (2012).

Some remarks about the above properties of fBm follow. It is seen from (a) and (b) that fBm $\{B_H(t)\}$ with $0 < H < 1$ is a self-similar process with stationary increments, but the increments are dependent when $H \neq 1/2$. It is noted from (c) through (f) that fBm has different properties depending on the value of H. In particular, (c) indicates that the increment process $\{\Delta B_H(j)\}$ is a stationary long-memory process with positive correlation when $1/2 < H < 1$.

The infinitesimal rule described in (d) holds for $1/2 < H < 1$. This comes from Gripenberg and Norros (1996), who proved that

$$\mathrm{E}\left(\int_R f(s)\,dB_H(s) \int_R g(t)\,dB_H(t)\right) = \int_R \int_R f(s)\,g(t)\,\phi(s,t)\,ds\,dt, \tag{8.29}$$

where

$$\phi(s,t) = H(2H-1)\,|s-t|^{2H-2} \quad (s,t \in R = (-\infty,\infty))\,, \tag{8.30}$$

and $f(s)$ and $g(t)$ are deterministic functions such that $\| f \|_H^2 < \infty$ and $\| g \|_H^2 < \infty$ with

$$\| f \|_H^2 = \int_R \int_R f(s) f(t) \phi(s,t)\, ds\, dt.$$

It follows (Exercise 8.2.6) that, when $1/2 < H < 1$,

$$\begin{aligned} \mathrm{E}\Big(B_H(s) B_H(t) \Big) &= \mathrm{E}\left(\int_0^s dB_H(u) \int_0^t dB_H(v) \right) \\ &= \int_0^s \int_0^t \phi(u,v)\, du\, dv \\ &= \frac{1}{2}\left[s^{2H} + t^{2H} - |s-t|^{2H} \right]. \end{aligned} \tag{8.31}$$

Note that this last equality is not valid for $0 < H \le 1/2$ since the kernel $\phi(u,v)$ is not integrable in that range.

The quadratic variation property of $B_H(t)$ also depends on H, as is described in (e). We have

$$\sum_{j=1}^n B_H(t_{j-1}) \left(B_H(t_j) - B_H(t_{j-1}) \right) = \frac{1}{2} \left(B_H^2(1) - \sum_{j=1}^n \left(B_H(t_j) - B_H(t_{j-1}) \right)^2 \right).$$

Thus it follows from (8.28) that the usual definition of the integral based on the Riemann sum leads us to

$$\begin{aligned} \int_0^1 B_H(t)\, dB_H(t) &= \mathrm{plim}_{n\to\infty} \sum_{j=1}^n B_H(t_{j-1}) \left(B_H(t_j) - B_H(t_{j-1}) \right) \\ &= \begin{cases} \text{not integrable} & (0 < H < 1/2), \\ \frac{1}{2}\left(B_H^2(1) - 1 \right) & (H = 1/2), \\ \frac{1}{2} B_H^2(1) & (1/2 < H < 1). \end{cases} \end{aligned} \tag{8.32}$$

The nonsemimartingale property described in (f) differentiates between the cases of $H = 1/2$ and $H \neq 1/2$. A simplified definition of a semimartingale is given by

Definition 8.2 (Semimartingale) A continuous-time process $S(t)$ is called a *semimartingale* if it is decomposed as $S(t) = A(t) + M(t)$, where $A(t)$ is a process of finite variation and $M(t)$ is a martingale.

Any martingale is a semimartingale, so that Bm is a semimartingale. The more general and involved definition of a semimartingale may be found in

Klebaner (2012). An important fact about $B_H(t)$ with $H \neq 1/2$ is that it is neither a semimartingale nor a martingale, whereas $B_H(t)$ with $H = 1/2$ is a martingale. Some other examples of semimartingales follow (Klebaner, 2012).

(1) $S(t) = W^2(t)$ is a semimartingale because $S(t) = A(t) + M(t)$, where $A(t) = t$ is a finite variation process and $M(t) = W^2(t) - t$ is a martingale. Alternatively, we can express $W^2(t)$ as

$$W^2(t) = 1 + 2 \int_0^t W(s)\, dW(s),$$

which shows that $W^2(t)$ is a semimartingale.

(2) $S(t) = N(t)$, where $N(t)$ is a Poisson process with rate λ, is a semimartingale since $N(t)$ itself is a finite variation process. Alternatively, we have $S(t) = \lambda t + N(t) - \lambda t$, where $A(t) = \lambda t$ is a finite variation process and $M(t) = N(t) - \lambda t$ is a martingale.

(3) A solution process of the Itô stochastic differential equation, that is, the process $Y(t)$ that satisfies

$$dY(t) = \mu(Y(t), t)\, dt + \sigma(Y(t), t)\, dW(t),$$

is a semimartingale. For instance, the ordinary O–U process is a semimartingale because

$$Y(t) = e^{-ct} \int_0^t e^{cs}\, dW(s) = A(t) + W(t),$$

where

$$A(t) = -ce^{-ct} \int_0^t e^{cs} W(s)\, ds$$

is a finite variation process.

The nonsemimartingale nature of fBm is reflected in the above properties (b) through (e). As a consequence the Itô calculus developed for martingales or semimartingales is inapplicable. Various approaches to overcoming this difficulty have been suggested. See, for details, Biagini, Hu, Øksendal, and Zhang (2008) and Nourdin (2012).

In any case, we have recognized that the process $\{B_H(t)\}$ behaves quite differently depending on the value of H. In particular, the case of $H \in (0, 1/2)$ needs to be taken care of because of the unbounded quadratic variation.

Exercises for Section 8.2

Exercise 8.2.1 For the stationary long-memory process $\{v_j\}$ defined by

$$v_j = (1-L)^{-d}\varepsilon_j, \quad 0 < d < 1/2, \quad \{\varepsilon_j\} \sim \text{i.i.d.}(0,1),$$

prove that

$$\lim_{n\to\infty} \frac{1}{n^{2d+1}} \text{Var}\left(\sum_{j=1}^{n} v_j\right) = \frac{\Gamma(1-2d)}{(2d+1)\Gamma(1+d)\Gamma(1-d)}.$$

Exercise 8.2.2 For fGn $\{\Delta B_H(j)\}$ with $H \neq 1/2$, prove that

$$\begin{aligned}\text{Cov}\big(\Delta B_H(j),\, \Delta B_H(j+h)\big) &= \frac{1}{2}\left[|h+1|^{2H} + |h-1|^{2H} - 2|h|^{2H}\right] \\ &= H(2H-1)|h|^{2H-2} + O\big(|h|^{2H-3}\big).\end{aligned}$$

Exercise 8.2.3 For fGn $\{\Delta B_H(j)\}$ with $0 < H < 1/2$, prove that

$$\sum_{h=-\infty}^{\infty} \text{Cov}\,(\Delta B_H(j),\, \Delta B_H(j+h)) = 0.$$

Exercise 8.2.4 Using the expression for $B_H(t)$ given in (8.24), prove that

$$\text{Var}\,(B_H(t) - B_H(s)) = |s-t|^{2H} \qquad (H \in (0,1)).$$

Exercise 8.2.5 Prove that, for $H \in (0,1)$,

$$\text{plim}_{n\to\infty} \frac{1}{n^{1-2H}} \sum_{j=1}^{n} \big(B_H(t_j) - B_H(t_{j-1})\big)^2 = 1 \qquad (t_j = j/n),$$

which yields

$$\text{plim}_{n\to\infty} \sum_{j=1}^{n} \big(B_H(t_j) - B_H(t_{j-1})\big)^2 = \begin{cases} \infty & (0 < H < 1/2), \\ 1 & (H = 1/2), \\ 0 & (1/2 < H < 1). \end{cases}$$

Exercise 8.2.6 Prove that, for $H \in (1/2,1)$,

$$H(2H-1)\int_0^s \int_0^t |u-v|^{2H-2}\, du\, dv = \frac{1}{2}\left[s^{2H} + t^{2H} - |s-t|^{2H}\right].$$

8.3 Fractional Unit Root Distribution

In this section we consider the unit root distribution asssociated with the model discussed in the last section, which is

$$y_j = \rho y_{j-1} + v_j, \quad (1-L)^d v_j = \varepsilon_j, \quad \{\varepsilon_j\} \sim \text{i.i.d.}\left(0, \sigma_\varepsilon^2\right), \tag{8.33}$$

where $y_0 = 0$, the true value of ρ is unity, $0 < d < 1/2$, and $\{v_j\}$ is the stationary long-memory process given by

$$v_j = (1-L)^{-d}\varepsilon_j = \sum_{\ell=0}^{\infty} \psi_\ell\, \varepsilon_{j-\ell}, \quad \psi_\ell = \frac{\Gamma(\ell+d)}{\Gamma(d)\Gamma(\ell+1)}. \tag{8.34}$$

Let $\hat{\rho}$ be the LSE of ρ in the model (8.33). Then we have

$$\hat{\rho} = \sum_{j=2}^{n} y_{j-1}y_j \Big/ \sum_{j=2}^{n} y_{j-1}^2, \quad \hat{\rho} - 1 = (A_n - B_n) \Big/ \sum_{j=2}^{n} y_{j-1}^2, \tag{8.35}$$

where, putting $d = H - 1/2$,

$$A_n = \frac{1}{2}\left(y_n^2 - y_1^2\right) = O_p\left(n^{2d+1}\right) = O_p(n^{2H}),$$
$$B_n = \frac{1}{2}\sum_{j=2}^{n} v_j^2 = O_p(n), \quad \sum_{j=2}^{n} y_{j-1}^2 = O_p\left(n^{2d+2}\right) = O_p\left(n^{2H+1}\right).$$

It follows from the FCLT described in (8.23) and the CMT that

$$n(\hat{\rho} - 1) = \frac{1}{\sigma_\varepsilon^2 n^{2H}}(A_n - B_n) \Big/ \frac{1}{\sigma_\varepsilon^2 n^{2H+1}} \sum_{j=2}^{n} y_{j-1}^2$$
$$= \left(\frac{1}{2} Y_n^2(1) + o_p(1)\right) \Big/ \frac{1}{n}\sum_{j=2}^{n} Y_n^2\left(\frac{j-1}{n}\right) \Rightarrow Q_H,$$

$$Q_H = \frac{1}{2} B_H^2(1) \Big/ \int_0^1 B_H^2(t)\, dt \qquad (1/2 < H < 1). \tag{8.36}$$

The distribution of Q_H in (8.36) is called the *fractional unit root distribution*, and $P(Q_H < x)$ can be computed if the associated c.f. can be derived. However, the derivation turns out to be hard. Even the c.f. of the statistic

$$S_H = \int_0^1 B_H^2(t)\, dt \overset{\mathcal{D}}{=} \int_0^1\int_0^1 K_H(s,t)\, dW(s) dW(t) \overset{\mathcal{D}}{=} \sum_{n=1}^{\infty} \frac{1}{\lambda_n(H)} Z_n^2 \tag{8.37}$$

is hard to derive, where $\{Z_n\} \sim \text{NID}(0,1)$ and $0 < \lambda_1(H) < \lambda_2(H) < \cdots$ are eigenvalues of

$$K_H(s,t) = \text{Cov}\Big(B_H(s), B_H(t)\Big) = \frac{1}{2}\left[s^{2H} + t^{2H} - |s-t|^{2H}\right].$$

The derivation of the FD of $K_H(s,t)$ is difficult because the seemingly simple integral equation

$$f(t) = \lambda \int_0^1 K_H(s,t) f(s)\, ds$$

cannot be converted into any plausible differential equation unless $H = 1/2$. In fact, when $H \neq 1/2$, we have

$$\begin{aligned}
f(t) &= \frac{\lambda}{2}\left[\int_0^1 s^{2H} f(s)\, ds + t^{2H}\int_0^1 f(s)\, ds - \int_0^t (t-s)^{2H} f(s)\, ds \right.\\
&\qquad \left. - \int_t^1 (s-t)^{2H} f(s)\, ds\right],\\
f'(t) &= \lambda H\left[t^{2H-1}\int_0^1 f(s)\, ds - \int_0^t (t-s)^{2H-1} f(s)\, ds \right.\\
&\qquad \left. + \int_t^1 (s-t)^{2H-1} f(s)\, ds\right],\\
f''(t) &= \lambda H(2H-1)\left[t^{2H-2}\int_0^1 f(s)\, ds - \int_0^t (t-s)^{2H-2} f(s)\, ds \right.\\
&\qquad \left. - \int_t^1 (s-t)^{2H-2} f(s)\, ds\right].
\end{aligned}$$

Because of the existence of the term $|s-t|^{2H}$ in the kernel, it is impossible to obtain any plausible differential equation equivalent to the original integral equation.

It is also impossible to apply Girsanov's theorem (see Theorem 7.4). Instead of the O–U process, we need to consider the *fractional O–U (fO–U) process* given by

$$dY_H^{\alpha}(t) = \alpha Y_H^{\alpha}(t)\, dt + dB_H(t), \quad Y_H^{\alpha}(0) = 0, \quad t \in [0,T]. \tag{8.38}$$

fBm is the case where $\alpha = 0$, whereas the ordinary O–U process is the case where $H = 1/2$. The fractional version of Girsanov's theorem was established in Kleptsyna, Le Breton, and Roubaud (2000), which is described as follows. Define the auxiliary process

$$dY_H^\beta(t) = \beta Y_H^\beta(t)\, dt + dB_H(t), \qquad Y_H^\beta(0) = 0, \quad t \in [0, T]. \tag{8.39}$$

Let $\big(C[0,T], \mathcal{B}(C)\big)$ be the measurable space of continuous functions on $[0, T]$ with the associated σ-field generated under the supremum norm. Let $\mu_{Y_H^\alpha}$ and $\mu_{Y_H^\beta}$ be the probability measures on $\big(C[0,T], \mathcal{B}(C)\big)$ induced by $\{Y_H^\alpha(t)\}$ and $\{Y_H^\beta(t)\}$, respectively, by the relation

$$\mu_{Y_H^\alpha}(A) = P\Big(\omega : Y_H^\alpha \in A\Big), \quad \mu_{Y_H^\beta}(A) = P\Big(\omega : Y_H^\beta \in A\Big), \quad A \in \mathcal{B}(C).$$

Then Kleptsyna, Le Breton, and Roubaud (2000) proved that the measures $\mu_{Y_H^\alpha}$ and $\mu_{Y_H^\beta}$ are equivalent and that the Radon–Nikodym derivative evaluated at $Y_H^\gamma \in C[0,T]$, where $dY_H^\gamma(t) = \gamma Y_H^\gamma(t)\, dt + dB_H(t)$ with $Y_H^\gamma(0) = 0$, is given by

$$\begin{aligned} \frac{d\mu_{Y_H^\alpha}}{d\mu_{Y_H^\beta}}\left(Y_H^\gamma\right) = \exp\Bigg[&\frac{\beta^2-\alpha^2}{2}\int_0^T \left(Q_H^\gamma(t)\right)^2 dv_H(t) \\ &-(\beta-\alpha)\int_0^T Q_H^\gamma(t)\, dZ_H^\gamma(t)\Bigg]. \end{aligned} \tag{8.40}$$

Here $v_H(t) = a_H t^{2-2H}$, $a_H = \Gamma(3/2 - H)/\left(2H\,\Gamma(H+1/2)\,\Gamma(3-2H)\right)$, and

$$\begin{aligned} Q_H^\gamma(t) &= \frac{d}{dv_H(t)}\int_0^t \frac{(s(t-s))^{1/2-H}}{\kappa_H} Y_H^\gamma(s)\, ds \\ &= \frac{1}{4(1-H)a_H}\left\{t^{2H-1} Z_H^\gamma(t) + \int_0^t s^{2H-1}\, dZ_H^\gamma(s)\right\}, \end{aligned} \tag{8.41}$$

$$Z_H^\gamma(t) = \kappa_H^{-1}\int_0^t (s(t-s))^{1/2-H}\, dY_H^\gamma(s), \tag{8.42}$$

with $\kappa_H = 2H\,\Gamma(H+1/2)\,\Gamma(3/2-H)$. Note that, when $H = 1/2$, it holds that $v_H(t) = t$ and $Q_H^\gamma(t) = Z_H^\gamma(t) = Y_H^\gamma(t)$, so that the Radon–Nikodym derivative for the fO–U process reduces to that for the O–U process.

It is shown in Kleptsyna and Le Breton (2002) that the sample paths of the process $Q_H^\gamma(t)$ in (8.41) belong to $L^2([0,T], dv_H)$, and the process $Z_H^\gamma(t)$ in (8.42) is a Gaussian semimartingale with the decomposition

$$\begin{aligned} Z_H^\gamma(t) &= \gamma\int_0^t Q_H^\gamma(s)\, dv_H(s) + M_H(t), \\ M_H(t) &= \kappa_H^{-1}\int_0^t (s(t-s))^{1/2-H}\, dB_H(s), \end{aligned} \tag{8.43}$$

where $M_H(t)$ is a martingale with $\mathrm{Cov}(M_H(s), M_H(t)) = a_H \min(s,t)$, which will be examined in detail in the next section. Note here that, when $H = 1/2$, $M_H(t)$ reduces to $B_{1/2}(t) = W(t)$.

The Radon–Nikodym derivative in (8.40) is composed of a deterministic function $v_H(t)$ and two stochastic processes $Q_H^\gamma(t)$ and $Z_H^\gamma(t)$, which are totally different from the original process $Y_H^\alpha(t)$, although they are linear functionals of $Y_H^\gamma(t)$. Putting $\alpha = 0$ and $T = 1$, it follows from (8.40) that

$$\begin{aligned}
\mathrm{E}\left(e^{\theta S_H}\right) &= \mathrm{E}\left[\exp\left\{\theta \int_0^1 \left(Y_0^\alpha(t)\right)^2 dt\right\}\right] \\
&= \mathrm{E}\left[\exp\left\{\theta \int_0^1 \left(Y_H^\beta(t)\right)^2 dt\right\} \frac{d\mu_{Y_H^0}}{d\mu_{Y_H^\beta}}\left(Y_H^\beta\right)\right] \\
&= \mathrm{E}\left[\exp\left\{\theta \int_0^1 \left(Y_H^\beta(t)\right)^2 dt + \frac{\beta^2}{2}\int_0^1 \left(Q_H^\beta(t)\right)^2 dv_H(t)\right.\right. \\
&\qquad \left.\left. -\beta \int_0^1 Q_H^\beta(t)\, dZ_H^\beta(t)\right\}\right].
\end{aligned}$$

It is impossible to make the above expression simpler with any choice of β. Thus the fractional version of Girsanov's theorem is of no help in the derivation of the fractional unit root distribution.

However, it is quite useful for computing the distribution of the MLE of α in (8.38). Let us define the likelihood for $\{Y_H(t)\}$ in (8.39) by the Radon–Nikodym derivative $d\mu_{Y_H^\alpha}(Y_H^\alpha)/d\mu_{Y_H^0}$ obtained from (8.40). Then the MLE $\tilde{\alpha}$ of α is given by

$$\tilde{\alpha} = \frac{\displaystyle\int_0^T Q_H^\alpha(t)\, dZ_H^\alpha(t)}{\displaystyle\int_0^T \left(Q_H^\alpha(t)\right)^2 dv_H(t)} = \frac{U(T,H,\alpha)}{V(T,H,\alpha)}, \tag{8.44}$$

whereas the LSE $\hat{\alpha}$ of α is given by

$$\hat{\alpha} = \frac{\displaystyle\int_0^T Y_H^\alpha(t)\, dY_H^\alpha(t)}{\displaystyle\int_0^T \left(Y_H^\alpha(t)\right)^2 dt}.$$

As was mentioned above, it is hard to compute the distribution of the LSE $\hat{\alpha}$. However, it is possible to compute that of the MLE $\tilde{\alpha}$. For this purpose, we consider the joint m.g.f. of $U(T,H,\alpha)$ and $V(T,H,\alpha)$ in (8.44). It follows from the fractional version of Girsanov's theorem that

$$
\begin{aligned}
m(\theta_1,\theta_2) &= \mathrm{E}\left[\exp\left\{\theta_1\int_0^T Q_H^\alpha(t)\,dZ_H^\alpha(t)+\theta_2\int_0^T \left(Q_H^\alpha(t)\right)^2 dv_H(t)\right\}\right]\\
&= e^{-(\theta_1+\alpha-\beta)T/2}\mathrm{E}\left[\exp\left\{\frac{\theta_1+\alpha-\beta}{4(1-H)a_H}\,Z_H^\beta(T)\int_0^T t^{2H-1}\,dZ_H^\beta(t)\right\}\right],
\end{aligned}
$$

where $\beta = \sqrt{\alpha^2-2\theta_2}$. Then we can proceed to the computation of the associated c.f. of the MLE. We do not pursue the matter further here. See Kleptsyna and Le Breton (2002) and Bishwal (2008) for its final derivation. Tanaka (2013, 2015) presented graphs of probability densities of the MLE and discussed the MLE-based test on α.

The eigenvalues of $K_H(s,t)$ in (8.37) also remain to be derived, although Bronski (2003) obtained

$$
\frac{1}{\lambda_n(H)} = \frac{\Gamma(2H+1)\,\sin\pi H}{(n\pi)^{2H+1}} + o\left(n^{-(2H+1)}\right) \qquad (n\to\infty).
$$

Since we need values of $1/\lambda_n(H)$ for small n to compute the distribution, the above expansion is not helpful for this purpose. Assuming that the multiplicity of every eigenvalue is unity, Theorem 2.8 suggests an upper bound B_2 of $1/\lambda_1(H)$ given by

$$
\begin{aligned}
B_2 &= \sqrt{\int_0^1\int_0^1 K_H^2(s,t)\,ds\,dt} = \sqrt{\sum_{n=1}^\infty \frac{1}{\lambda_n^2(H)}}\\
&= \sqrt{\frac{4H+3}{2(4H+1)(4H+2)} - \frac{\Gamma^2(2H+1)}{\Gamma(4H+3)}},
\end{aligned}
$$

where

$$
\frac{1}{\lambda_1(H)} < B_2 < M = \sum_{n=1}^\infty \frac{1}{\lambda_n(H)} = \int_0^1 K_H(t,t)\,dt = \frac{1}{2H+1}.
$$

Table 8.1 reports B_2, M, and B_2/M for various values of H. It can be seen that the upper bound B_2 decreases as H becomes large, which implies that $1/\lambda_1(H)$ decreases with H, whereas the contribution of $1/\lambda_1(H)$ to the total sum increases as H becomes large. However, it is found that the distribution of S_H is not monotone, so that it cannot be approximated by $Z^2/\lambda_1(H)$.

In the next two sections we suggest a martingale approximation to the distributions of S_H in (8.37) and Q_H in (8.36).

Table 8.1 *Upper bounds B_2 of $1/\lambda_1(H)$ for various values of H.*

H	0.1	0.2	0.3	0.5	0.7	0.8	0.9
B_2	0.472	0.457	0.442	0.408	0.375	0.360	0.346
M	0.833	0.714	0.625	0.500	0.417	0.385	0.357
B_2/M	0.567	0.640	0.707	0.816	0.901	0.937	0.970

8.4 Approximation to Fractional Brownian Motion

The last section revealed that the derivation of the distribution associated with quadratic functionals of fBm $\{B_H(t)\}$ is difficult. In this section we approximate nonsemimartingale fBm by a martingale process. For this purpose we utilize an important result obtained by Norros, Valkeila, and Virtamo (1999).

Let us consider the zero-mean Gaussian process $\{M_H(t)\}$ defined in (8.43) as

$$M_H(t) = \kappa_H^{-1} \int_0^t (u\,(t-u))^{1/2-H}\, dB_H(u), \tag{8.45}$$

where H is assumed to be $H \in (0,1)$ and $\kappa_H = 2H\,\Gamma(H+1/2)\,\Gamma(3/2-H)$. The process $\{M_H(t)\}$ is called a fundamental martingale with

$$\mathrm{Cov}\big(M_H(s), M_H(t)\big) = a_H(\min(s,t))^{2-2H}, \quad a_H = \frac{\Gamma(3/2-H)}{2H\,\Gamma(H+1/2)\,\Gamma(3-2H)}.$$

In fact, it follows from (8.29) that $\mathrm{Cov}(M_H(s), M_H(t))$ is equal to

$$\begin{aligned} &\kappa_H^{-2}\mathrm{E}\left(\int_0^s \int_0^t (uv(s-u)(t-v))^{1/2-H}\, dB_H(u)\, dB_H(v)\right) \\ &= \kappa_H^{-2} H(2H-1) \int_0^s \int_0^t (uv(s-u)(t-v))^{1/2-H}\, |u-v|^{2H-2}\, du\, dv. \end{aligned}$$

Using the formula (Norros, Valkeila, and Virtamo, 1999)

$$\int_0^1 (x(1-x))^{-\alpha}\, |x-y|^{2\alpha-1}\, dx = B(\alpha, 1-\alpha) \quad (0<\alpha,\, y<1), \tag{8.46}$$

where $B(p,q)$ is the beta function, we obtain, for $0<u<s<t<1$,

$$\begin{aligned} \int_0^t (v(t-v))^{1/2-H} |u-v|^{2H-2}\, dv &= \int_0^1 (x(1-x))^{1/2-H} \left|x - \frac{u}{t}\right|^{2H-2} dx \\ &= \Gamma(H-1/2)\,\Gamma(3/2-H). \end{aligned}$$

We also have

$$\int_0^s (u(s-u))^{1/2-H} du = s^{2-2H} \int_0^1 (x(1-x))^{1/2-H} dx = \frac{\Gamma^2(3/2-H)}{\Gamma(3-2H)} s^{2-2H}.$$

Thus, using these results, it holds that, for $s < t$,

$$\begin{aligned}\text{Cov}\,(M_H(s),\ M_H(t)) &= \frac{H(2H-1)\,\Gamma(H-1/2)\,\Gamma^3(3/2-H)}{\kappa_H^2\,\Gamma(3-2H)}\,s^{2-2H} \\ &= a_H\ (\min(s,t))^{2-2H}.\end{aligned}$$

The process $\{M_H(t)\}$ is a zero-mean Gaussian martingale called the *fundamental martingale*. Note that $M_H(t) = W(t)$ when $H = 1/2$. Here are some properties of $M_H(t)$ with $0 < H < 1$ (Exercises 8.4.1 and 8.4.2).

(a) $\text{Cov}\,(M_H(s), M_H(t)) = a_H\ (\min(s,t))^{2-2H}$.

(b) $\text{Cov}(B_H(s), M_H(t)) = s, \quad \text{Cov}(B_H(s), M_H(t) - M_H(s)) = 0 \quad (s < t)$.

(c) $\text{Corr}(B_H(t), M_H(t)) = 1/\sqrt{a_H}$.

(d) $M_H(t) \overset{\mathcal{D}}{=} N_H(t) = b_H \int_0^t u^{1/2-H}\, dW(u), \quad b_H = \sqrt{(2-2H)\,a_H}$.

(e) $\text{Corr}(B_H(t), N_H(t)) = \dfrac{\Gamma^2(3/2-H)}{\Gamma(2-2H)} \Big/ \sqrt{a_H}$.

The covariance structure in (a) leads to the martingale nature of $M_H(t)$. The property (b) implies that the increments of $M_H(t)$ are independent of $B_H(s)$ for $s < t$. The correlation property (c) is of most interest. It will be found that the correlation of $B_H(t)$ and $M_H(t)$ is quite high when $H > 0.4$, which implies that $B_H(t)$ may be well approximated by $M_H(t)$. The property (d) describes that $M_H(t)$ has the same distribution as $N_H(t)$ defined by a simple integral with respect to $W(t)$. Thus, not $M_H(t)$ but $N_H(t)$ may be useful for approximating the quadratic functional of $B_H(t)$. The correlation of $B_H(t)$ and $N_H(t)$ described in (e) does not depend on t, but it is not high, as is presented in what follows.

Figure 8.2 is concerned with the properties (c) and (e) mentioned above and plots the correlation of $B_H(t)$ and $M_H(t)$ (solid line) and the correlation of $B_H(t)$ and $N_H(t)$ (broken line) as a function of H $(0 < H < 1)$. It is seen that the former correlation is very close to 1 for $H > 0.4$. In fact, it is greater than 0.99 for $H > 0.4$, which is, of course, equal to 1 when $H = 1/2$. The high correlation may be partly explained from the definition of $M_H(t)$ in (8.45), which is the integral of the U-shaped weight function with respect to $B_H(u)$ $(u \le t)$, although the reason why it is very high for $H > 0.4$ is not clear. However, the latter correlation is not high. It decreases as H moves away from $H = 1/2$, but it will be shown later that $N_H(t)$ is useful for approximating the distribution associated with fBm.

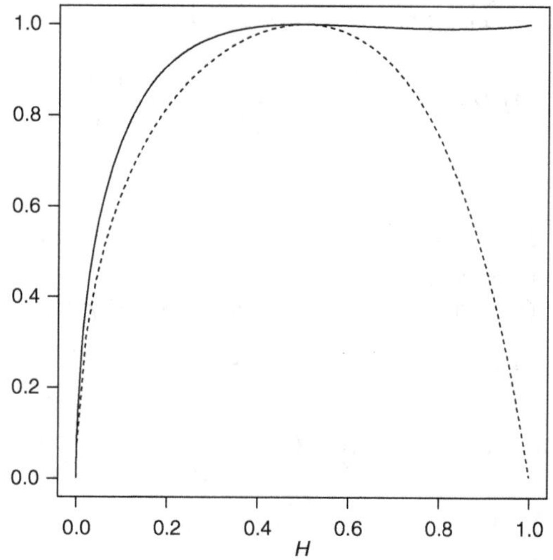

Figure 8.2 $\mathrm{Corr}(B_H(t), M_H(t))$ and $\mathrm{Corr}(B_H(t), N_H(t))$ for $0 < H < 1$.

From the above discussions it may be justified that, when $H > 1/2$, $B_H(t)$ can be approximated by $\gamma\, M_H(t)$, where γ is chosen so that

$$\mathrm{E}\big(B_H^2(t)\big) = t^{2H} = \gamma^2\, \mathrm{E}\big(M_H^2(t)\big) = \gamma^2 a_H t^{2-2H},$$

which yields $\gamma = t^{2H-1}/\sqrt{a_H}$. Then $B_H(t)$ can further be approximated by

$$\begin{aligned} C_H(t) &= t^{2H-1}\, N_H(t)/\sqrt{a_H} \\ &= \sqrt{2(1-H)}\, t^{2H-1} \int_0^t u^{1/2-H}\, dW(u). \end{aligned} \tag{8.47}$$

Note that, when $H = 1/2$, the approximation is exact. In fact, it holds that $C_H(t) = B_H(t) = W(t)$ for $H = 1/2$. It is ensured that $\mathrm{E}(B_H(t)) = \mathrm{E}(C_H(t)) = 0$ and $\mathrm{Var}(B_H(t)) = \mathrm{Var}(C_H(t)) = t^{2H}$. Note, however, that the covariance structure of $C_H(t)$ is different from that of $B_H(t)$ since

$$\mathrm{Cov}\,(C_H(s), C_H(t)) = (st)^{2H-1}\,(\min(s,t))^{2-2H}\,. \tag{8.48}$$

As a first approximation we take up the statistic S_H defined in (8.37) as

$$S_H = \int_0^1 B_H^2(t)\, dt \overset{\mathcal{D}}{=} \int_0^1 \int_0^1 \frac{1}{2}\left[s^{2H} + t^{2H} - |s-t|^{2H}\right] dW(s)\, dW(t).$$

We now consider (Exercise 8.4.3), as an approximation of S_H,

$$\begin{aligned} T_H &= \int_0^1 C_H^2(t)\,dt \\ &= \frac{2(1-H)}{4H-1}\int_0^1\int_0^1 \left[1-(\max(s,t))^{4H-1}\right](s\,t)^{1/2-H}\,dW(s)\,dW(t) \\ &\stackrel{\mathcal{D}}{=} \int_0^1\int_0^1 (st)^{2H-1}\,(\min(s,t))^{2-2H}\,dW(s)\,dW(t). \end{aligned} \tag{8.49}$$

Of course, when $H=1/2$, it holds that $S_H=T_H$.

Let us derive the FD of the kernels in (8.49). We first work with the kernel defined by

$$K_H(s,t) = \frac{2(1-H)}{4H-1}\left[1-(\max(s,t))^{4H-1}\right](s\,t)^{1/2-H}. \tag{8.50}$$

Putting $f(t)=t^{H-1/2}h(t)$, the integral equation

$$f(t) = \lambda\int_0^1 K_H(s,t)f(s)\,ds$$

leads us to the following differential equation and two boundary conditions:

$$h''(t) + \left(2(1-H)\lambda t^{2H-1} - \frac{2H(2H-1)}{t^2}\right)h(t) = 0, \tag{8.51}$$

$$h(1)=0, \quad \lim_{t\to 0}\left((2H-1)t^{-2H}h(t)+t^{1-2H}h'(t)\right) = 0. \tag{8.52}$$

Using the fact that

$$\left((2H-1)t^{-2H}h(t)+t^{1-2H}h'(t)\right)' = -2(1-H)\lambda h(t),$$

it can be shown that the above differential equation with the two boundary conditions is equivalent to the original integral equation (Exercise 8.4.4). The differential equation in (8.51) is a special case of Bessel's equation defined in (2.56) as

$$f''(t) - \frac{2\alpha-1}{t}f'(t) + \left(\beta^2\gamma^2 t^{2\gamma-2} + \frac{\alpha^2-\nu^2\gamma^2}{t^2}\right)f(t) = 0, \tag{8.53}$$

and its general solution is given in (2.57) as

$$f(t) = t^{\alpha}\left(c_1 J_\nu\left(\beta t^{\gamma}\right) + c_2 J_{-\nu}\left(\beta t^{\gamma}\right)\right),$$

where $J_\nu(z)$ is the Bessel function of the first kind defined in (2.58). Thus the general solution to (8.51) is given by

$$h(t) = t^{1/2}\left(c_1 J_\nu\left(\eta(\lambda)t^{H+1/2}\right) + c_2 J_{-\nu}\left(\eta(\lambda)t^{H+1/2}\right)\right),$$

where $\eta(\lambda) = \sqrt{2(1-H)\lambda}/(H+1/2)$ and $\nu = (2H-1/2)/(H+1/2)$. From the two boundary conditions in (8.52), we have the homogeneous equation in $\boldsymbol{c} = (c_1, c_2)'$

$$M(\lambda)\,\boldsymbol{c} = \begin{pmatrix} J_\nu(\eta(\lambda)) & J_{-\nu}(\eta(\lambda)) \\ 1 & 0 \end{pmatrix} \boldsymbol{c} = \boldsymbol{0}.$$

Following from $|M(\lambda)| = -J_{-\nu}(\eta(\lambda))$ and using the fact that

$$J_{-\nu}(z) = \frac{(z/2)^{-\nu}}{\Gamma(1-\nu)} \prod_{n=1}^{\infty} \left(1 - \frac{z^2}{a_n^2}\right),$$

where $0 < a_1 < a_2 < \cdots$ are positive zeros of $J_{-\nu}(z)$, we obtain, as the FD of $K_H(s,t)$ in (8.50),

$$D(\lambda) = \Gamma(1-\nu) J_{-\nu}(\eta(\lambda)) \bigg/ \left(\frac{\eta(\lambda)}{2}\right)^{-\nu}. \tag{8.54}$$

We can also work with the kernel

$$L_H(s,t) = (st)^{2H-1} \left(\min(s,t)\right)^{2-2H} \tag{8.55}$$

and obtain the same FD as that of $K_H(s,t)$ (Exercise 8.4.5).

The above discussions are summarized in the following theorem (Tanaka, 2014).

Theorem 8.3 *The FD $D(\lambda)$ of the kernels $K_H(s,t)$ in (8.50) and $L_H(s,t)$ in (8.55) is given by (8.54), and the distribution function of T_H in (8.49) can be computed following*

$$P(T_H < x) = \frac{1}{\pi} \int_0^\infty \mathrm{Re}\left[\frac{1 - e^{-i\theta x}}{i\theta} (D(2i\theta))^{-1/2}\right] d\theta. \tag{8.56}$$

Figure 8.3 presents graphs of the probability densities of T_H for various values of H. The density of $T_{1/2}$ coincides with that of $S_{1/2}$, but the others are approximations to the density of S_H. Note that both S_H and T_H converge to $\chi^2(1)/3$ as $H \to 1$, which becomes monotone and diverges at 0, unlike the densities for $H < 1$.

To see the closeness of the distributions of T_H to S_H, we compare the first two moments of S_H and T_H. It holds (Exercise 8.4.6) that

$$\mathrm{E}(S_H) = \mathrm{E}(T_H) = \int_0^1 t^{2H}\, dt = \frac{1}{2H+1},$$

Table 8.2 *Comparison of variances of S_H and T_H.*

H	0.5	0.6	0.7	0.8	0.9	0.95	1.0
$\mathrm{Var}(S_H)$	1/3	0.3065	0.2819	0.2598	0.2399	0.2308	2/9
$\mathrm{Var}(T_H)$	1/3	0.3030	0.2778	0.2564	0.2381	0.2299	2/9

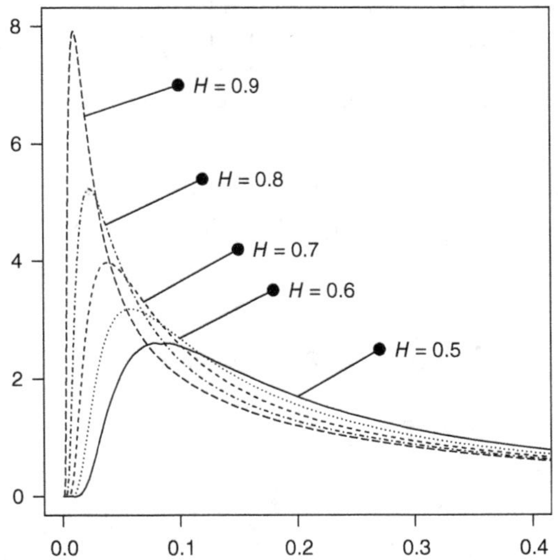

Figure 8.3 Approximate densities of $\int_0^1 B_H^2(t)\,dt$.

$$\mathrm{Var}(S_H) = \frac{1}{2}\int_0^1\int_0^1 \left(s^{2H} + t^{2H} - |s-t|^{2H}\right)^2 ds\,dt$$

$$= \frac{4H+3}{(4H+1)(4H+2)} - \frac{2\Gamma^2(2H+1)}{\Gamma(4H+3)},$$

$$\mathrm{Var}(T_H) = \left.\frac{d^2\mathrm{E}\left(e^{\theta T_H}\right)}{d\theta^2}\right|_{\theta=0} - \mathrm{E}^2(T_H)$$

$$= \frac{4H+5}{3(2H+1)^2} - \frac{1}{(2H+1)^2} = \frac{2}{3(2H+1)}.$$

Table 8.2 reports $\mathrm{Var}(S_H)$ and $\mathrm{Var}(T_H)$ for various values of H, where the case for $H = 1$ is also presented as a limiting case. It can be seen that, as a whole, the variances of the approximate densities coincide with the true ones up to the second decimal places. There is no evidence that the approximation is poor, but more details remain to be investigated.

Table 8.3 *Percentage points of the distribution of T_H.*

H	0.01	0.05	0.1	0.5	0.9	0.95	0.99
0.5	0.0345	0.0565	0.0765	0.2905	1.1958	1.6557	2.7875
0.6	0.0246	0.04160	0.0577	0.2502	1.1181	1.5582	2.6407
0.7	0.0169	0.0296	0.0423	0.2175	1.0521	1.4744	2.5131
0.8	0.0106	0.0196	0.0292	0.1909	0.9952	1.4015	2.4008
0.9	0.0053	0.0108	0.0174	0.1694	0.9456	1.3374	2.3011
0.95	0.0029	0.0064	0.0115	0.1601	0.9231	1.3081	2.2551
1.0	0.00005	0.0013	0.0053	0.1516	0.9018	1.2805	2.2116

Table 8.3 reports percentage points of the distribution of T_H for the same values of H as in Table 8.2. It is anticipated that the density will behave quite differently as H approaches 1.

Exercises for Section 8.4

Exercise 8.4.1 Prove that, for $s \le t$,

$$\mathrm{Cov}(B_H(s), M_H(t)) = s,$$

where $\{B_H(s)\}$ is fBm and $\{M_H(t)\}$ is the fundamental martingale.

Exercise 8.4.2 For the process $\{N_H(t)\}$ defined by

$$N_H(t) = b_H \int_0^t u^{1/2-H}\, dW(u), \quad b_H = \sqrt{\frac{\Gamma(3/2-H)}{2H\,\Gamma(H+1/2)\,\Gamma(2-2H)}},$$

prove that

$$\mathrm{Corr}(B_H(t), N_H(t)) = \frac{\Gamma^2(3/2-H)}{\Gamma(2-2H)} \Big/ \sqrt{a_H},$$

where $a_H = \Gamma(3/2-H)/\,(2H\,\Gamma(H+1/2)\,\Gamma(3-2H))$.

Exercise 8.4.3 Show that the following relations hold:

$$\begin{aligned} T_H &= \int_0^1 C_H^2(t)\, dt \\ &= \frac{2(1-H)}{4H-1} \int_0^1 \int_0^1 \left[1 - (\max(s,t))^{4H-1}\right] (s\,t)^{1/2-H}\, dW(s)\, dW(t) \\ &\overset{\mathcal{D}}{=} \int_0^1 \int_0^1 (st)^{2H-1}\, (\min(s,t))^{2-2H}\, dW(s)\, dW(t), \end{aligned}$$

where

$$C_H(t) = \sqrt{2(1-H)}\, t^{2H-1} \int_0^t u^{1/2-H}\, dW(u).$$

Exercise 8.4.4 Show that the integral equation

$$f(t) = \lambda \int_0^1 K_H(s,t) f(s)\, ds$$

with

$$K_H(s,t) = \frac{2(1-H)}{4H-1}\left(1 - (\max(s,t))^{4H-1}\right)(s\,t)^{1/2-H}$$

is equivalent to the following differential equation with two boundary conditions:

$$h''(t) + \left(2(1-H)\lambda t^{2H-1} - \frac{2H(2H-1)}{t^2}\right) h(t) = 0,$$

$$h(1) = 0, \quad \lim_{t\to 0}\left((2H-1)t^{-2H} h(t) + t^{1-2H} h'(t)\right) = 0,$$

where $h(t) = f(t)/t^{H-1/2}$.

Exercise 8.4.5 Show that the FD of $L_H(s,t) = (st)^{2H-1}(\min(s,t))^{2-2H}$ is given by

$$D_H(\lambda) = \Gamma(1-\nu)\left(\frac{\eta}{2}\right)^{\nu} J_{-\nu}(\eta),$$

where $\eta = \sqrt{2(1-H)\lambda}/(H+1/2)$ and $\nu = (2H-1/2)/(H+1/2)$.

Exercise 8.4.6 For the statistics

$$S_H = \int_0^1 B_H^2(t)\, dt, \quad T_H = \int_0^1 C_H^2(t)\, dt,$$

where $C_H(t)$ is given in Exercise 8.4.3, show that

$$\mathrm{Var}(S_H) = \frac{4H+3}{(4H+1)(4H+2)} - \frac{2\Gamma^2(2H+1)}{\Gamma(4H+3)},$$

$$\mathrm{Var}(T_H) = \frac{2}{3(2H+1)}.$$

8.5 Approximation to the Fractional Unit Root Distribution

In this section we deal with the statistic Q_H defined in (8.36) as

$$Q_H = \frac{1}{2} B_H^2(1) \Big/ \int_0^1 B_H^2(t)\, dt,$$

which gives the fractional unit root distribution. The exact density of Q_H remains to be computed, but we attempt to compute approximate densities of Q_H for various values of H. As was discussed in the previous section, the approximation is based on replacing the fBm $B_H(t)$ by $C_H(t)$ defined in (8.47).

8.5.1 The Fredholm Determinant for the Approximate Distribution

Following Tanaka (2014), let us consider the approximation of Q_H by

$$R_H = \frac{1}{2} C_H^2(1) \Big/ \int_0^1 C_H^2(t)\, dt = \frac{U_H}{V_H}, \tag{8.57}$$

where $1/2 \le H < 1$ and

$$\begin{aligned} U_H &= \frac{1}{2}\left(\int_0^1 s^{1/2-H}\, dW(s)\right)^2 \\ &= \frac{1}{2}\int_0^1\int_0^1 (s\,t)^{1/2-H}\, dW(s)\, dW(t), \end{aligned} \tag{8.58}$$

$$\begin{aligned} V_H &= \int_0^1 t^{4H-2}\left(\int_0^t u^{1/2-H}\, dW(u)\right)^2 dt \\ &= \frac{1}{4H-1}\int_0^1\int_0^1 \left[1-(\max(s,t))^{4H-1}\right] \\ &\qquad \times (s\,t)^{1/2-H}\, dW(s)\, dW(t). \end{aligned} \tag{8.59}$$

Then we deal with the following approximate distribution:

$$\begin{aligned} &P(R_H < x) \\ &\quad = P\left(xV_H - U_H > 0\right) = P\left(\int_0^1\int_0^1 K_H(s,t;x)\, dW(s)\, dW(t) > 0\right), \end{aligned}$$

where

$$K_H(s,t;x) = \frac{x}{4H-1}\left[1-(\max(s,t))^{4H-1}\right](s\,t)^{1/2-H} - \frac{1}{2}(s\,t)^{1/2-H}. \tag{8.60}$$

Let us derive the FD of $K_H(s,t;x)$. We proceed in much the same way as in the last section. Putting $f(t) = t^{H-1/2}h(t)$, the integral equation with the kernel $K_H(s,t;x)$ takes the following form:

$$t^{2H-1}h(t) = \lambda \left[\frac{x}{4H-1} \left\{ \int_0^1 h(s)\,ds - t^{4H-1} \int_0^t h(s)\,ds \right.\right.$$
$$\left.\left. - \int_t^1 s^{4H-1} h(s)\,ds \right\} - \frac{1}{2} \int_0^1 h(s)\,ds \right],$$

which leads to

$$h''(t) + \left(\lambda x t^{2H-1} - \frac{2H(2H-1)}{t^2} \right) h(t) = 0. \tag{8.61}$$

The differential equation (8.61) is a special case of Bessel's equation in (8.53), and the general solution is given by

$$h(t) = t^{1/2} \left(c_1 J_\nu \left(\kappa t^{H+1/2} \right) + c_2 J_{-\nu} \left(\kappa t^{H+1/2} \right) \right),$$

where $\nu = (2H - 1/2)/(H + 1/2)$ and $\kappa = \sqrt{\lambda x}/(H + 1/2)$. It can be shown (Exercise 8.5.1) that the above integral equation is equivalent to the differential equation (8.61) with two boundary conditions

$$(2H - 1 - 2x)h(1) + h'(1) = 0, \quad \lim_{t \to 0} G(t) = 0, \tag{8.62}$$

where

$$G(t) = (2H-1)t^{-2H} h(t) + t^{1-2H} h'(t). \tag{8.63}$$

In fact, the equivalence may be proved by noting that

$$\lim_{t \to 0} G(t) = 0, \quad G(1) = 2xh(1), \quad G'(t) = -\lambda x h(t).$$

Then it follows (Exercise 8.5.2) that the two boundary conditions in (8.62) give us the following homogeneous equation in $\boldsymbol{c} = (c_1, c_2)'$:

$$M(\lambda)\,\boldsymbol{c} = \begin{pmatrix} b & -\left(2xJ_{-\nu}(\kappa) + \sqrt{\lambda x} J_{1-\nu}(\kappa)\right) \\ 1 & 0 \end{pmatrix} \boldsymbol{c} = \boldsymbol{0},$$

where b is some constant. Let the positive zeros of $J_{-\nu}(\kappa)$ and $J_{1-\nu}(\kappa)$ be $0 < a_1 < a_2 < \cdots$ and $0 < b_1 < b_2 < \cdots$, respectively. Then it follows from Watson (1958) that

$$J_{-\nu}(\kappa) = \frac{(\kappa/2)^{-\nu}}{\Gamma(1-\nu)} \prod_{n=1}^{\infty} \left(1 - \frac{\kappa^2}{a_n^2} \right), \quad J_{1-\nu}(\kappa) = \frac{(\kappa/2)^{1-\nu}}{\Gamma(2-\nu)} \prod_{n=1}^{\infty} \left(1 - \frac{\kappa^2}{b_n^2} \right).$$

Thus we have

$$\begin{aligned}
|M(\lambda)| &= 2xJ_{-\nu}(\kappa) + \sqrt{\lambda x}\,J_{1-\nu}(\kappa) \\
&= \frac{(\kappa/2)^{-\nu}}{\Gamma(1-\nu)}\left[2x\prod_{n=1}^{\infty}\left(1-\frac{\kappa^2}{a_n^2}\right) + \frac{\kappa/2}{1-\nu}\sqrt{\lambda x}\prod_{n=1}^{\infty}\left(1-\frac{\kappa^2}{b_n^2}\right)\right] \\
&= \frac{2x(\kappa/2)^{-\nu}}{\Gamma(1-\nu)}\left[\prod_{n=1}^{\infty}\left(1-\frac{\kappa^2}{a_n^2}\right) + \frac{\lambda}{4(1-H)}\prod_{n=1}^{\infty}\left(1-\frac{\kappa^2}{b_n^2}\right)\right],
\end{aligned}$$

and we obtain, as the FD of $K_H(s,t;x)$,

$$\begin{aligned}
D_H(\lambda;x) &= \prod_{n=1}^{\infty}\left(1-\frac{\kappa^2}{a_n^2}\right) + \frac{\lambda}{4(1-H)}\prod_{n=1}^{\infty}\left(1-\frac{\kappa^2}{b_n^2}\right) \\
&= \Gamma(1-\nu)\left(J_{-\nu}(\kappa) + \frac{\lambda}{2}\frac{J_{1-\nu}(\kappa)}{\sqrt{\lambda x}}\right)\Big/\left(\frac{\kappa}{2}\right)^{-\nu}, \qquad (8.64)
\end{aligned}$$

where $\nu = (2H-1/2)/(H+1/2)$ and $\kappa = \sqrt{\lambda x}/(H+1/2)$.

Then Imhof's (1961) formula gives us the following theorem.

Theorem 8.4 *The FD $D_H(\lambda;x)$ of the kernel $K_H(s,t;x)$ in (8.60) is given by (8.64), and the distribution function of R_H in (8.57) can be computed following*

$$\begin{aligned}
P(R_H < x) &= P\left(\frac{1}{2}C_H^2(1)\Big/\int_0^1 C_H^2(t)\,dt < x\right) \\
&= \frac{1}{2} + \frac{1}{\pi}\int_0^{\infty}\frac{1}{\theta}Im\left[(D_H(2i\theta;x))^{-1/2}\right]d\theta.
\end{aligned}$$

Figure 8.4 plots probability densities of R_H for various values of H ($\geq 1/2$). These are approximate densities of the original fractional unit root distribution Q_H, although the density for $H = 1/2$ is exact and is the same as that shown in Figure 8.1 for $g = 0$. The densities have positive support and diverge at the origin, and are shifted to the right as H becomes large. These densities have an interesting moment property, which we shall discuss in the next section.

Figure 8.5 draws the distribution functions of R_H (solid line) for $H = 0.5$ and $H = 0.8$ together with the corresponding simulated distributions of $Q_{H,n}$ (broken line), where

$$Q_{H,n} = n\frac{y_n^2/2}{\sum_{j=2}^{n} y_{j-1}^2}, \quad y_j = y_{j-1} + v_j, \quad (1-L)^{H-1/2}v_j = \varepsilon_j, \quad y_0 = 0.$$

Note that $Q_{H,n} \Rightarrow Q_H$ as $n \to \infty$. The simulated distribution of $Q_{H,n}$ was produced with $\{\varepsilon_j\} \sim$ NID(0, 1) from $n = 1000$ and number of replications $= 10{,}000$. Note that $R_H = Q_H$ for $H = 0.5$, so that $Q_{H,n} \Rightarrow R_H$ as $n \to \infty$

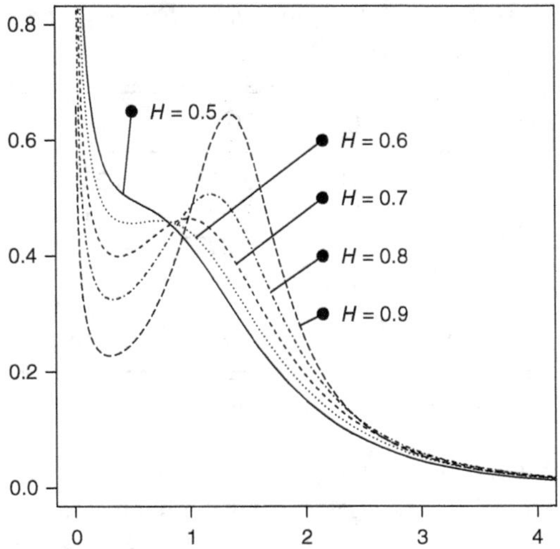

Figure 8.4 Approximate densities of $\frac{1}{2}\, B_H^2(1) \Big/ \int_0^1 B_H^2(t)\, dt$.

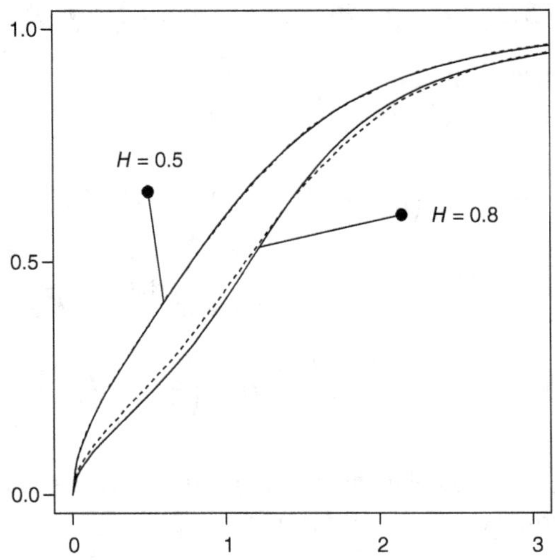

Figure 8.5 Fractional unit root distributions.

when $H = 0.5$. Allowing for this fact, the conjecture presents itself that the approximate distribution for $H = 0.8$ performs reasonably well and captures the characteristics of the true distribution.

Table 8.4 reports percentage points, means, and standard deviations of the distributions of R_H for various values of H. It can be seen that the percentage

Table 8.4 *Percentage points of the approximate fractional unit root distribution.*

H	0.01	0.05	0.5	0.95	0.99	Mean	SD
0.5	0.00066	0.0136	0.7812	2.8137	4.2541	1	0.944
0.6	0.00076	0.0188	0.8981	2.9770	4.5130	1.1	1.019
0.7	0.0011	0.0273	1.0245	3.0973	4.7184	1.2	1.063
0.8	0.0018	0.0439	1.1624	3.1475	4.8367	1.3	1.070
0.9	0.0039	0.0908	1.3155	3.0474	4.7634	1.4	1.001
0.95	0.0082	0.1729	1.4007	2.8401	4.5144	1.45	0.888

points in the lower tail increase as H becomes large, whereas those in the upper tail do not change much as H varies. It is also of interest to note that $\mathrm{E}(R_H) = H + 1/2$. This holds for any $H \in [1/2, 1)$, which will be proved in the next section.

8.5.2 An Interesting Moment Property

Here we compute the expectation of $R_H = U_H/V_H$ in (8.57). For this purpose let us define the joint m.g.f. of U_H and $-V_H$ by

$$\begin{aligned} m_H(\theta_1, -\theta_2) &= \mathrm{E}\left(e^{\theta_1 U_H - \theta_2 V_H}\right) = \left[D_H\left(-2\theta_1; \theta_2/\theta_1\right)\right]^{-1/2} \\ &= \left[\Gamma(1-\nu)\left(\frac{\sqrt{-2\theta_2}}{2H+1}\right)^{\nu}\left\{J_{-\nu}\left(\frac{\sqrt{-2\theta_2}}{H+1/2}\right)\right.\right. \\ &\qquad \left.\left. - \frac{\theta_1}{\sqrt{-2\theta_2}} J_{1-\nu}\left(\frac{\sqrt{-2\theta_2}}{H+1/2}\right)\right\}\right]^{-1/2}, \end{aligned}$$

where $D_H(\lambda; x)$ is the FD defined in (8.64) and $\nu = (2H - 1/2)/(H + 1/2)$. Following the formula in (7.15), we compute

$$\mathrm{E}(R_H) = \mathrm{E}\left(\frac{U_H}{V_H}\right) = \int_0^\infty \left.\frac{\partial m_H(\theta_1, -\theta_2)}{\partial \theta_1}\right|_{\theta_1 = 0} d\theta_2. \tag{8.65}$$

Putting $\xi = \sqrt{-2\theta_2}/(H + 1/2)$, we have

$$m_H(\theta_1, -\theta_2) = \left[\Gamma(1-\nu)\left(\frac{\xi}{2}\right)^{\nu}\left(J_{-\nu}(\xi) - \frac{\theta_1}{(H+1/2)\xi} J_{1-\nu}(\xi)\right)\right]^{-1/2},$$

so that we obtain

$$\begin{aligned}\left.\frac{\partial m_H(\theta_1,-\theta_2)}{\partial\theta_1}\right|_{\theta_1=0} &= \frac{1}{2}\left(\Gamma(1-\nu)\right)^{-1/2}\frac{(\xi/2)^{-\nu/2}}{(H+1/2)\xi}J_{1-\nu}(\xi)\left(J_{-\nu}(\xi)\right)^{-3/2}\\ &= \frac{H+1/2}{2}\left[\Gamma(1-\nu)\left(\frac{\xi}{2}\right)^{\nu}J_{-\nu}(\xi)\right]^{-3/2}\\ &\quad\times\left\{\frac{\Gamma(1-\nu)}{2(H+1/2)^2}\left(\frac{\xi}{2}\right)^{\nu-1}J_{1-\nu}(\xi)\right\}.\end{aligned}$$

This last factorization is necessary because we use the change of variables rule to execute integration in (8.65). More specifically, we consider the change of variables from $\theta_2>0$ to u given by

$$\begin{aligned}u &= \Gamma(1-\nu)\left(\frac{\xi}{2}\right)^{\nu}J_{-\nu}(\xi) = \Gamma(1-\nu)\sum_{k=0}^{\infty}\frac{(-\xi^2/4)^k}{k!\,\Gamma(k+1-\nu)}\\ &= 1+\Gamma\left(\frac{1-H}{H+1/2}\right)\sum_{k=1}^{\infty}\frac{(2\theta_2)^k}{k!\,(2H+1)^{2k}\,\Gamma\left(k+\dfrac{1-H}{H+1/2}\right)},\end{aligned}$$

where the second equality follows from the definition of the Bessel function given in (2.58). It is clear that we have $1<u<\infty$ for $0<\theta_2<\infty$. Using the relation ((2.127); Bowman, 2010)

$$\frac{d}{dz}\left(z^{\nu}J_{-\nu}(z)\right) = -z^{\nu}J_{1-\nu}(z), \tag{8.66}$$

it can be shown (Exercise 8.5.3) that

$$\begin{aligned}du &= d\left(\Gamma(1-\nu)\left(\frac{\xi}{2}\right)^{\nu}J_{-\nu}(\xi)\right)\\ &= \frac{\Gamma(1-\nu)}{2(H+1/2)^2}\left(\frac{\xi}{2}\right)^{\nu-1}J_{1-\nu}(\xi)\,d\theta_2.\end{aligned}$$

Thus we finally arrive at

$$\begin{aligned}\mathrm{E}(R_H) &= \int_0^{\infty}\left.\frac{\partial m(\theta_1,-\theta_2)}{\partial\theta_1}\right|_{\theta_1=0}d\theta_2\\ &= \frac{H+1/2}{2}\int_0^{\infty}\left[\Gamma(1-\nu)\left(\frac{\xi}{2}\right)^{\nu}J_{-\nu}(\xi)\right]^{-3/2}\\ &\quad\times\left\{\frac{\Gamma(1-\nu)}{2(H+1/2)^2}\left(\frac{\xi}{2}\right)^{\nu-1}J_{1-\nu}(\xi)\right\}d\theta_2\\ &= \frac{H+1/2}{2}\int_1^{\infty}u^{-3/2}\,du = H+\frac{1}{2},\end{aligned}$$

which establishes

Theorem 8.5 *The mean of $R_H = U_H/V_H$ is given by*

$$E(R_H) = E\left(\frac{U_H}{V_H}\right) = E\left(\frac{1}{2}C_H^2(1) \Big/ \int_0^1 C_H^2(t)\,dt\right) = H + \frac{1}{2}. \tag{8.67}$$

The above theorem yields a very simple expression for $\mathrm{E}(U_H/V_H)$, which motivates us to compute, from (8.58) and (8.59),

$$\mathrm{E}(U_H) = \frac{1}{2}\int_0^1 t^{1-2H}\,dt = \frac{1}{4(1-H)},$$
$$\mathrm{E}(V_H) = \frac{1}{4H-1}\int_0^1 \left(1 - t^{4H-1}\right) t^{1-2H}\,dt = \frac{1}{2(1-H)(2H+1)},$$

so that

$$\mathrm{E}(R_H) = \mathrm{E}\left(\frac{U_H}{V_H}\right) = \frac{\mathrm{E}(U_H)}{\mathrm{E}(V_H)} = \frac{1}{4(1-H)} \Big/ \frac{1}{2(1-H)(2H+1)} = H + \frac{1}{2}.$$

This implies that the expectation of the ratio is equal to the ratio of expectations, which occurs if R_H is independent of V_H. However, this is not the case. The similar situation occurred for g-integrated Bm $\{F_g(t)\}$ discussed in Section 8.1, where it holds that

$$\mathrm{E}(R(g)) = \mathrm{E}\left(\frac{1}{2}F_g^2(1) \Big/ \int_0^1 F_g^2(t)\,dt\right) = g + 1.$$

Here g is a positive integer that corresponds to $g = H - 1/2$ if g is allowed to be fractional.

It is conjectured that the same property will hold for the original fractional unit root distribution, that is,

$$\mathrm{E}(Q_H) = \mathrm{E}\left(\frac{B_H^2(1)/2}{\int_0^1 B_H^2(t)\,dt}\right) = \frac{\mathrm{E}\left(B_H^2(1)/2\right)}{\mathrm{E}\left(\int_0^1 B_H^2(t)\,dt\right)} \tag{8.68}$$
$$= \frac{1/2}{1/(2H+1)} = H + \frac{1}{2}.$$

Exercises for Section 8.5

Exercise 8.5.1 Putting $f(t) = t^{H-1/2}h(t)$, show that the integral equation

$$f(t) = \lambda \int_0^1 K_H(s,t;x) f(s)\,ds,$$

where

$$K_H(s,t;x) = \frac{x}{4H-1}\left(1-(\max(s,t))^{4H-1}\right)(s\,t)^{1/2-H} - \frac{1}{2}(s\,t)^{1/2-H},$$

is equivalent to the following differential equation with two boundary conditions:

$$h''(t) + \left(\lambda x t^{2H-1} - \frac{2H(2H-1)}{t^2}\right) h(t) = 0,$$

$$(2H-1-2x)h(1) + h'(1) = 0,$$

$$\lim_{t\to 0} G(t) = 0, \quad G(t) = (2H-1)t^{-2H}h(t) + t^{1-2H}h'(t).$$

Exercise 8.5.2 Show that the differential equation with the two boundary conditions given in Exercise 8.5.1 yields

$$M(\lambda)\,\boldsymbol{c} = \begin{pmatrix} b & -\left(2xJ_{-\nu}(\kappa) + \sqrt{\lambda x}J_{1-\nu}(\kappa)\right) \\ 1 & 0 \end{pmatrix} \boldsymbol{c} = \boldsymbol{0},$$

where b is some constant.

Exercise 8.5.3 Prove that

$$\begin{aligned}\frac{2^{\nu/2}}{(2H+1)\sqrt{\Gamma(1-\nu)}} &\int_0^\infty \xi^{-\nu/2-1} J_{1-\nu}(\xi)\,(J_{-\nu}(\xi))^{-3/2}\,dx \\ &= H + 1/2,\end{aligned}$$

where $\xi = \sqrt{-2x}/(H+1/2)$ and $\nu = (2H-1/2)/(H+1/2)$ with $1/2 < H < 1$.

References

Anderson, T. W. (1971). *The Statistical Analysis of Time Series*. Wiley, New York.

Anderson, T. W. and Darling, D. A. (1952). Asymptotic theory of certain 'goodness of fit' criteria based on stochastic processes. *Annals of Mathematical Statistics*, **23**, 193–212.

Biagini, F., Hu, Y., Øksendal, B., and Zhang, T. (2008). *Stochastic Calculus for Fractional Brownian Motion and Applications*. Springer, London.

Billingsley, P. (1999). *Convergence of Probability Measures* (Second Edition). Wiley, New York.

Bishwal, J. P. N. (2008). *Parameter Estimation in Stochastic Differential Equations*. Springer, Berlin.

Bobkoski, M. J. (1983). Hypothesis testing in nonstationary time series, PhD Thesis. University of Wisconsin.

Bowman, F. (2010). *An Introduction to Bessel Functions*. Dover, New York.

Bronski, J. C. (2003). Asymptotics of Karhunen–Loève eigenvalues and tight constants for probability distributions of passive scalar transport. *Communications in Mathematical Physics*, **238**, 563–582.

Chan, N. H. and Wei, C. Z. (1988). Limiting distributions of least squares estimates of unstable autoregressive processes. *Annals of Statistics*, **16**, 367–401.

Courant, R. and Hilbert, D. (1953). *Methods of Mathematical Physics, Vol. I.* Wiley, New York.

Darling, D. A. (1955). The Cramér–Smirnov test in the parametric case. *Annals of Mathematical Statistics*, **26**, 1–20.

Davydov, Y. (1970). The invariance principle for stationary processes. *Theory of Probability and Its Applications*, **15**, 487–498.

Dickey, D. A. and Fuller, W. A. (1979). Distribution of the estimators for autoregressive time series with a unit root. *Journal of the American Statistical Association*, **74**, 427–431.

Donsker, M. D. (1951). An invariance principle for certain probability limit theorems. *Memoires of the American Mathematical Society*, **6**, 1–12.

Durbin, J. (1973). Weak convergence of the sample distribution function when parameters are estimated. *Annals of Statistics*, **1**, 279–290.

Engle, R. F. and Granger, C. W. J. (1987). Co-integration and error correction: Representation, estimation, and testing. *Econometrica*, **55**, 251–276.

Evans, G. B. A. and Savin, N. E. (1981). Testing for unit roots: 1. *Econometrica*, **49**, 753–779.

Ferguson, T. S. (1967). *Mathematical Statistics: A Decision Theoretic Approach*. Academic Press, New York.

Giraitis, L., Koul, H., and Surgailis, D. (2012). *Large Sample Inference for Long Memory Processes*. Imperial College Press, London.

Girsanov, I. V. (1960). On transforming a certain class of stochastic processes by absolutely continuous substitution of measures. *Theory of Probability and Its Applications*, **5**, 285–301.

Gradshteyn, I. S. and Ryzhik, I. M. (1965). *Table of Integrals, Series, and Products* (Fifth Edition). Academic Press, New York.

Gripenberg, G. and Norros, I. (1996). On the prediction of fractional Brownian motion. *Journal of Applied Probability*, **33**, 400–410.

Helland, I. S. (1982). Central limit theorems for martingales with discrete or continuous time. *Scandinavian Journal of Statistics*, **9**, 79–94.

Hochstadt, H. (1973). *Integral Equations*. Wiley, New York.

Hosking, J. R. M. (1981). Fractional differencing. *Biometrika*, **68**, 165–176.

Imhof, J. P. (1961). Computing the distribution of quadratic forms in normal variables. *Biometrika*, **48**, 419–426.

Jazwinski, A. H. (1970). *Stochastic Processes and Filtering Theory*. Academic Press, New York.

Jin, S. and Li, W. V. (2015). Expectation of the limiting distribution of the LSE of a unit root process. *Statistica Sinica*, **25**, 529–536.

Kac, M., Kiefer, J., and Wolfowitz, J. (1955). On tests of normality and other tests of goodness of fit based on distance methods. *Annals of Mathematical Statistics*, **26**, 189–211.

Kariya, T. (1980). Locally robust tests for serial correlation in least squares regression. *Annals of Statistics*, **8**, 1065–1070.

Klebaner, F. C. (2012). *Introduction to Stochastic Calculus with Applications* (Third Edition). Imperial College Press, London.

Kleptsyna, M. L. and Le Breton, A. (2002). Statistical analysis of the fractional Ornstein–Uhlenbeck type process. *Statistical Inference for Stochastic Processes*, **5**, 229–248.

Kleptsyna, M. L., Le Breton, A., and Roubaud, M. C. (2000). Parameter estimation and optimal filtering for fractional type stochastic systems. *Statistical Inference for Stochastic Processes*, **3**, 173–182.

Kolmogorov, A. N. (1940). Wienersche Spiralen und einige andere interessante Kurven im Hilbertschen Raum. *Doklady Akademii Nauk SSSR*, **26**, 115–118.

Kuo, H. -H. (2006). *Introduction to Stochastic Integration*. Springer, New York.

Liptser, R. S. and Shiryaev, A. N. (2001a). *Statistics of Random Processes I: General Theory* (Second Edition). Springer-Verlag, New York.

Liptser, R. S. and Shiryaev, A. N. (2001b). *Statistics of Random Processes II: Applications* (Second Edition). Springer-Verlag, New York.

Loève, M. (1977). *Probability Theory I* (Fourth Edition). Springer-Verlag, New York.

Loève, M. (1978). *Probability Theory II* (Fourth Edition). Springer-Verlag, New York.

MacNeill, I. B. (1974). Tests for change of parameter at unknown times and distributions of some related functionals on Brownian motion. *Annals of Statistics*, **2**, 950–962.

Mandelbrot, B. B. and Van Ness, J. W. (1968). Fractional Brownian motions, fractional Browninan noises and applications. *SIAM Review*, **10**, 422–437.

Marinucci, D. and Robinson, P. M. (1999). Alternative forms of fractional Brownian motion. *Journal of Statistical Inference and Planning*, **80**, 111–122.

Nabeya, S. (1989). Asymptotic distributions of test statistics for the constancy of regression coefficients under a sequence of random walk alternatives. *Journal of the Japan Statistical Society*, **19**, 23–33.

Nabeya, S. (1992). Limiting moment generating function of Cramér–von Mises–Smirnov goodness of fit statistics under null and local alternatives. *Journal of the Japan Statistical Society*, **22**, 113–122.

Nabeya, S. and Perron, P. (1994). Local asymptotic distributions related to the AR(1) model with dependent errors. *Journal of Econometrics*, **62**, 229–264.

Nabeya, S. and Tanaka, K. (1988). Asymptotic theory of a test for the constancy of regression coefficients against the random walk alternative. *Annals of Statistics*, **16**, 218–235.

Nabeya, S. and Tanaka, K. (1990a). A general approach to the limiting distribution for estimators in time series regression with nonstable autoregressive errors. *Econometrica*, **58**, 145–163.

Nabeya, S. and Tanaka, K. (1990b). Limiting powers of unit-root tests in time-series regression. *Journal of Econometrics*, **46**, 247–271.

Norros, I., Valkeila, E., and Virtamo, J. (1999). An elementary approach to a Girsanov formula and other analytical results on fractional Brownian motions. *Bernoulli*, **5**, 571–587.

Nourdin, I. (2012). *Selected Aspects of Fractional Brownian Motion*. Springer, Berlin.

Nyblom, J. and Mäkeläinen, T. (1983). Comparisons of tests for the presence of random walk coefficients in a simple linear model. *Journal of the American Statistical Association*, **78**, 856–864.

Perron, P. (1991). A continuous-time approximation to the unstable first-order autoregressive model: The case without an intercept. *Econometrica*, **59**, 211–236.

Pettitt, A. N. (1976). A two-sample Anderson–Darling rank statistic. *Biometrika*, **63**, 161–168.

Phillips, P. C. B. (1987a). Time series regression with a unit root. *Econometrica*, **55**, 277–301.

Phillips, P. C. B. (1987b). Towards a unified asymptotic theory for autoregression. *Biometrika*, **74**, 535–547.

Phillips, P. C. B. and Magdalinos, T. (2007). Limit theory for moderate deviations from a unit root. *Journal of Econometrics*, **136**, 115–130.

Rao, C. R. (1973). *Linear Statistical Inference and Its Applications* (Second Edition). Wiley, New York.

Rutherford, D. E. (1946). Some continuant determinants arising in physics and chemistry. *Proceedings of the Royal Society of Edinburgh*, **A-62**, 229–236.

Soong, T. T. (1973). *Random Differential Equations in Science and Engineering*. Academic Press, New York.

Sowell, F. (1990). The fractional unit root distribution. *Econometrica*, **58**, 495–505.

Sukhatme, S. (1972). Fredholm determinant of a positive definite kernel of a special type and its application. *Annals of Mathematical Statistics*, **43**, 1914–1926.

Tanaka, K. (1983). Non-normality of the Lagrange multiplier statistic for testing the constancy of regression coefficients. *Econometrica*, **51**, 1577–1582.

Tanaka, K. (1990a). The Fredholm approach to asymptotic inference on nonstationary and noninvertible time series models. *Econometric Theory*, **6**, 411–432.

Tanaka, K. (1990b). Testing for a moving average unit root. *Econometric Theory*, **6**, 433–444.

Tanaka, K. (1996). *Time Series Analysis: Nonstationary and Noninvertible Distribution Theory*. Wiley, New York.

Tanaka, K. (2013). Distributions of the maximum likelihood and minimum contrast estimators associated with the fractional Ornstein–Uhlenbeck process. *Statistical Inference for Stochastic Processes*, **16**, 173–192.

Tanaka, K. (2014). Distributions of quadratic functionals of the fractional Brownian motion based on a martingale approximation. *Econometric Theory*, **30**, 1078–1109.

Tanaka, K. (2015). Maximum likelihood estimation for the non-ergodic fractional Ornstein–Uhlenbeck process. *Statistical Inference for Stochastic Processes*, **18**, 315–332.

Tanaka, K. (2017). *Time Series Analysis: Nonstationary and Noninvertible Distribution Theory* (Second Edition). Wiley, New York.

Tanaka, K. (2023). Extensions of Darling's formula for the Fredholm determinant. *Unpublished Manuscript*.

Taqqu, M. S. (1975). Weak convergence to fractional Brownian motion and to the Rosenblatt process. *Zeitschrift für Wahrscheinlichkeitstheorie und Verwandte Gebiete*, **31**, 287–302.

Watson, G. N. (1958). *A Treatise on the Theory of Bessel Functions* (Second Edition). Cambridge University Press, Cambridge.

Watson, G. S. (1961). Goodness-of-fit tests on a circle. *Biometrika*, **48**, 109–114.

Whittaker, E. T. and Watson, G. N. (1958). *A Course of Modern Analysis* (Fourth Edition). Cambridge University Press, Cambridge.

Author Index

Subject Index

For EU product safety concerns, contact us at Calle de José Abascal, 56–1°, 28003 Madrid, Spain or eugpsr@cambridge.org.

www.ingramcontent.com/pod-product-compliance
Lightning Source LLC
LaVergne TN
LVHW011528200926
848484LV00001B/188

* 9 7 8 1 0 0 9 5 6 6 9 9 5 *